ROUTLEDGE HANDBOOK OF MARINE GOVERNANCE AND GLOBAL ENVIRONMENTAL CHANGE

This comprehensive handbook provides a detailed and unique overview of current thinking about marine governance in the context of global environmental change.

Many of the most profound impacts of global environmental change, and climate change in particular, will occur in the oceans. It is vital that we consider the role of marine governance in adapting to and mitigating these impacts. This comprehensive handbook provides a thorough review of current thinking about marine environmental governance, including law and policy, in the context of global environmental change. Initial chapters describe international law, regimes, and leadership in marine environmental governance, in the process considering how existing regimes for climate change and the oceans should and can be coordinated. This is followed by an exploration of the role of non-state actors, including scientists, nongovernmental organisations, and corporations. The next section includes a collection of chapters highlighting governance schemes in a variety of marine environments and regions, including coastlines, islands, coral reefs, the open ocean, and regional seas. Subsequent chapters examine emerging issues in marine governance, including plastic pollution, maritime transport, sustainable development, environmental justice, and human rights.

Providing a definitive overview, the *Routledge Handbook of Marine Governance and Global Environmental Change* is suitable for advanced students in marine and environmental governance, environmental law and policy, and climate change, as well as practitioners, activists, stakeholders, and others concerned about the world's oceans and seas.

Paul G. Harris is the Chair Professor of Global and Environmental Studies at the Education University of Hong Kong. He is the author of six books and editor of 20 volumes on climate change and global environmental politics, policy, and justice.

ROUTLEDGE HANDBOOK OF MARINE GOVERNANCE AND GLOBAL ENVIRONMENTAL CHANGE

Edited by
Paul G. Harris

LONDON AND NEW YORK

Cover image: Hong Kong Beach, 2021 © Paul G. Harris

First published 2022
by Routledge
4 Park Square, Milton Park, Abingdon, Oxon OX14 4RN

and by Routledge
605 Third Avenue, New York, NY 10158

Routledge is an imprint of the Taylor & Francis Group, an informa business

British Library Cataloguing-in-Publication Data
A catalogue record for this book is available from the British Library

Library of Congress Cataloging-in-Publication Data
A catalog record has been requested for this book

ISBN: 978-1-138-55591-4 (hbk)
ISBN: 978-1-032-00459-4 (pbk)
ISBN: 978-1-315-14974-5 (ebk)

DOI: 10.4324/9781315149745

Typeset in Bembo
by codeMantra

CONTENTS

FIGURES

TABLES

CONTRIBUTORS

Amanda Alva is a research assistant in the Conservation Governance Lab, School of Forestry and Wildlife, Auburn University.

Elia Elisa Cia Alves is Associate Professor in the International Relations Department and the Graduate Program in Public Management and International Cooperation, Federal University of Paraiba, Brazil.

Emma Avoyan is a postdoctoral researcher in the Department of Geography, Planning and Environment, Radboud University.

Luciana Fernandes Coelho is a PhD candidate at the Global Ocean Institute, World Maritime University.

Erik van Doorn is a research associate at the Walther Schücking Institute for International Law, Kiel University.

Kelly Dunning is Assistant Professor and director of the Conservation Governance Lab, School of Forestry and Wildlife, Auburn University.

Pedro Fidelman is Associate Professor and Principal Research Fellow at the Centre for Policy Futures, University of Queensland.

Jennicca Gordon is the Carbon Officer at the John Muir Trust and LLM graduate in Global Environment and Climate Change Law at the University of Edinburgh.

Lars H. Gulbrandsen is Professor and Research Director at the Fridtjof Nansen Institute, Lysaker, Norway.

Paul G. Harris is the Chair Professor of Global and Environmental Studies at the Education University of Hong Kong.

Janos Hennicke is Head of Division 3.1 at the Foundation of International Marine Nature Conservation, German Federal Agency for Nature Conservation, Vilm.

Benjamin Hofmann is a postdoctoral researcher in environmental social sciences at Eawag, the Swiss Federal Institute of Aquatic Science and Technology.

Rozemarijn J. Roland Holst is Assistant Professor in Public International Law, Utrecht University.

Marko Joas is Professor in Public Administration, and Dean of the Faculty of Social Sciences, Business and Economics, at Åbo Akademi University.

Greg Johnson is a research assistant in the Conservation Governance Lab, School of Forestry and Wildlife, Auburn University.

Shailly Kedia is a fellow at the Energy and Resources Institute, India.

E. Carina H. Keskitalo is Professor of Political Science in the Department of Geography, Umeå University, Sweden.

Anastasia Kuswardani is a researcher at the Marine Research Centre, Ministry of Marine Affairs and Fisheries, Indonesia.

Dmitry L. Lajus is Associate Professor in the Department of Ichthyology and Hydrobiology, Faculty of Biology, St. Petersburg State University, Russia.

Judith van Leeuwen is Assistant Professor in the Environmental Policy Group, Wageningen University, the Netherlands.

Mitchell Lennan is Affiliate Researcher in the Strathclyde Centre for Environmental Law and Governance, University of Strathclyde, Scotland.

Nele Matz-Lück is Co-Director of the Walther Schücking Institute for Public International Law and Professor of Public International Law at Kiel University.

Hugh McDonald is a fellow at the Ecologic Institute, Berlin.

Jason Monios is Professor of Maritime Logistics at Kedge Business School, Marseille.

Kapil Narula is Senior Researcher at the Institute for Environmental Sciences, University of Geneva.

Colin von Negenborn is a research associate in the Department of Philosophy and at the Walther Schücking Institute for International Law, Kiel University.

Gabriela A. Oanta is Associate Professor of Public International Law and holder of the Jean Monnet Chair 'European Union law of the sea', University of A Coruña, Spain.

Konrad Ott is Chair for Philosophy and Ethics of the Environment at Kiel University.

Tim Packeiser is Senior Policy Advisor in Ocean Governance at WWF Germany.

Achmad Poernomo is Senior Lecturer at the Jakarta Technical University of Fisheries, and Senior Research Fellow, Center for Southeast Asian Studies, Jakarta.

Tony George Puthucherril is Professor at the Jindal Global Law School, O.P. Jindal Global University, India.

Arne Riedel is Senior Fellow at the Ecologic Institute, Berlin.

Henrik Ringbom is Head of Research at Åbo Akademi University, and Professor II at the Scandinavian Institute of Maritime Law at the University of Oslo.

Luís Paulo Santana is undertaking graduate studies in the Department of Anthropology, Federal University of Pernambuco, Brazil.

Anemoon Soete is a post-doctoral researcher in the Department of European, Public and International Law, Ghent University.

Otto Spijkers is Professor in the China Institute of Boundary and Ocean Studies and the Research Institute of Environmental Law, Wuhan University.

Andrea Quirino Steiner is an associate professor in the Department of Political Science, Federal University of Pernambuco, Brazil.

Jan van Tatenhove is Professor of Marine Governance and Maritime Spatial Planning at Aalborg University and Professor at Van Hall Larenstein University of Applied Sciences.

Nata Tavonvunchai is a PhD candidate at the School of Biological and Marine Sciences, University of Plymouth.

Nina Tynkkynen is Professor in Environmental Governance and Policy, Faculty of Social Sciences, Business and Economics, Åbo Akademi University.

Alice B.M. Vadrot is Associate Professor in International Relations and the Environment, Department of Political Science, University of Vienna.

Joanna Vince is Senior Lecturer in the Politics and International Relations Program, School of Social Sciences, University of Tasmania.

Tony R. Walker is Associate Professor in the School for Resource and Environmental Studies, Dalhousie University.

Nicole Wienrich is Research Associate in Ocean Governance and Arctic Governance at the Institute for Advanced Sustainability Studies, Potsdam.

Chuxiao Yu is an associate research professor at the Research Institute of Environmental Law, Wuhan University.

Małgorzata Zachara-Szymańska is an associate professor of international studies at Jagiellonian University in Krakow, Poland.

PART 1

Introduction

1
THE GROWING CHALLENGE FOR MARINE GOVERNANCE
Global environmental change

Paul G. Harris

The places that humans inhabit day to day are experiencing the effects of global environmental change. Many of those effects, such as the impacts of global warming and other manifestations of climate change on land-based ecosystems and human communities, are increasingly reported in the daily news. They include unusually widespread and massive wildfires, prolonged droughts, intense cyclonic storms, and heavy rains, flooding, and landslides, and much more. Other widescale environmental occurrences on land, such as deforestation and habitat destruction, diminution of biodiversity and agricultural productivity, omnipresent plastic rubbish, fetid waterways, and the spread of pathogens due to human activities, are also featured every day in news reports around the world. But what of the environments that few people experience directly on a regular basis, such as sparsely inhabited mountain ranges and remote polar landscapes? Widescale environmental changes are underway in those places, too, but they are less likely to receive much attention in the news. As the proverb intones, 'out of sight, out of mind'. Too often, this is also the case with environmental changes underway in the world's oceans and seas, which are profoundly affected by climate change, notably by warming and acidification, and by other impacts, such as widespread pollution and intensive extraction of marine living resources, the latter manifested in formerly vibrant fisheries in which sought-after species are now commercially extinct. It is not that millions of humans are not affected by these changes; they are, profoundly. It is that they are often those people who are, like the high seas, often invisible to those actors with the power to do something about it.

Global environmental change is thus creating new and profound challenges for the governance of marine environments and the people who depend on them for their well-being, and often for their survival. More attention needs to be directed toward the changes that are underway, and growing, and the associated challenges that result. This handbook tries to do just that, in the process offering a unique treatment of marine environmental governance in the Anthropocene – the current geological epoch characterised by unprecedented anthropogenic impacts on the earth and its ecosystems (see Biermann 2022). The chapters that follow explore challenges for governing the marine environment that arise from global-scale (or nearly global-scale) environmental changes. They survey many of the key issues arising from global environmental change and the multifarious actors and institutions involved in governing oceans and seas. Devoting more attention to the impacts of global environmental

DOI: 10.4324/9781315149745-2

change on marine environmental governance is justified because humanity is extremely reliant on the oceans for resources and ecosystem services. As global environmental change intensifies, so, too, do the threats to those resources and services. Effective marine governance becomes more important, and more challenging, as time passes. With all of this in mind, this book is intended to be a useful resource for researchers, students and others interested in marine governance in the current era of accelerating global environmental change.

Global environmental change: challenges for oceans and seas

For the purposes of this book, global environmental change is defined broadly to include changes to environments that are occurring over wide geographic areas, such as ocean warming and changes to ocean chemistry as a result of carbon pollution, as well as those changes that are happening in many geographically removed places simultaneously, such as agricultural runoff leading to ocean dead zones (eutrophication) and decimation of fish numbers and varieties due to overfishing or habitat destruction in many locations across the world's ocean. Global environmental changes make marine governance more difficult: they make existing problems worse, thus increasing current governance challenges, and they create new challenges altogether, creating a need for not only more effective marine environmental governance but also completely new types and layers of governance. The result is often that existing institutions and mechanisms for managing the marine environment are found to be lacking in their capabilities, and increasingly that existing institutions and mechanisms do not exist and need to be created from scratch, or that uncoordinated existing marine governance regimes and schemes need to be synchronised with regimes and schemes from other issue areas. For example, fisheries regimes may need to be coordinated with regimes for regulating agricultural runoff or, as several chapters in this book make clear, the law of the sea regime may need to be coordinated with the climate change regime.

The environmental changes that are challenging marine governance are as varied as the oceans and seas. They include overexploitation of fish and other marine resources; widespread eutrophication caused by the runoff of pollutants from land environments; plastic pollution so extensive in marine ecosystems that it justifies its own term – the 'plastisphere' (Thomas 2021); coastal over-development and habitat destruction; intensifying shipping traffic in busy seaways and new ship traffic in relatively remote and vulnerable sea regions, among many other changes. The greatest challenge to marine ecosystems and their management is climate change (see Harris 2019). In 2019, the Intergovernmental Panel on Climate Change (IPCC) produced a special report describing the impacts of climate change on the oceans (and the cryosphere) (Intergovernmental Panel on Climate Change 2019). The list of those impacts is long and worrying. In its latest report on the science of climate change (Intergovernmental Panel on Climate Change 2021), the IPCC declared that human influence on the oceans is unequivocal, resulting in 'widespread and rapid changes' in marine and other biospheres (Intergovernmental Panel on Climate Change 2021: SPM-5). Ocean warming has already resulted in extensive coral bleaching and changes to fisheries. The heat content of the oceans and global mean sea-surface temperatures have increased. The oceans have warmed about 1°C since the second half of the nineteenth century, with much of that increase in recent decades (Intergovernmental Panel on Climate Change 2021: SPM-5). Marine heatwaves have become more common. Global mean sea level has risen substantially, partly due to the melting of land ice, such as that of the Greenland Ice Sheet, with major implications for coastal communities, low-lying countries, and small-island states. The extent of Arctic sea ice – its range throughout the year and its thickness – has decreased markedly, and other changes to polar marine ecosystems are increasing, not least in the Southern Ocean around Antarctica. Acidification and other changes to ocean chemistry, such as increasing salinity and decreasing oxygen content of seawater, are affecting marine organisms. Changes to both ocean temperatures and salinity, among other factors, are affecting ocean currents, turnover, and upwelling.

Marine governance amidst global environmental change

All of these manifestations of global environmental change in oceans and seas, and indeed many other manifestations that are not mentioned here, present new challenges for *governing* the marine environment. Marine environmental governance can be broadly conceived of as marine policies and actions that prevent or limit adverse environmental outcomes and promote desirable ones (cf. Delmas and Young 2009: 6). Governance has always been difficult in the marine environment (see, e.g., Grip 2017), but global environmental change adds new layers of difficulty that cannot be ignored. This is especially true with respect to the most profound of all types of global environmental change: climate change. As the Intergovernmental Panel on Climate Change has observed, 'Impacts of climate-related changes in the ocean … increasingly challenge current governance efforts to develop and implement adaptation responses from local to global scales, and in some cases pushing them to their limits' (Intergovernmental Panel on Climate Change 2019: 29). Consequently, it is imperative to strengthen those and other governance efforts that are directly and indirectly related to the world's oceans and seas. How have governments, communities and other actors concerned about the marine environment responded to global environmental changes? How will, and how should, they do so in the future? The contributors to this book aim to answer these and many related questions. Their chapters are organised into sections that respectively focus on marine environmental governance with respect to (a) international law, regimes, and institutions (Part 2 of the book); (b) non-state actors (Part 3); (c) particular environments and regions (Part 4); and (d) emerging issues (Part 5). The following subsections highlight some of the ideas presented in each chapter.

International law, regimes, and leadership in marine environmental governance

The substantive chapters begin in Part 2 of the book, which focuses on international law, regimes, and institutions in marine governance amidst global environmental change. In Chapter 2, Erik van Doorn provides a foundation for most subsequent chapters with a detailed description of the law of the sea. He argues that marine governance is very largely premised on the law of the sea, which evolved through centuries to become customary international law. The importance of clarifying the law of the sea and adapting it to modern needs resulted in negotiation and agreement of the 1982 United Nations Convention on the Law of the Sea (UNCLOS). Combined with existing practices, UNCLOS has subsequently affected the use of oceans and seas as the Anthropocene has become more evident there. The question is whether UNCLOS and the law of the sea more broadly are up to the task as global environmental changes become more intense, more widespread, and more influential in the marine environment. The likely outcome for marine environmental governance may be a combination of creative interpretation of existing law and creation of new formal and informal ('soft') law agreements and the evolution of practices by states, all of which are taken up in greater detail in several other chapters in this book.

One of the big challenges for marine environmental governance amidst global environmental change is the imperfect fit between regimes developed for ocean management and those created to manage global climate change. The nexus between these two regimes – or, more accurately, sets of regimes – is taken up by Rozemarijn J. Roland Holst in Chapter 3. Holst points out that, despite the many connections between oceans and climate – the bulk of carbon pollution and indeed heat from global warming have been absorbed by the oceans, for example – oceans and seas were, at least until very recently, largely absent from international negotiations on combatting climate change. Where oceans have found their way into international efforts to address climate change are in countries' so-called Nationally Determined Contributions – their action pledges – to the 2015 Paris Agreement on Climate Change. While the oceans regime is adaptive enough to incorporate considerations of climate change, Holst argues that cumulative impacts of climate change expose the weaknesses of approaching the management of different forms of ocean pollution independently. To be sure, there are opportunities to overcome this sectoral approach within both the oceans and climate regimes, but coordination is still lacking. Neither regime is up to the task of effectively managing ocean challenges exacerbated by climate change, so 'purposeful coordination' will become increasingly necessary and urgent.

By focusing on the problem of ocean acidification, in Chapter 4, Jennicca Gordon further explicates the important relations between the oceans and climate regimes, in the process especially highlighting the legal connections. Due to emissions of carbon dioxide from the burning of fossil fuels, changes in ocean chemistry are occurring at an accelerating rate. In particular, as the oceans absorb that carbon, they become acidic; today, they are almost one-third more acidic than 200 years ago. Impacts on marine organisms and ecosystems are increasingly apparent, ranging from adverse effects on plankton to reductions in shellfish harvests. Gordon argues that ocean acidification is creating specific new challenges for international law. While extant multilateral environmental agreements can serve as the basis for marine governance in this context, it is still unclear which should take the lead and what kind of new rules will need to be agreed upon to do so effectively. An obvious problem is that the cause of this problem – carbon emissions governed by the climate change regime – and the impacts – acidification of oceans and seas – are the subjects of different regimes. Gordon draws on the experience of the Convention on Biological Diversity to suggest avenues for strengthening the oceans and climate regimes so that they can more effectively respond to acidification. Connections across all three regimes – oceans, climate, and biodiversity – point to avenues for more effective marine governance of this aspect of global environmental change.

Marine governance will be conceived by some people as addressing the management challenges of the world's great oceans, but often it is about regional seas with proximity to large human populations. The latter are the objects of analysis in Chapter 5, by Luciana Fernandes Coelho and Nata Tavonvunchai. Coelho and Tavonvunchai maintain that regional seas programmes – formalised agreements among littoral states, often under the auspices of the United Nations or other international organisations – are important regimes for ocean governance at the regional level. In particular, in addition to their normal focus on addressing traditional forms of ocean pollution, regional seas programmes provide opportunities for storing carbon within marine habitats. They can do this by shifting focus towards ecosystem-based management of marine environments, in the process protecting those habitats, such as seagrass beds and mangroves, which absorb and retain on the order of one-half of all carbon stored in marine (including coastal) ecosystems. Coelho and Tavonvunchai argue that preserving these 'blue carbon' ecosystems can contribute to the management objectives of both the oceans and climate regimes. A number of regional seas programmes have adopted ecosystem-based management schemes, the most prominent examples being marine protected areas and integrated coastal zone management. These schemes have the effect of promoting blue carbon storage without necessarily trying to do so specifically. This implies that more conscious efforts to protect, and eventually enlarge, blue carbon ecosystems would result in both more effective marine governance amidst climate change and more effective governance of the climate change problem itself, thanks to regional seas.

Adding to Coelho and Tavonvunchai's analysis of blue carbon and regional seas programmes, in Chapter 6, Gabriela A. Oanta explores the marine ('blue') dimensions of the European Union's (EU) European Green Deal, a nearly continent-wide initiative to realise environmental sustainability in general and carbon neutrality in particular. Towards that end, the European Green Deal has implied and explicit marine components – implied because maritime considerations matter for the programme's wider objectives and explicit because it includes specific maritime sectoral policies and the union's Integrated Maritime Policy. Oanta argues that the European Green Deal should be developed in ways that comport with the EU's policies that affect oceans and seas. In particular, she calls for policies and laws that move the union beyond its established Integrated Maritime Policy towards achieving coordinated climate-related objectives through the European Green Deal, the European Climate Law, and promotion of the broader blue economy.

Whether efforts to address the marine challenges of global environmental change are likely to be successful will, to a substantial extent, be a function of leadership by actors and institutions. Their strategies for leadership are examined in Chapter 7 by Małgorzata Zachara-Szymańska. Zachara-Szymańska defines leadership in terms of multilevel activity directed towards building capacities to act. She argues that leadership is central to marine environmental governance in identifying problems and building trust among actors that, without that leadership, might face difficulties in working collectively. Leadership, or the lack of it, can be exercised by individuals, such as national leaders, corporate executives, and activists, with roles in the regimes and institutions that formulate and implement marine environmental governance. Through their leadership, they can promote new rules and measures for action, build trust among the entities that must carry out action, and address fragmentation among those entities. Zachara-Szymańska explores several dimensions of leadership: structural leadership among states and international organisations, which is often premised on the distribution of material power; cognitive leadership, which is focused on how knowledge is created, distributed and transformed, in the process affecting conceptual frameworks that guide or at least affect policies; and relational leadership, a dimension that highlights the important interactions among actors that are central to whether marine environmental governance will be effective.

Non-state actors in marine environmental governance

International law and regimes are absolutely key to marine environmental governance. For example, the law of the sea has applied for centuries out of necessity, and as maritime affairs among states became more complex, it was codified and clarified in UNCLOS. But understanding marine environmental governance, not least in the context of global environmental change, also requires that we consider the many roles played by non-state actors. The chapters in Part 3 aim to do this, starting with Chapter 8 by Alice B.M. Vadrot. Vadrot's chapter focuses on experts and the scientific knowledge that they bring to bear in marine environmental governance. Her chapter introduces a number of prominent features of expertise and science in marine governance. Indeed, without experts and the ideas that they bring to the process, it would be all but impossible to apprehend the environmental challenges of marine governance, including those manifested in global environmental change. At the same time, however, experts and the science that they produce are often 'shaped' by political, legal, and economic contexts. In short, much as other ideas in marine environmental governance may be 'constructed' by powerful social and other forces, so, too, may be the experts who produce the science that informs governance policy be influenced by those same forces.

Civil society, including nongovernmental organisations, publics, and individuals, are the subject of Chapter 9, by Andrea Quirino Steiner, Elia Elisa Cia Alves, and Luís Paulo Santana. The role of civil society in environmental governance generally increased markedly from the early 1990s as democracy took root in more countries (some of which have since experienced retrenchment towards less democratic, more authoritarian forms of rule). The growing influence of civil society has similarly been the case in marine governance. As Steiner, Alves and Santana note, civil society operates at all levels of governance – including the local, the regional, and the international – through a wide variety of actors – from individuals and nongovernmental organisations to social movements and informal networks. Civil society actors operationalise their influence on marine environmental governance through various means, including diplomacy, advocacy, education, and provision of technical information. They have done so in a wide range of issues areas, such as management of fisheries, conservation of biological diversity, 'policing' activities on the high seas, and surveying – and sometimes cleaning up – pollution. Civil society actors have been successful in influencing marine environmental governance on many occasions; at others, they have been less successful. Through case studies, Steiner, Alves, and Santana get to the bottom of how and why.

In Chapter 10, E. Carina H. Keskitalo, Dmitri L. Lajus, and Lars H. Gulbrandsen draw on the experience of certification by the Marine Stewardship Council (MSC) in Russia to illuminate 'private' marine environmental governance. Private governance comprises non-state actors playing roles in achieving policy outcomes. As Chapter 10 notes, private governance became prominent in forest stewardship over the last several decades, with nongovernmental 'certification' programmes used to identify the origins and environmental impact of timber products. Similar programmes have been applied to marine certification, notably in certifying the sustainability of fisheries. Doing this has not been without its difficulties, for example with respect to the location of authority and coordination among actors, and the impacts of climate change at sea are magnifying difficulties by exacerbating conflict among actors and making the past an unreliable guide to the future. Focusing especially on the MSC's efforts to implement its certification initiatives in Russia, Keskitalo, Lajus and Gulbrandsen's chapter demonstrates that effective marine governance, including in its private forms, will require improved understanding of the local and regional context in which it is applied.

Extending the analysis of private marine environmental governance amidst climate change, in Chapter 11, Nata Tavonvunchai explores blue carbon financing by private-sector investors. As Tavonvunchai notes, the gap between available funding for marine conservation and actual funding for it is enormous. Public-sector financing, whether from national governments or international organisations, is – and almost certainly will remain – inadequate. Consequently, increasing levels of private-sector financing will be essential if marine conservation is to be successful. Tavonvunchai's chapter looks in particular at efforts to sequester carbon in marine sediments and biomass, notably in saltmarshes, mangrove forests, and seagrass beds. Exploiting these forms of blue carbon storage has some potential attraction to private investors, including insurers, asset managers, investment funds, and multinational corporations. A variety of funding models have been attempted and implemented, with 'blue carbon credits' being the most common. However, many blue carbon projects are relatively unattractive to private investors, while others that are attractive have met with varying degrees of success. To increase and improve outcomes that lead to marine conservation, greater emphasis is needed to value natural assets and to make investments that are centred on local communities.

Governing marine environments and regions

Part 4 of the book expands our perspective on marine environmental governance by way of several studies that examine particular marine environments and regions. In Chapter 12, Greg Johnson, Amanda Alva, and Kelly Dunning look at coastal environments associated with some of the communities and countries that are most dependent on the seas. Johnson, Alva, and Dunning are particularly interested in understanding decision-making related to coastal and marine governance that impacts communities and environments with high sensitivity and exposure to hazards associated with global environmental change. While they want to know what happens at the national level, they also explore the role of women, indigenous people, and other vulnerable populations in low-income parts of Africa, Asia, the Caribbean, and Latin America. Their analysis shows that substantial progress is being made to realise policies that effectively address marine pollution, overfishing, and other ecological stresses. This is being achieved through participatory, flexible institutions. That said, they also find that governance capacity and policy enforcement can be lacking and are made more challenging by the impacts of climate change, not least sea level rise and worsening tropical storms.

In Chapter 13, Tony George Puthucherril provides further examination of coastal and near-shore habitats, in the process describing the role of 'interactive' marine environmental governance amidst global change. Puthucherril builds on arguments from the previous chapter to remind us that coastal and near-shore environments are increasingly threatened due to human activities, such as pollution, extraction of resources, acidification of seawater and rising sea levels. New approaches to marine governance are needed to address these and other shoreline and near-shore environmental threats. Puthucherril argues that interactive governance and integrated coastal zone management are critical in this respect. Interactive governance is a process whereby problems are addressed through interactions among civil, public, and private stakeholders. Integrated coastal zone management is an iterative process involving informed participation and cooperation among actors. Using case studies from a number of countries, the chapter shows that both interactive governance and integrated coastal zone management provide promising avenues for more effectively protecting the integrity of coastal and near-shore environments in the long term.

Islands are among the geographic features of the world's oceans and seas that are most vulnerable to global environmental change, especially the consequences of global warming and climate change. In Chapter 14, Anemoon Soete looks at islands' vulnerabilities associated with rising seas, vulnerable shorelines, and territorial integrity. As Soete notes, the impacts felt by islands are both local and global: they are experienced locally, but they are intimately connected to issues that are affecting the global oceans, such as overexploitation of fisheries and illegal, unreported, and unregulated fishing. Most profoundly, islands are 'canaries in the coalmine' of climate change: they are directly impacted by changes to the seas and, especially in the case of low-lying islands, the threats that they face from climate change are potentially existential. These impacts are already being felt and will undoubtedly grow much worse in the coming decades, raising profound questions of fairness given that the most vulnerable island countries are among those that have contributed the least to causing climate change. As the chapter reveals, these communities are already trying to address the many growing challenges that are being exacerbated by global environmental change. Their experiences point to the need for changes in extant practices of marine environmental governance and very likely entirely new approaches specifically suited to a future of continuous change.

Many of the world's island countries are ringed by, and often formed by, coral reefs, which are among the ecosystems that are suffering the most severely from climate change. Chapter 15, by Pedro Fidelman, is devoted to understanding some of the governance challenges associated with coral reefs, in the process highlighting the importance of interactions among the ecological and social aspects of marine governance. Fidelman points out that coral reefs around the world's oceans are degrading quickly as a consequence of global climate change and local factors, such as water pollution and overexploitation. Innovative marine governance is required to reverse the degradation, where possible, and to minimise the decline where it is not. That governance is so far lacking. Using Australia's Great Barrier Reef as a case study, Fidelman explores 'social-ecological reflexivity', an idea that he argues has the potential to bolster coral reef management and increase its chances of success. Combining social-ecological reflexivity with aspects of institutional analysis, he proposes a conceptual framework that illuminates the circumstances and conditions under which innovative governance for coral reefs affected by global change might be realised.

Many of the coral reefs that are threatened by climate change are found in Southeast Asia, the subject of Chapter 16, by Anastasia Kuswardani and Achmad Poernomo. Kuswardani and Poernomo examine fisheries and aquaculture in the region, which together are the primary sources of dietary protein for the majority of the people living there, and which make up large parts of the economies of the region's countries. Climate change has severely impacted Southeast Asian fisheries and aquaculture, in turn adversely affecting the livelihoods of people in fishing villages and beyond. Responses to the adverse changes to fisheries and aquaculture in the region vary from country to country, although there are some vehicles for cooperation, for example in the context of the Association of Southeast Asian Nations. Cooperation among the region's countries is vital given their many shared borders and marine ecosystems. Chapter 16 reviews the nature of the threat to fisheries and aquaculture in the region and explores the means by which countries there are managing the impacts of climate change.

Shifting hemispheres, in Chapter 17 Marko Joas, Henrik Ringbom, and Nina Tynkkynen look at the impact of global environmental change on the Baltic Sea. As a consequence of very complex ecological and societal characteristics, the Baltic Sea is a case study of multilevel and 'network-type' systems of marine environmental governance. As such, it can offer not only lessons for improved local management but also experiences that may be transferrable to other regions and ecosystems. In addition to describing the threats that global environmental change presents to the Baltic Sea, Joas, Ringbom, and Tynkkynen examine regulatory institutions for the sea and explore the evolution of the region's multilevel environmental governance system. They ask how approaches to marine environmental governance for this regional sea are challenged by climate change combined with existing ecological threats, notably eutrophication caused by pollutants entering the sea from the surrounding riparian system. Their chapter highlights the potential utility and adaptability of existing management systems for regional seas in a future characterised by global environmental change.

In contrast, in Chapter 18, Emma Avoyan and Jan van Tatenhove draw in existing experiences with managing the impacts of global environmental change on the Black Sea to show where building on current practice may be less fruitful. The Black Sea is one of the most polluted of the world's seas, a situation that is being exacerbated by the effects of climate change. International cooperation among the littoral countries will be essential if the pollution aspects of decline are to be addressed. However, just that sort of cooperation is sorely lacking due to geopolitical disagreements, imbalances in resources and lack of willingness among the governments of surrounding countries. Effective marine governance will require major institutional reform and a reassessment by local governments of the importance of having a Black Sea that is environmentally healthy.

Moving farther from the tropics, Chapters 19 and 20 look at challenges to marine environmental governance in polar seas, beginning with an analysis by Nicole Wienrich, Hugh McDonald, Arne Riedel, Tim Packeiser, and Janos Hennicke of the challenges of extreme change in the Arctic and Southern Oceans. In both regions, warming seas, declining sea ice and acidification of waters are adversely affecting the distribution of marine species. Although the specific long-term impacts on fisheries are uncertain, the general increasing pace and scale of change appear to be inevitable. Due to the loss of sea ice, human activities in polar seas, such as shipping, fishing, tourism, and exploration for (and extraction of) hydrocarbons, are increasing and will accelerate in the future. Chapter 19 describes existing governance frameworks for polar seas and notes that ecosystem-based approaches are being promoted by the Arctic Council and the Antarctic Commission for the Conservation of Antarctic Marine Living Resources. Existing barriers to effective conservation of polar marine ecosystems include the desire among actors to exploit newly accessible resources, a lack of international cooperation, and a shortage of scientific information. Wienrich *et al.* argue that the use of more collaborative processes and the implementation of new management schemes, including area-based management measures, may help to overcome those barriers.

Chapter 20 by Benjamin Hofmann builds on the preceding chapter by focusing on the management of increased shipping activities in the Arctic Ocean, which is changing dramatically due to regional warming and declines in the extent and thickness of sea ice. Increased shipping, which has been made possible by climate change, poses new threats to Arctic Ocean ecosystems, including the growing danger of accidental spillage of fuel oil and emissions of black carbon – itself a contributor to climate change – into the atmosphere. In response to these threats, traditional state-based governance approaches, notably in the context of the International Maritime Organisation (IMO), are being supplemented by newer forms of transnational governance that involve nongovernmental and corporate actors. For example, the latter have pledged to end gradually the use of the most polluting fuels and to forgo trans-Arctic shipping along some routes. Hofmann argues that this combination of traditional approaches to marine environmental governance with newer innovative approaches can serve as an example for governing other marine environments affected by global change. In particular, he advocates innovative governance involving non-state actors in the region, which in turn can stimulate traditional state-oriented institutions to bring their resources to bear on addressing the impacts of global environmental change.

The final chapter in Part 4 of the book moves from littoral and regional seas to marine environments far from shores. In Chapter 21, Mitchell Lennan explores pelagic ecosystems and the expansive open oceans that cover half of Earth's surface. These areas are far from immune to the impacts of global environmental change, as exemplified by ocean warming and acidification. Those impacts, combined with others, such as deoxygenation, pollution, and overexploitation, undermine pelagic ecosystems. As those ecosystems suffer, they diminish the ability of the oceans to compensate for humanity's past and ongoing carbon dioxide emissions. Lennan describes the existing governance regime for the high seas, which he characterises as fragmented and ad hoc. He uses an analysis of regional fisheries management organisations to illustrate governance efforts on the high seas. These organisations have the potential to aid in adapting to the movement of fisheries, which is likely to arise due to climate-induced environmental changes, through cooperative management and designation of marine protected areas. A legally binding agreement on protecting biodiversity beyond national jurisdictions would, Lennan argues, enhance the potential for more effective environmental governance on the high seas as the impacts of global environmental change become more pronounced.

Emerging issues in environmentally sustainable marine governance

Part 5 of the book shifts from a geographic orientation to a perspective that looks at particular challenges for marine environmental governance amidst global environmental change. In Chapter 22, Judith van Leeuwen, Tony R. Walker, and Joanna Vince examine uncertainty and 'multiplicity' in the massive and growing challenge for global marine governance of plastic pollution. They note that plastic pollution in oceans and seas is having long-term impacts on marine organisms and ecosystems, yet there are a few effective strategies for dealing with that pollution. Importantly, much of the problem originates on land; about four-fifths of the plastics in the world's oceans come from land-based sources. Responding to this problem has involved many different actors, both state and non-state, but attempts at effective governance have, for the most part, failed. In their chapter, van Leeuwen, Walker, and Vince highlight ongoing debates regarding the best means by which to manage marine plastic pollution effectively, and they describe the multiplicity of governance schemes across sectors, scales, and types of actors. They argue that these schemes have the potential to become more effective through experimentation, learning, and critical monitoring leading to transformative changes in the ways that plastics are produced, used, and discarded.

Building on Chapter 20's discussion of shipping in the Arctic Ocean, in Chapter 23, Judith van Leeuwen and Jason Monios look at maritime commerce and transport more broadly, in the process highlighting the imperfect match between the policies of the IMO and the challenges posed by climate change to the world's oceans. As the principal regulator of international maritime transport, the IMO is an obvious venue for devising policies for addressing climate-related challenges for shipping. However, as van Leeuwen and Monios argue, the IMO is not up to the task for a number of reasons. It has always advocated a staid approach to addressing environmental concerns related to shipping, relying, for example, on existing technologies rather than pushing members to adopt new ones. Although the IMO has called for improvements in the energy efficiency of ships, it has not embraced the decarbonisation of maritime transport. Indeed, greenhouse gas emissions from shipping continue to increase, and even the IMO does not expect them to fall before 2050. Because the IMO has lacked initiative and consequently lacks legitimacy in this issue area, actions to limit climate-changing pollution from shipping have fallen to often-fragmented schemes implemented by regional organisations, ports, and private actors. If van Leeuwen and Monios are right, pressure will build to coordinate global efforts to address the climatic impacts of shipping, and the IMO may eventually be forced to rethink its relatively go-slow approach.

When we think of large human-made objects floating on the world's oceans, we naturally imagine ships. However, as the impacts of climate change on the oceans, not least sea level rise, become more prominent in the future, it is likely that whole communities will move to floating cities. These will often comprise large platforms floating on the sea surface, enabling them to rise as seas rise. A number of countries are considering the development of floating cities as strategies for adapting to climate change. The legal and regulatory issues that arise from the development of floating cities are taken up in Chapter 24, by Otto Spijkers and Chuxiao Yu. As Spijkers and Yu describe it, the development of floating cities is a realistic strategy for adapting to climate change, especially its impacts upon low-lying littoral communities. The technology for floating cities will be relatively easy to work out, but less simple may be the legal, regulatory, and international political considerations. Chapter 24 considers how and whether existing international law will be able to guide the governance of future floating cities. It evaluates how to classify floating cities in the context of international law and examines the consequences of those classifications, particularly in the context of UNCLOS.

Floating cities are one potential route to development in the marine context as the impacts of global environmental change manifest themselves. More broadly, we should consider the wider role that oceans and seas play in *sustainable* development. That is the question taken up in Chapter 25, by Shailly Kedia and Kapil Narula. Their chapter focuses on the United Nations' Sustainable Development Goals (SDGs), and the oceans-oriented SDG14 in particular. SDG14, the only one of 17 SDGs that is directly related to the world's oceans, aims to 'conserve and sustainably use the oceans, seas and marine resources for sustainable development' through realisation of numerous sustainability targets. The relationship between sustainable development and oceans is two-way. The oceans are threatened by environmentally unsustainable activities globally leading to ocean pollution, collapse of fisheries, ocean warming, acidification, and eutrophication, as noted in several of the chapters in this book. Alternatively, the oceans are vehicles for realising environmentally sustainable development. Put simply, environmentally sustainable development can help to protect oceans, and healthy oceans can help in the achievement of sustainable development. As Kedia and Narula show, implementing SDG14 will require proactively addressing the oceanic causes and impacts of global environmental change. Doing this is difficult due to gaps in marine environmental governance and a shortage of capacity – technology, funding, and so forth – in many of the countries that are most reliant on successful implementation of SDG14.

The final chapter in Part 5 and the penultimate chapter of the book considers the normative aspects of marine environmental governance, in the process highlighting the vitally important roles of ethics, justice and human rights, all of which will become more important and more challenged as environmental change is felt more forcefully around the world. In Chapter 26, Konrad Ott, Colin von Negenborn, and Nele Matz-Lück argue that the ethics of marine environmental change should provide the grounding for political and policy responses. They identify what they consider to be vital 'ethical building blocks' and consider how these might be translated into legal practice in the marine context. They ask whether human rights approaches and viewing the oceans as a legal entity have utility in marine environmental governance. Their analysis leads them to conclude that the notion of a 'covenant' of leading states, within the context of UNCLOS, has the potential to take seriously claims for environmental justice while having sufficient political viability to shape marine environmental governance amidst global environmental change. The initiative that they advocate, which they believe is implicit in SDG14, involves a 'global initiative for ocean recovery and restoration', aggressive cuts in carbon dioxide and other pollutants, expansion of marine protected areas and protection of coastal ecosystems, and the sustainable regulation of fisheries. Doing all of these things will promote ethics, justice and human rights while also realising the objectives discussed by other contributors to this book.

Conclusion

Taken together, the chapters reveal a picture of marine environmental governance in the new era of global environmental change. They highlight longstanding challenges of managing marine ecosystems, resources, and relationships with human users of the sea, whether they are individuals, civic organisations, or nation-states. More to the point, they show, first, how global environmental change is increasing and exacerbating those challenges and, second, whether existing mechanisms for marine environmental governance are and will be able to cope with that change. In most cases, they conclude that those mechanisms are not up to the task, indeed that they will sometimes be obstacles to more effective marine environmental governance in the future. At the very least, in nearly all cases existing mechanisms for marine environmental governance will need to evolve. Many of the chapters are helpful in that they point us to what it will take for that to happen. As such, the chapters in this book offer valuable lessons about marine environmental governance today and in the future – a future that will be characterised by environmental change of global proportions. An important question is whether the solutions that are identified and proposed can and will take root. Based on what the contributors to this book have determined and argued, the future prospects for effective marine governance look decidedly mixed, at best, given the mounting challenges of global environmental change. Might the experience of trying to develop an effective global regime to combat climate change offer any clues about the prospects for marine environmental governance? That is a question taken up in the final chapter.

References

Biermann, F. (2022) 'Earth system governance: world politics in the post-environmental age', in P.G. Harris (ed), *Routledge Handbook of Global Environmental Politics*, 2nd ed. Routledge, London, pp. 283–94.

Delmas, M.A. and Young, O.R. (2009) 'Introduction: new perspectives on governance for sustainable development', in M.A. Delmas and O.R. Young (eds), *Governance for the Environment: New Perspectives*. Cambridge University Press, Cambridge, pp. 4–40.

Grip, K. (2017) 'International marine environmental governance: a review', *Ambio* vol 46, pp. 413–427.

Harris, P.G. (ed) (2019) *Climate Change and Ocean Governance: Politics and Policy for Threatened Seas*. Cambridge University Press, Cambridge.

Intergovernmental Panel on Climate Change. (2019) *Special Report on the Ocean and Cryosphere in a Changing Climate*, https://www.ipcc.ch/srocc/

Intergovernmental Panel on Climate Change. (2021) *Climate Change 2021: The Physical Science Basis*. Cambridge University Press, Cambridge.

Thomas, R. (2021) 'Welcome to the 'plastisphere': the synthetic ecosystem evolving at sea', *The Guardian*, 11 August, https://www.theguardian.com/environment/2021/aug/11/welcome-to-the-plastisphere-the-synthetic-ecosystem-evolving-at-sea

PART 2

International law, regimes, and leadership in marine environmental governance

2
THE GLOBAL OCEANS REGIME
The law of the sea and beyond

Erik van Doorn

Law is very often reactive. As an instrument of policymakers, regulation aims to react to an aspect of society that needs improvement or encouragement. This is not at all different for international law. The challenge with international law is that, differing to a certain extent from a national legal system, the entities that make the rules, those that are supposed to obey these rules and those that enforce the very same rules are traditionally the same: states. There is thus no official hierarchy in the international legal system, and this might create problems regarding enforcement. In simple terms, international law of the sea deals with who gets what in the ocean (Rothwell and Stephens 2016). The body of international law of the sea consists of a wide range of international legal instruments. The cradle of these instruments is most often within an international organisation, such as the United Nations, its specialised agencies or other intergovernmental organisations. The instruments that contain rules concerning the law of the sea might be either binding or non-binding. These are all factors that play a role in the assessment of the law of the sea being able to consider all kinds of changes, among them global environmental change.

There are generally two ways in which law, and specifically international law of the sea, can cope with global environmental change. Either it is possible to deal with the change within the already existing legal framework or it is necessary to adapt the existing body of rules. Lawyers tend to prefer the former option because the traditional idea is that it is the duty of policymakers to change regulation or create new rules, akin to the latter option. The content of the adaptation of rules needs to be distinguished, however, from the possibility of adaptation in the first place. Whereas the former is indeed in the realm of policymakers, the latter is for lawyers to analyse. It is not the aim of this chapter to assess the value of norms on an ethical scale but rather to scrutinise the legal possibilities to deal with global environmental change (see Chapter 26). In order to do so, the following section of this chapter sets out the framework of the law of the sea with an emphasis on the 1982 United Nations Convention on the Law of the Sea (LOSC) as its backbone. The subsequent section sketches the Anthropocene and more specifically planetary boundaries as potential forms of global environmental change that the law of the sea will have to address in order not to become outdated. The chapter then dives into the stability of the law of the sea considering global environmental change and the feasibility of adaptation. The conclusion picks up the threads and summarises how international law of the sea copes with global environmental change.

DOI: 10.4324/9781315149745-4

The law of the sea framework

Before the law of the sea was codified in international treaties in the twentieth century, at least two millennia of legal interaction with the oceans had passed. Since '[i]nternational law at any given time represents an equilibrium between opposing pressures', this is also applicable to the development of the law of the sea (Oxman 2006, 850). The latter's entire history 'has been one of oscillation between freedom and restriction' (Rayfuse and Warner 2008, 408). More precisely, the history of the international law of the sea is one of the constant struggles between the extension of coastal sovereignty on the one hand and freedom of the seas on the other hand (O'Connell 1982, 1).

Emergence of the law of the sea

At the beginning of the seventeenth century, the Dutch fleet captured a Portuguese galleon in the Strait of Malacca and sold it as a prize in Amsterdam. A young lawyer, Huig de Groot, defended the Dutch (see Anand 1983, 77–89). He energetically argued on the basis of both ancient Greek texts and writings from sixteenth-century Spanish and Italian lawyers why states cannot and should not claim large parts of the ocean: 'the sea can in no way become the private property of any one because nature not only allows but enjoins its common use' (Grotius 1608/1916, 30). De Groot's idea for freedom of the seas is based on major, nowadays unrealistic, assumptions. First, no one is effectively able to occupy marine areas. Second, marine resources are inexhaustible. Third, one form of use might interfere with other uses only to a minimal extent (Koers 1973, 17). Fourth, possible negative human impacts on the quality of the marine environment would be negligible (Schrijver 1988, 93). With time, all of these assumptions clearly turned out to be incorrect.

Both the increased exploitation of the ocean and the onsets of global environmental change have been drivers to slowly move away from an absolute freedom-of-the-seas idea. Until the negotiations of the LOSC, most developments that diminished the freedom of the sea were due to the increased use of the ocean. It was only near the end of the twentieth century that global environmental change challenged the very foundation that had supported the law of the sea over centuries. Increased military use was the first incentive for coastal states to nibble away at the freedom of the sea. Van Bijnkershoek proposed in the eighteenth century that coastal states would be able to exercise control over a belt of ocean adjacent to their coasts the breadth of which 'extends as far as cannon will carry; for that is as far as we seem to have both command and possession' (Van Bijnkershoek 1744/1923, 364 [44 in the translation]). It took another two centuries before the next large encroachment upon the freedom of the seas occurred. Just after the Second World War, many American states began to claim rights over their continental shelves and the fishing grounds right above them (Treves 2015, 10–12; Proelss 2017, 411–415).

In the 1960s, a polarisation occurred between industrialised states with the technology to enlarge their reach over the ocean and thus increase territoriliasation, and those without the technological capacity. The Maltese delegation at the United Nations proposed in 1967 that the ocean floor beyond national jurisdiction be declared the common heritage of humankind, supported overwhelmingly by other former colonies and newly industrialising states. Proponents of the common heritage of humankind made clear that the idea is contrary to property and therewith also to territorial aspirations of states (Pardo and Christol 1983, 657). The Maltese proposal finally led to the third United Nations Conference on the Law of the Sea, which concluded in 1982 with the adoption of the LOSC (Morin 2017). The last president of the Conference proclaimed at this occasion that the negotiators 'achieved [the] fundamental objective of producing a comprehensive constitution for the oceans which will stand the test of time' (Koh 1982, xxxiii). So far, he has been at least close to the truth. The LOSC divides the ocean into different zones and subsequently creates a regulatory framework for issues that permeate the very boundaries that it sets.

Maritime zones

All maritime zones are normally measured from the low-water baseline (LOSC, Art. 5). Bays, reefs, and river mouths might lead to deviations from this default option (LOSC, Arts. 10, 6, and 9, respectively). In cases 'where the coastline is deeply indented and cut into, or if there is a fringe of islands along the coast in its immediate vicinity', a coastal state may also draw straight baselines (LOSC, Art. 7). Since international law does not define terms like 'deeply indented and cut into' or 'immediate vicinity', there is some flexibility for coastal states to draw these lines. Global environmental change will lead to, for instance, sea-level rise or the erosion of coral reefs, which may have a direct effect on the legitimacy of the baseline (see Chapters 13 and 15). All waters landward from the baseline, for instance, rivers and lakes, are part of the state's internal waters (LOSC, Art. 8(1)). Coastal states themselves will need to publish their baselines (LOSC, Arts. 5 and 16). Global environmental change may force coastal states to adjust their charts that contain the official baselines but the incentive to do so might be lacking (Lathrop 2015, 78).

Changing baselines can consequently cause changes to the boundaries of a coastal state's maritime zones. Since the rights of coastal states differ for every maritime zone, global environmental change might lead to situations where coastal states gain or lose rights in certain areas of the ocean. Coastal states can claim a belt of ocean with a breadth of up to 12 nautical miles, measured from the baseline, as territorial sea (LOSC, Art. 3). This territorial sea falls under the sovereignty of the coastal state with, as its most significant exception, the customary right of innocent passage by ships from all states (LOSC, Art. 17). Innocent passage refers to 'traversing the sea' (LOSC, Art. 18(1)(a)) in a manner 'not prejudicial to the peace, good order or security of the coastal State' (LOSC, Art. 19(1)). Coastal states shall not interfere with this passage (LOSC, Art. 24).

Archipelagic states enjoy in their archipelagic waters rights that are akin to those in a territorial sea (LOSC, Arts. 46 and 49). Yet, third states do have a right of archipelagic sea-lane passage (LOSC, Arts. 53 and 54). Whether a feature qualifies as an island or just as a rock is another matter. The latter can, by definition in the law of the sea, not 'sustain human habitation or economic life' of its own (LOSC, Art. 121). There is no definition of what this might mean. Global environmental change, when it arrives in the form of a rising ocean or dramatic changes in temperature, might have an impact on whether a certain geographic feature can sustain human life by itself and thus whether it can qualify as an island (see Chapter 14). A rock can have its own territorial sea but no exclusive economic zone or continental shelf, whereas an island can have the latter two maritime zones as well. Entitlement to an exclusive economic zone or continental shelf might have a significant impact on a state's ability to derive economic benefits from the ocean.

The claims from 1945 onwards of coastal states over large parts of their continental shelves and the fishing grounds above them lead to two new regimes. The sovereign rights over the natural resources of the continental shelf were the most direct consequence of these coastal states' claims (LOSC, Art. 77). These rights extend to 200 nautical miles from the coastal state's baseline (LOSC, Art. 76). If a coastal state establishes rights to a continental shelf beyond 200 nautical miles because the geological continental shelf extends beyond that distance, it will have to share the benefits from its exploration in this area (LOSC, Art. 82). The coastal state does not have to claim its continental shelf (LOSC, Art. 77(3)). Since the regime for the continental shelf covers the ocean floor, the regulations concerning the exclusive economic zone are mainly of interest for the water column. In their exclusive economic zones, coastal states have sovereign rights over the exploration and exploitation of natural resources as well as over other economic activities such as water, tidal, and wind energy (LOSC, Art. 56(1)(a)). Furthermore, they have jurisdiction over the protection of the marine environment, marine scientific research, artificial islands, installations, and structures (LOSC, Art. 56(1)(b)). Yet, 'the coastal State shall have due regard to the rights and duties of other States' (LOSC, Art. 56(2)). Except for the exceptions that give the coastal state exclusive rights, the freedoms of the high seas apply in the exclusive economic zone (LOSC, Art. 58).

De Groot's freedoms of the sea apply to the ocean beyond the continental shelf and the exclusive economic zone. The LOSC lists six of these freedoms, which are not exclusive: freedom of navigation, overflight, fishing, scientific research, to lay submarine cables and pipelines, and to construct artificial islands and other installations (Art. 87). The ocean floor beyond the continental shelf is the common heritage of humankind, following Malta's proposal in 1967. Fishing on the high seas only accounts for about 10% of global marine catches but the freedom of fishing in this part of the ocean is no longer absolute (FAO 2020, 94; LOSC, Arts. 116–119; see Chapter 21). So, whereas the living resources beyond national jurisdiction are traditionally subject to the freedom of the sea, the mineral resources of the ocean floor beyond national jurisdiction are the common heritage of humankind (LOSC, Arts. 136, 1(1)(1), and 133). One would think that other genetic resources than fish would also fall under the freedom of the seas but since these resources were only discovered in the deep ocean in the late 1970s, the regulations concerning their harvest are still under negotiation.

Overarching marine issues subject to international regulation

Beyond the division of maritime zones, there are issues that spread throughout the ocean without regard to human-made boundaries. Marine pollution is one of them. States generally agree that they have an obligation to protect and preserve the marine environment (LOSC, Art. 192). This obligation mainly materialises in the duty to prevent, reduce, and control marine pollution (LOSC, Art. 194). Marine pollution is defined as 'the introduction by man, directly or indirectly, of substances or energy into the marine environment, including estuaries, which results or is likely to result in such deleterious effects as harm to living resources and marine life, hazards to human health, hindrance to marine activities, including fishing and other legitimate uses of the sea, impairment of quality for use of seawater and reduction of amenities' (LOSC, Art. 1(1)(4)). This definition is broad, which shows on more than one occasion. Marine pollution is not limited to direct introduction but also comprises indirect introduction and thus includes marine pollution from land, from vessels, from installations and devices and through the atmosphere (LOSC, Arts. 194(3) and 207–212). It deals not only with substances but also with energy. The latter secures that light, heat and sound could also qualify as marine pollution.

States should do everything in their power to prevent the spread of marine pollution to parts of the ocean under the jurisdiction of other states or beyond these (LOSC, Art. 194(2)). Global and regional cooperation, also through international organisations, shall lead to further international rules, standards and recommended practices and procedures for the protection of the marine environment (LOSC, Art. 197; see Chapter 5). The competent international organisation for when it comes to vessel-based pollution is the International Maritime Organization (LOSC, Art. 211; see Chapter 23). Enforcement of the rules that are applicable to vessels can take place by flag states, port states, and coastal states, depending on the location of the vessel (LOSC, Arts. 217–220). The duty to cooperate through an organisation is even stronger for states that border an enclosed or semi-enclosed sea (LOSC, Art. 123(b); Oral 2015; see Chapters 17–19).

Furthermore, states should notify each other immediately when there is the likeliness of being affected by the pollution (LOSC, Art. 197). In order to monitor any pollution, states shall endeavour 'as far as practicable' the observation, surveillance, measurement, evaluation, and analysis of pollution risks and effects (LOSC, Art. 204). In addition, a state that has reasonable grounds to believe that activities 'may cause substantial pollution or significant and harmful changes to the marine environment ... shall, as far as practicable, assess the potential effects of such activities' (LOSC, Art. 206). The term 'practicable' in both provisions appears to give the state in question a significant amount of discretion but the International Court of Justice considers the duty to conduct an environmental impact assessment to be customary international law (International Court of Justice 2010, para. 204; Harrison 2017, 31–35). It follows that states shall promote studies, scientific research programmes, and exchange of information concerning marine pollution (LOSC, Art. 200; Hubert 2015, 316–319 and 322). This duty is again explicit for states surrounding enclosed or semi-enclosed seas (LOSC, Art. 123(b)). Newly industrialising states shall receive support to do so (LOSC, Art. 202). The right to conduct marine scientific research is irrespective of the geographical location of a state (LOSC, Art. 238). There is a duty to actively promote the publication and dissemination of information and knowledge (LOSC, Art. 244).

Legal aspects of the marine Anthropocene

With no regard to whether the management of humankind's interactions with its marine environment succeeds, humankind has nowadays such an impact on our planet that scientists suggested to call the current geological era the 'Anthropocene' (Vidas et al. 2015, 5–7). Such a new geological era has the potential to shake up the foundations of international law and thus international law of the sea and marine governance in a broader sense. Law, generally reactive, benefits from stability in both international relations and the natural environment context in which it operates. Both the territorial integrity of states and the changing geopolitical situation because of global environmental change will thus have an effect on the functioning of international law (Vidas et al. 2015, 4). The substantial amounts of water on the surface of the Earth, particularly its great oceans, is one of the main reasons for a well-regulated and stable climate (Vidas et al. 2015, 7 and 8). One could actually trace humankind's impact on the planet back to the moment when the sea level stabilised (Vidas et al. 2015, 11). At that moment, 'the oceans were still essentially pristine, and climate and sea levels were essentially stable (barring fluctuations, usually regional, in such phenomena as floods and droughts)' (Vidas et al. 2015, 12). Therefore, the question is how to keep the living conditions for humankind also stable enough in the future.

To keep the Earth liveable for current and future generations, a group of scientists proposed the idea of defining planetary boundaries (Rockström et al. 2009a, 2009b). They identified nine processes of the Earth system for which they tried to quantify a boundary, some of which have a clear marine component. Staying within these nine planetary boundaries would assure 'a safe operating space for humanity' (Rockström et al. 2009a, 472). The framework should not lead to the conclusion that the planetary boundaries are strictly separate from each other but rather the contrary. The interdependence of the boundaries is such that a change in one process of the Earth system might have an effect on another process, either large or small (Rockström et al. 2009a, 474, 2009b). The nine processes of the Earth system with planetary boundaries are stratospheric ozone depletion, atmospheric aerosol loading, climate change, ocean acidification, change in land use, freshwater use, biochemical flows, novel entities, and integrity of the biosphere (Steffen et al. 2015).

At first sight, stratospheric ozone depletion and atmospheric loading appear to lack an obvious connection with marine governance. Ozone depletion seems furthest away (e.g., although it may adversely affect phytoplankton) from the ocean but a decrease of ozone-depleting substances like chlorofluorocarbons in the stratosphere might in the long-term mean that there will be an increase of these substances in the ocean (Wang et al. 2021). Interactions between the lower atmosphere and the upper ocean connect atmospheric aerosol loading also with the ocean (Brooks and Thornton 2018). Yet, both these processes have so far no direct bearing on marine governance nor the other way around. This is very different concerning the processes of climate change and acidification of the ocean (see Chapters 3 and 4).

Change in land use, freshwater use, biogeochemical flows and novel entities all have a clear terrestrial component. Yet, as the vast majority of marine pollution comes from land, the relationship of these processes with the ocean becomes obvious (Harrison 2017, 64 and 65). The use of fertilisers and agricultural run-off are components of the global cycles of nitrogen and phosphorus; the two elements that quantify biogeochemical flows in the planetary boundaries framework. For the phosphorus cycle, the amount of phosphorus that reaches the ocean through freshwater systems quantifies the planetary boundary (Steffen et al. 2015). Yet the regulation of this kind of pollution in the LOSC is relatively weak. States shall only 'endeavour' the harmonisation of their policies and minimisation of 'the release of toxic, harmful or noxious substances' shall occur 'to the fullest extent possible' (LOSC, Art. 207). The latter might sound positive but what threshold it exactly encapsulates is for the states to decide. The 1995 Global Programme of Action for the Protection of the Marine Environment from Land-based Activities, adopted under the auspices of the United Nations Environment Programme, tries to fill in the framework of the LOSC. The Global Programme itself is however non-binding, so it is up to states to implement the measures that it contains. There remain weaknesses in this approach, but the process is an ongoing one (Harrison 2017, 70–73).

The process of novel entities is a broadening of what was the process of chemical pollution in earlier versions of the framework. Chemical pollution from land is often regulated in conventions that are not specifically applied to the marine environment but do have an effect on it. Explicitly mentioned are the persistent organic pollutants and the 2001 Stockholm Convention as its international regulatory framework (Rockström et al. 2009b). The Stockholm Convention aims at the elimination of certain persistent organic pollutants, which also might pose a threat to marine ecosystems. The Convention furthermore restricts import and export, and regulates permits for trade. It sets relatively high environmental standards and includes a precautionary approach (Harrison 2017, 73 and 74). The 2013 Minamata Convention on Mercury takes a similar approach (Minamata Convention on Mercury; Harrison 2017, 74 and 75). The Convention is named after the Japanese city where thousands of people suffered from mercury poisoning due to industrial run-off and the consumption of contaminated marine fish. Plastics comprise a further category of novel entities, one that is not yet subject to a specific global regime as such (see Chapter 22). Plastics can also contribute to climate change as a process, thus establishing a link between the planetary boundaries for novel entities and climate change (Stoett and Vince 2019).

The general obligation to protect and preserve the marine environment qualifies the sovereign right of states to exploit their natural resources (LOSC, Art. 193). Global environmental change and exploitation of marine resources mutually affect each other. The usage of the resources of the continental shelf, mainly oil and gas, has an obvious effect on our planet globally. The exploitation of fossil fuels itself is energy-intensive and has consequently an effect on our climate, the burning of these fuels even more so (see Chapter 20). The exploitation of the resources in the continental shelf is predominantly subject to the national laws of the coastal state because it is the coastal state that has the exclusive usage right over these natural resources (Harrison 2017, 209–217).

This is slightly different for marine living resources because of the simple fact that they are mobile. Overfishing might decrease marine biological diversity. Although unlikely to occur (Hannesson 2006, 44–47), the extinction of fish species would contribute to the loss of genetic biodiversity (Steffen et al. 2015). The other way around, global environmental change in the form of climate change heats up the ocean, which leads fish stocks to move either to colder or to warmer parts of the ocean depending on the preference of the species. With fish stocks migrating in and out of maritime zones of different states, international cooperation is thus required for the conservation and management of marine living resources (see Chapter 16). Yet the LOSC only supplies rules that apply either in the exclusive economic zone or in the high seas (Arts. 61–67, 116–120 and 123(a)).

To bridge the gap between regulations for the exclusive economic zones on the one hand and the high seas on the other hand, compatibility of measures needs to become a goal. For this, states negotiated an implementation agreement to the LOSC in the first half of the 1990s: the Agreement for the Implementation of the Provisions of the United Nations Convention on the Law of the Sea of 10 December 1982 Relating to the Conservation and Management of Straddling Fish Stocks and Highly Migratory Fish Stocks (FSA). The ecosystem approach, although not explicitly referred to in the Fish Stocks Agreement, could have been the main driver behind the rules on compatibility (FSA, Art. 7; Orrego Vicuña 1999, 78). The holistic approach to the ocean as an ecosystem emphasises the interactions between marine species but also forms the basis of the interactions between the legal regimes within and beyond national jurisdiction (Orrego Vicuña 1999, 176). States shall ensure that any 'measures do not result in harmful impact on the living marine resources as a whole' (FSA, Art. 7(2)(f)). The ecosystem approach might help to prevent further loss of functional biodiversity (Steffen et al. 2015). It helps as well that states' obligations towards the marine environment include specifically the protection and preservation of 'rare and fragile ecosystems as well as the habitat of depleted threatened or endangered species and other forms of marine life' (LOSC, Art. 194(5)). In addition, the International Tribunal for the Law of the Sea confirmed that the provisions in the LOSC on the protection and preservation of the marine environment also apply to fisheries (ITLOS 1999, para. 70).

Much research, now and in the future, will be necessary to establish where the planetary boundaries exactly lie and whether they possibly shift. Part of this research will be marine scientific research. This is true for potentially all nine boundaries, not the least for the two core processes: climate change and biodiversity (Steffen et al. 2015). Together with biogeochemical flows, these processes happen to be the two whose planetary boundaries have already been crossed (Rockström et al. 2009a, 473 and 474, 2009b). It is questionable whether the current regulations in the LOSC can fully cater for the new research and the instruments and platform that are necessary to carry it out (Hubert 2015, 329–331).

Stability and adaptation

The question now is whether the international law of the sea can effectively deal with ongoing global environmental change. For many of the processes that form the foundation of the concept of planetary boundaries, the LOSC provides the framework for the regulation of the oceanic component of human interactions with these processes. This framework does provide a certain flexibility (Boyle 2005; Barrett 2016). Beyond this flexibility, there is the option of amending the Convention (Arts. 312–316). Yet this has never happened and is unlikely to occur in the future because one objection would be enough to reject an amendment (LOSC, Art. 313(2); Boyle 2005, 564). Furthermore, the proposal of one amendment could open the gates for a flood of amendments, which state parties might not want to risk. Any rejected amendment might also interfere with the development of the particular norm as customary international law (Barrett 2016, 16). Parties to the LOSC, however, may conclude agreements that modify or even suspend the application of the Convention's provisions (Art. 311(3)). In the 1990s, state representatives negotiated two so-called 'implementation agreements' to the LOSC (one to implement Part XI in 1994 and the 1995 FSA). However, 'the politically prudent label of an "implementing agreement" is a euphemism for the word "amendment" which would have been more correct from the legal point of view' (Scovazzi 2000, 125). A third implementation agreement is on its way concerning gaps in the law in the sea for biological diversity beyond national jurisdiction. To keep the LOSC flexible enough to deal with aspects of global environmental change, this appears an attractive possibility (Buga 2015, 52–55).

With a look at the sources of the law of the sea, amendments and the adoption of implementation agreements are within the realm of written, conventional international law, but there are many more ways to further develop or even update the norms in the law of the sea. The LOSC, for example, was, to a large extent, a codification of customary international law (Bernhardt 1989, 321–326). The practice of states can also modify the Convention's provisions and can develop through the institutions that the LOSC created (Harrison 2015). Yet the Meeting of the State Parties to the LOSC has a fairly limited scope (LOSC, Art. 319(2)(e); Harrison 2015, 387–390). The United Nations General Assembly has a more significant role to play through the adoption of its annual resolutions on the law of the sea and sustainable fisheries but also through the meeting of the Open-Ended Informal Consultative Process on Ocean and Law of the Sea (Harrison 2017, 289–296).

The resolutions of the United Nations General Assembly amount to soft law as do many more decisions of international organisations within the United Nations system. In the field of fisheries, the Food and Agriculture Organization adopted many instruments, starting with the Code of Conduct on Responsible Fisheries in 1995. International plans of actions and guidelines followed (Boyle 2005, 572–574). Thus under the influence of soft law, 'subtle evolutionary changes in existing treaties may come about through the process of interpretation' (Boyle 2005, 574). Probably not even living up to possible definitions of soft law, the 2015 Sustainable Development Goals can nonetheless achieve effects of evolutionary interpretation akin to those of soft law (see Chapter 25).

Conclusion

Over the centuries, maybe even millennia, the law of the sea evolved to regulate human interactions with the marine environment. It did so often in response to global changes of a technological nature. The further people could go out on the ocean, the more interaction with the ocean evidently needed to be regulated. When technology, be that in the form of sailing vessels or equipment for the exploitation of resources, became available to more people, the need for regulation further increased. These developments culminated in the negotiation of the LOSC in the second half of the twentieth century, whose main achievements are not only the division of the ocean in maritime zones but also providing a framework for activities that straddle the very boundaries of these maritime zones. Global environmental change in the Anthropocene is now the main driver for the law of the sea as it currently stands.

Yet, the law of the sea has also had an indirect impact on global environmental change itself. Not only in the idea of the freedom of the seas but how the increases in coastal sovereignty over parts of the ocean have allowed humankind to take out immense amounts of both living and non-living resources on the one hand and the introduction of vast quantities of harmful substances no longer useful in the terrestrial environment on the other hand. Both these processes of extraction and pollution have contributed to global environmental changes, particularly in recent decades. Notwithstanding, the law of the sea has not only been a result and a cause of global environmental change. It can also be a remedy. The ocean has links to all the nine processes that form the basis of the planetary boundary framework, and so does the law of the sea. The law of the sea can thus play a role in contributing the humankind's efforts not to cross more than three boundaries and reduce the pressure on the planet through all of these processes. If it turns out that the current regulations are not fit for this purpose, there exist many mechanisms to further develop the law of the sea not only through new agreements but also through soft law and means of state practice.

Acknowledgements

The author would like to acknowledge the support of Kiel Marine Science and the European Union's Horizon 2020 project 'EuroSea' (grant number 862626). Christa Marandino, Jurgita Ovadnevaite, and Birgit Quack were very helpful in the supply of material concerning the interactions between the ocean and the atmosphere. Scarlett Sett Díaz deserves many thanks for the thorough proofreading.

References

Agreement Relating to the Implementation of Part XI of the United Nations Convention on the Law of the Sea of 10 December 1982 (1994) *International Legal Materials*, vol 33, pp. 1309–1327.

Agreement for the Implementation of the Provisions of the United Nations Convention on the Law of the Sea of 10 December 1982 Relating to the Conservation and Management of Straddling Fish Stocks and Highly Migratory Fish Stocks (1995) *International Legal Materials*, vol 34, pp. 1542–1580.

Anand, R.P. (1983) *Origin and Development of the Law of the Sea. History of International Law Revisited*. Martinus Nijhoff, The Hague etc.

Barrett, J. (2016) 'The UN Convention on the Law of the Sea: A "Living" Treaty?', in J. Barrett and R. Barnes (eds) *Law of the Sea: UNCLOS as a Living Treaty*. British Institute of International and Comparative Law, London.

Bernhardt, R. (1989) 'Custom and Treaty in the Law of the Sea', *Recueil des Cours*, vol 205 (1987), pp. 247–340.

Bijnkershoek, C. van (1744/1923) 'De dominio maris dissertatio', in *Opera Minora*, 2nd edition (translation by R. van Deman Magoffin). Oxford University Press, New York.

Boyle, A. (2005) 'Further Development of the Law of the Sea Convention: Mechanisms for Change', *International and Comparative Law Quarterly*, vol 54, pp. 563–584.

Brooks, S.D. and Thornton, D.C.O. (2018) 'Marine Aerosols and Clouds', *Annual Review of Marine Science*, vol 10, pp. 20.1–20.25.

Buga, I. (2015) 'Between Stability an Change in the Law of the Sea Convention: Subsequent Practice, Treaty Modification, and Regime Interaction', in D.R. Rothwell et al. (eds) *The Oxford Handbook of the Law of the Sea*. Oxford University Press, Oxford.

Food and Agriculture Organization of the United Nations. (2020) *The State of World Fisheries and Aquaculture 2020*, FAO, Rome.

Grotius, H. (1608/1916) *Mare liberum sive de iure quod Batavis competit ad indicana commercia. Dissertatio* (translation by R. van Deman Magoffin). Oxford University Press, New York.

Hannesson, R. (2006) *The Privatization of the Oceans*. MIT Press, Cambridge and London.

Harrison, J. (2015) 'The Law of the Sea Convention Institutions', in D.R. Rothwell et al. (eds) *The Oxford Handbook of the Law of the Sea*. Oxford University Press, Oxford.

Harrison, J. (2017) *Saving the Oceans through Law: The International Legal Framework for the Protection of the Marine Environment*. Oxford University Press, Oxford.

Hubert, A.-M. (2015) 'Marine Scientific Research and the Protection of the Seas and Oceans', in R. Rayfuse (ed) *Research Handbook on International Marine Environmental Law*. Edward Elgar, Cheltenham and Northampton.

International Court of Justice. (2010) 'Pulp Mills on the River Uruguay (Argentina v. Uruguay) Judgment of 20 April 2010', *I.C.J. Reports*, pp. 14–107.

International Tribunal for the Law of the Sea. (1999) 'Southern Bluefin Tuna (New Zealand v. Japan; Australia v. Japan) Provisional Measures Order of 27 August 1999', *ITLOS Reports*, pp. 280–301.

Koers, A.W. (1973) *International Regulation of Marine Fisheries: A Study of Regional Fisheries Organizations*. Fishing News, West Byfleet and London.

Koh, T.T.B. (1982) *A Constitution for the Oceans*, available at http://www.un.org/depts/los/convention_agreements/texts/koh_english.pdf

Lathrop, C.G. (2015) 'Baselines', in D.R. Rothwell et al. (eds) *The Oxford Handbook of the Law of the Sea*. Oxford University Press, Oxford.

Minamata Convention on Mercury. (2016) *International Legal Materials*, vol 55, pp. 582–616.

Morin, M. (2017) 'Les 50 ans de l'allocution d'Arvid Pardo aux Nations Unies', *Annuaire de Droit Maritime et Océanique*, vol XXXV, pp. 13–34.

O'Connell, D.P. (1982) *The International Law of the Sea*, vol I (edited by I.A. Shearer). Clarendon Press, Oxford.

Oral, N. (2015) 'Forty Years of the UNEP Regional Seas Programme: From Past to Future', in R. Rayfuse (ed), *Research Handbook on International Marine Environmental Law*. Edward Elgar, Cheltenham and Northampton.

Orrego Vicuña, F. (1999) *The Changing International Law of High Seas Fisheries*. Cambridge University Press, Cambridge.

Oxman, B.H. (2006) 'The Territorial Temptation: A Siren Song at Sea', *American Journal of International Law*, vol 100, pp. 830–851.

Pardo, A. and Christol, C.Q. (1983) 'The Common Interest: Tension between the Whole and the Parts', in R.St.J. Macdonald and D.M. Johnston (eds) *The Structure and Process of International Law: Essays in Legal Philosophy Doctrine and Theory*. Martinus Nijhoff Publishers, The Hague etc.

Proelss, A. (2017) 'Article 55 Special Legal Regime of the Exclusive Economic Zone', in A. Proelss (ed) *United Nations Convention on the Law of the Sea: A Commentary*. C.H. Beck, München.

Rayfuse, R. and Warner, R. (2008) 'Securing a Sustainable Future for the Oceans beyond National Jurisdiction: The Legal Basis for an Integrated Cross-Sectoral Regime for High Seas Governance for the 21st Century', *International Journal of Marine and Coastal Law*, vol 23, pp. 399–421.

Rockström et al., J. (2009a) 'A Safe Operating Space for Humanity', *Nature*, vol 461, pp. 472–475.

Rockström et al., J. (2009b) 'Planetary Boundaries: Exploring the Safe Operating Space for Humanity', *Ecology & Society*, vol 14(2), article 32.

Rothwell, D.R. and Stephens, T. (2016) *The International Law of the Sea*, 2nd edition. Hart, Oxford and Portland.

Schrijver, N.J. (1988) 'Permanent Sovereignty Over Natural Resources Versus the Common Heritage of Mankind: Complementary or Contradictory Principles of International Economic law?', in P. de Waart et al. (eds) *International Law and Development*. Martinus Nijhoff, Dordrecht etc.

Scovazzi, T. (2000) 'The Evolution of International Law of the Sea: New Issues, New Challenges', *Recueil des Cours*, vol 286, pp. 39–243.

Steffen et al., W. (2015) 'Planetary Boundaries: Guiding Human Development on a Changing Planet', *Science*, vol 347, article 1259855.

Stockholm Convention on Persistent Organic Pollutants. (2001) *International Legal Materials*, vol 40, pp. 532–563.
Stoett, P. and Vince, J. (2019) 'The Plastic-Climate Nexus: Linking Science, Policy, and Justice', in P.G. Harris (ed) *Climate Change and Ocean Governance: Politics and Policy for Threatened Seas*. Cambridge University Press, Cambridge.
Treves, T. (2015) 'Historical Development of the Law of the Sea', in D.R. Rothwell et al. (eds) *The Oxford Handbook of the Law of the Sea*. Oxford University Press, Oxford.
United Nations Convention on the Law of the Sea. (1982) *International Legal Materials*, vol 21, pp. 1261–1354.
Vidas, D. et al. (2015) 'What Is the Anthropocene—And Why Is It Relevant for International Law?', *Yearbook of International Environmental Law*, vol 25, pp. 3–23.
Wang, P. et al. (2021) 'On the Effects of the Ocean on Atmospheric CFC-11 Lifetimes and Emissions', *Proceedings of the National Academy of Sciences of the United States of America*, vol 118, article e2021528118.

3

THE CLIMATE-OCEANS NEXUS

Oceans in the climate regime, climate in the oceans regime

Rozemarijn J. Roland Holst

From a scientific and Earth System perspective, the biophysical nexus between oceans and the global climate system is abundantly clear (Pörtner *et al.*, 2019). Yet, in terms of law and governance, this nexus is less pronounced in the relevant regimes and the interactions between them than one would expect. Climate change impacts on the oceans are pervasive and widely (although not evenly) dispersed: from sea-level rise, degradation of ecosystems and ecosystem services, to biodiversity loss, including changes to the abundance and distribution of commercially exploited species. Climate change, furthermore, has a main driver – CO_2 emissions – in common with another key challenge of global marine environmental change, namely ocean acidification (see Chapter 4). Climate change impacts have 'multiplier effects' alongside other anthropogenic threats to the marine environment, such as pollution and over-exploitation. As a result, climate change impacts on the oceans are central to a broader range of global governance challenges and interests beyond marine environmental protection per se, including maritime security and food security (Kaye, 2012; Kojima, 2018). The impacts of climate change on the ocean are thus trans-sectoral and cumulative. This is where the governance challenges posed by climate change reveal tensions with the existing legal framework for the oceans, which is traditionally premised on a sectoral approach to regulating the deleterious impacts of distinct human activities on the marine environment.

At the same time, the biophysical nexus means that oceans play a crucial role in mitigating the effects of global climate change on the Earth System as a whole. As the world's largest natural carbon sink, oceans capture around 25% of anthropogenic CO_2 emissions (Gattuso *et al.*, 2018: 337), and they have absorbed more than 90% of the world's excess heat since 1970 (Pörtner *et al.*, 2019: para. A.2). Ocean health is thus of central relevance to global climate change mitigation and adaptation strategies, as coordinated primarily under the umbrella of the 1992 United Nations Framework Convention on Climate Change (UNFCCC) and the 2015 Paris Agreement. Yet, oceans have infamously been 'the Cinderella of the UN climate negotiations' (Freestone, 2009: 383), and the climate change regime has been criticised for neglecting the oceans both as a major victim of the impacts of climate change and as a potential part of the solution (Galland, Harrould-Kolieb and Herr, 2012: 764).

This chapter will unpack the climate-oceans nexus by first discussing the ways in which oceans have started to make their way into the climate change regime, and secondly how climate change impacts fall within the scope of the oceans regime. In doing so, it will

DOI: 10.4324/9781315149745-5

highlight the (potential) legal avenues for enhanced interaction and coordination between the two and reveal notable gaps and limitations inherent in the respective regimes. It will show that neither can be expected to provide the full answer to the governance challenges posed by climate change impacts on the oceans. A multi-faceted yet synergistic approach to the interpretation, implementation, and further development of these two regimes is needed from the top-down and from the bottom-up.

Oceans in the climate change regime

The UNFCCC does what it says on the tin: it provides a 'framework', a general legal architecture, for international climate change cooperation, which leaves the progressive development of the broad principles and objectives set out therein to be negotiated and agreed upon in further instruments. The most recent of such instruments is the Paris Agreement (Bodansky, Brunnée and Rajamani, 2017). The general objective of the UNFCCC regime is stipulated in Article 2: 'stabilisation of greenhouse gas concentrations in the atmosphere at a level that would prevent dangerous anthropogenic interference with the climate system'. Guiding principles of the regime are the precautionary principle, the principle of intergenerational equity, common but differentiated responsibilities, and sustainable development (Article 3). The Paris Agreement introduced a more specific global average temperature target of 'well below 2°C above pre-industrial levels' while pursuing efforts to limit it to 1.5°C (Article 2(1)(a)). It contains an intricate mix of 'hard' and 'soft' obligations (Bodansky, 2016; Rajamani, 2016) across three central pillars of climate change action: mitigation (Articles 3–6), adaptation (Article 7) and loss and damage (Article 8). The main 'innovation' of the Paris Agreement was the introduction of a 'bottom-up' approach through which countries submit voluntary but progressively ambitious nationally determined contributions (NDCs) that set out their domestic mitigation and adaptation measures and targets (Article 4). The collective efforts under these NDCs will be subject to a global stocktaking process (Article 14), the first round of which is due to be completed in 2023. This approach is a break from the system contained in the 1997 Kyoto Protocol, which set out 'top down' binding reduction targets only for countries listed in Annex I of the UNFCCC. The first commitment period under the Kyoto Protocol ended in 2012 and the 2012 Doha amendment to the Kyoto Protocol containing a second commitment period never entered into force; hence its direct future relevance is comparatively limited. Finally, it is important to note that the UNFCCC regime as a whole does not cover greenhouse gas (GHG) emissions from shipping or aviation, as these are regulated by the International Maritime Organisation (IMO) and the International Civil Aviation Organisation (ICAO), respectively.

Oceans are not mentioned in the text of the Paris Agreement; only its preamble notes the 'importance of ensuring integrity of all ecosystems, including oceans'. In addition to the obligation to cut GHG emissions, Article 5 of the Paris Agreement contains the obligation to 'conserve and enhance' carbon sinks and reservoirs. This provision refers to Article 4(1)(d) of the UNFCCC which mentions oceans as an example of such natural sinks. This is the only explicit reference to oceans in the UNFCCC regime. Yet, as indicated above, in reality, the biophysical interrelationship between oceans and climate change goes far beyond the oceans' function as a sink. The 2019 IPCC *Special Report on Oceans and the Cryosphere in a Changing Climate* (IPCC SCROCC) reiterates that serious climate change mitigation efforts are also a precondition for effective adaptation to global climate change. In low-emissions-scenarios, the various risks posed by marine environmental change are reduced and more effective responses will be available, whereas in high-emissions-scenarios inevitable 'adaptation limits' will be encountered (Pörtner *et al.*, 2019: para. C.4.1–7).

It is thus the more remarkable that the climate-oceans nexus has only very recently started to receive some attention within the UNFCCC context. During the UNFCCC Conference of the Parties (COP) 23 in 2017, the 'Ocean Pathway' partnership was launched with the aim to increase the role of ocean considerations in UNFCCC processes and step-up action related to oceans and climate change, including as part of countries' NDCs and the global stocktaking process. This was subsequently lauded as 'a very positive development' by the UN Secretary-General in his 2018 Report on Oceans (UN Doc. A/73/368). The COP 25 in 2019 was dubbed to be 'the blue COP', but only a single decision was adopted that contains a minimal reference to the oceans: merely expressing 'appreciation and gratitude' for the IPCC SROCC and requesting the Subsidiary Body for Scientific and Technological Advice (SBSTA) to initiate an 'oceans dialogue' in which parties and non-party stakeholders are invited to submit inputs on how mitigation and adaptation action at the climate-oceans interface might be strengthened (COP Decision 1/CP.25). This 'dialogue' took place for the first time in 2020. The Glasgow Climate Pact adopted at COP 26 in 2021 requests the Chair of the SBSTA to organize and report on this dialogue annually from 2022 onwards. It furthermore invites the relevant work programmes and bodies constituted under the UNFCCC to 'consider how to integrate and strengthen ocean-based action in their existing mandates and workplans' and to report thereon (COP Decision 1/CP.26, paras 60–61). Although the wording of this decision is still very generic and no concrete targets or priority areas have been identified, it is the first time that a COP decision officially integrates the ocean across all areas of work. Future COPs will have to prove whether and to what extent this indeed marks the first step towards more detailed implementation and future decisions on oceans and climate change under the UNFCCC.

The need for an integrative approach to implementing obligations arising under the climate change and ocean regimes has been reiterated more vocally in other international fora, notably by the UN General Assembly in the yearly Resolution on Oceans of recent years that 'encourage' parties to ratify and implement the Paris Agreement, in recognition of the 'adverse impacts of climate change on the marine environment, marine biodiversity and sea level' (UNGA Res.75/239: para. 208). Also, the state parties to the Convention on Biological Diversity (CBD) repeatedly stress the need for 'integrated and co-beneficial implementation' of actions under the Paris Agreement, the CBD and other conventions in countries' NDCs and domestic climate action, 'taking into account the importance of ensuring the integrity and functionality of all ecosystems, including oceans, and the protection of biodiversity' (CBD COP Decision 14/5 (2018): para. 5).

While some form of purposeful institutional coordination to facilitate this on the part of the UNFCCC regime is currently still lacking, increasing attention for the climate-ocean nexus is nevertheless discernable from the bottom-up. A 2017 study analysed 161 NDCs submitted at that point, and found that 70% thereof referred to ocean issues, with 50 NDCs containing both ocean-related mitigation and adaptation measures (Gallo, Victor and Levin, 2017: 834). The countries most dramatically affected by climate change in the short term, in particular small island developing states, are notably the most active in pushing developments in recognition of the climate-oceans nexus, both unilaterally and within the UNFCCC context and in other international fora. Interestingly, across the board, ocean-inclusive NDCs paid significantly more attention to adaptation concerns, especially coastline impacts, ocean warming, and impacts on fisheries, than to ocean-based mitigation actions, such as ecosystem restoration, conservation, or offshore renewable energy developments (Gallo, Victor and Levin, 2017: 834). The latest NDC synthesis report of 48 new or updated NDCs submitted by 31 December 2020 appears to confirm this growing emphasis on adaptation.

Out of these adaptation measures, only 33% referred to the ocean, but more fundamentally concerning is the general finding that total GHG emissions reductions 'fall far short of what is required' (UN doc FCCC/PA/CMA/2021/2).

This means that also on the flip side of the mitigation coin we may expect growing attention to be paid to the oceans. In order for states to reach the targets under the Paris Agreement, strategies in addition to 'conventional' emissions reduction at the source may have to be explored, either to close the emissions gap or as a 'bridging technology'(Ghaleigh, 2016). Examples include negative-emissions solutions such as ocean-based carbon capture and storage (CCS) and – given the oceans' scale and natural capacity to absorb CO_2 – potentially even more controversial forms of 'geoengineering' or 'climate engineering' (Craik and Burns, 2016; McGee, Brent and Burns, 2018). Oceans are a particularly attractive CCS site. Existing empty offshore oil and gas fields can be used by means of technology that is already developed and employed by the oil and gas industry, and ocean-based CCS is thought to trigger fewer 'not in my back yard' objections than land-based storage sites (Bankes, 2019). While the climate change regime does not regulate such activities directly, the work of the IPCC appears to assume that the achievement of both 1.5°C and 2°C scenarios depends on the development of at least some of such strategies because emissions reduction at the source is likely to be insufficient (Masson-Delmotte *et al.*, 2018). At present, CCS is already part of several countries' NDCs (Gallo, Victor and Levin, 2017: 834). Whenever such negative-emissions technologies are deployed in the marine environment, the legal nexus with the oceans regime becomes clear again as any such activities are governed by the law of the sea and a number of more specialist regimes that will be discussed in the next section.

Climate change in the oceans regime

As the overarching legal framework for the oceans, the 1982 UN Convention on the Law of the Sea (UNCLOS) inevitably finds itself confronted with the governance challenges posed by climate change impacts on the oceans, despite the fact that at the time of its negotiation 'climate change' was not an issue on the minds of its drafters (Stephens, 2015; Boyle, 2016; Redgwell, 2019; see Chapter 2). Of particular relevance is UNCLOS Part XII on the protection and preservation of the marine environment. This part contains general obligations that were intended to evolve progressively over time in response to new developments, in particular through the adoption and implementation of subsequent more detailed rules and standards under external specialised regimes or on a regional basis (see Chapter 5). As we will see below, it is through these subsequent regulatory developments that certain aspects of climate change governance have made their way into the oceans regime. It should also be noted here that sea-level rise has direct implications for existing baselines and maritime entitlements as governed by UNCLOS (Veierud Busch, 2018). This aspect, however, falls outside the scope of this chapter.

Climate change mitigation obligations under the law of the sea and beyond

The obligations to protect and preserve the marine environment set out in Part XII of UNCLOS are 'general' obligations in the sense that they apply to all states at all times, regardless of where an activity takes place, thus including areas beyond national jurisdiction. The general obligation is stipulated in Article 192 and further elaborated in Article 194 and subsequent articles, which provide that states shall take 'measures to prevent, reduce and control pollution of the marine environment'. The general nature and wording of Part

XII are important for another reason, as it has proven an important avenue for evolutionary interpretation of states' obligations in relation to marine environmental protection in the light of global environmental change and subsequent legal developments. UNCLOS courts and tribunals have progressively interpreted the general obligations of Part XII as not being limited to 'pollution' in the strict sense of the word; they include the preservation of ecosystems and biodiversity, an interpretation in line with the emergence of subsequent regimes in international law, such as the CBD (see *Southern Bluefin Tuna* (1999) ITLOS Rep, 280, para. 70; *South China Sea Arbitration (Merits)* (2016) PCA case no. 2013–19, para. 945). Article 192 has furthermore been interpreted to cover both 'protection' from future damage, and 'preservation' in the sense of maintaining and improving the present condition, thus entailing both a positive obligation to take measures to protect and preserve the marine environment and a negative obligation not to degrade its current status (*South China Sea Arbitration (Merits)* (2016) PCA case no. 2013–19, para. 941).

For the impacts of GHGs on the marine environment specifically, it is important to note that the general obligation under Articles 192 and 194 covers pollution from 'any source', including land-based sources and pollution 'from or through the atmosphere (Articles 194(3)(a); 207 and 212). The term 'pollution of the marine environment' is broadly defined in UNCLOS as:

> the introduction by man, directly or indirectly, of substances or energy into the marine environment, including estuaries, which results or is likely to result in such deleterious effects as harm to living resources and marine life, hazards to human health, hindrance to marine activities, including fishing and other legitimate uses of the sea, impairment of quality for use of sea water and reduction of amenities.
>
> *(UNCLOS Article 1(1)(4))*

It is clear that the increased oceanic intake of GHGs like CO_2 (a 'substance') directly alters ocean chemistry and adds 'energy' into the marine environment that results in ocean warming, with all the known direct and indirect effects. That these effects are 'deleterious' not only for the marine environment but also for human health and other (economic) activities, is furthermore evident.

It is important to understand that the general obligation of states under Articles 192 and 194 is one of 'due diligence'; the content of which is informed by the other provisions of Part XII, and – importantly – other relevant rules of international law (*South China Sea Arbitration (Merits)* (2016) PCA case no. 2013–19, para. 941). 'Due diligence' is not an obligation of result that obliges states to prevent harm to the marine environment altogether, but rather an obligation of conduct that requires them to take those measures that are necessary to prevent or minimise harmful effects from activities under their jurisdiction and control, for example by conducting environmental impact assessments (EIA) (Article 206), applying the precautionary approach, using the best available technology, and ensuring a certain level of vigilance in the enforcement of applicable standards (*South China Sea Arbitration (Merits)* (2016) PCA case no. 2013–19). Accordingly, it can be argued that states are required to take active climate change mitigation measures in order to comply with their obligation of due diligence under UNCLOS. Yet, the standard of due diligence is a very general one and leaves unanswered the question as to what exact measures or what standard of mitigation is required.

In determining the content of this obligation of due diligence, the legal nexus with the climate change regime comes to the fore. The UNFCCC and the Paris Agreement are arguably relevant for interpreting and giving substance to the general obligations under

UNCLOS, in the same way as UNCLOS tribunals have previously referred to the CBD regime and the Convention on International Trade in Endangered Species (CITES) to interpret states' obligations under Article 194. On this basis, it can be argued that states should at least comply with their own (increasingly progressive) NDCs submitted under the Paris Agreement in order to meet the standard of care required as part of their due diligence obligation under UNCLOS. Such an integrative approach to the interpretation and implementation of obligations stemming from distinct regimes that may not be formally coordinated at the institutional level is not a novel concept in international law and governance (Trevisanut, Giannopoulos and Roland Holst, 2020). We also saw this approach stressed in the CBD's COP decision mentioned above. Similarly, the 2018 ILC Draft Guidelines on the Protection of the Atmosphere recommend that when states take measures to prevent, reduce, and control atmospheric pollution, relevant rules of international law, including the law of the sea, are 'interpreted and applied in order to give rise to a single set of compatible obligations, in line with the principles of harmonization and systemic integration' and Article 31(3)(c) of the Vienna Convention on the Law of Treaties (ILC Draft Guidelines on the Protection of the Atmosphere, UN Doc. A/73/10 (2018). Also, a proper application of the precautionary approach supports the argument that climate change mitigation measures are required as part of the obligation to protect the marine environment, especially, as noted above, because a low-emissions trajectory is a prerequisite for effective adaptation to marine environmental change.

Such an integrative approach is supported by the very character of, in particular, Part XII of UNCLOS, which has led commentators to describe it as a 'framework' or 'umbrella' Convention (Boyle, 2006; Redgwell, 2016). In many instances, the Convention purposefully defers to further cooperation between the parties, including on a regional basis, in order to elaborate, agree, and implement more detailed and progressive rules and standards for marine environmental protection consistent with the general principles set out in the Convention (see e.g., Articles 197, 237). A variety of provisions contained in Part XII furthermore contain so-called 'rules of reference', meaning that they refer to 'generally accepted international rules and standards' (GAIRS) contained in external instruments to give detailed content to general rights and obligations contained in the Convention. While these rules facilitate systemic integration and interpretation, they only do so in a sectorally limited sense, that is, per issue area for which they are provided. Not all rules of reference furthermore entail the same degree of normativity. Most positively, some provisions require states to cooperate on a global or regional basis to establish relevant GAIRS, and to adopt national pollution control measures that are 'no less effective' or that have 'at least the same effect' as these GAIRS (e.g., Article 208 on pollution from seabed activities and Article 210 on ocean dumping). In the 'weakest' form, other rules of reference merely require international standards to be 'taken into account' when states adopt national measures.

This sectoral approach to pollution regulation, coupled with different degrees of normativity, is relevant for the way in which emissions reduction standards make their way into the oceans regime, and explains why the emerging picture is a fragmented one. For GHG emissions from shipping, for example, Article 211 requires states to adopt measures that have 'at least the same effect' as the relevant GAIRS, which in this case are the relatively advanced and progressively stronger standards developed by the IMO under MARPOL. Yet, for land-based sources of pollution and pollution through or from the atmosphere, which together account for the bulk of climate-related stressors on the marine environment, the relevant rules in Articles 207 and 212 merely require states to 'take into account' GAIRS when taking measures to prevent, reduce, and control such pollution. For these non-ship sources of

GHG emissions, the mitigation standards and procedures adopted under the UNFCCC and Paris Agreement would arguably constitute the relevant GAIRS for the purposes of Articles 207 and 212. These rules of reference thereby extend the obligation to take (at least some) climate change mitigation measures to all parties to UNCLOS and requires them to take the Paris Agreement into account when doing so. To that extent, UNCLOS can be said to 'supplement' or reinforce the climate change regime (Redgwell, 2019: 449), but it cannot go above or beyond it. The measure of discretion afforded to states by the 'weak' rule of reference under Articles 207 and 212 makes any specific mitigation standards difficult to enforce via UNCLOS (Boyle, 2019).

Implementing ocean-based negative-emissions solutions

On the other side of the mitigation coin, only two ocean-based negative-emissions technologies have thus far reached more advanced stages of testing and development, and they have consequently also received regulatory attention: sub-seabed CCS and ocean fertilisation. The regulatory approaches adopted in respect of these two activities are briefly considered here because they provide another (indirect) link between the climate and ocean regimes. In both cases, regulatory action has been relatively proactive, which is noteworthy in a field where regulation is typically reactive.

For CCS, a permissive but regulated approach has been pursued, despite outstanding concerns about responsibility and liability for leakage and other potential risks to the marine environment (Rayfuse, 2012: 168–169). The storage of CO_2 in the marine environment falls under the scope of the London Dumping Convention (LC) and the London Dumping Protocol (LP) because the LP defines 'dumping' as 'deliberate disposal into the sea of wastes or other matter', as well as 'any storage of wastes or other matter in the seabed and subsoil thereof' (LP Article 1(4)(1)). This means that CCS qualifies as 'dumping', and it would thus be prohibited under the dumping regime and UNCLOS by virtue of the rule of reference in Article 210. This was generally perceived as an undesirable effect, and the parties to the LP adopted an amendment upon the LP's entry into force in 2006 to, first of all, confirm the applicability of the LP to CCS activities and, second, list CCS in sub-seabed geological formations (i.e., not in the water column) in Annex 1. This amendment entered into force in 2007 and as a result states may authorise the storage of CO_2 for CCS purposes, subject to certain permitting conditions and an assessment framework that regulates the risks to the marine environment (Res LP.1(1), and IMO Doc. LC/SG-CO2 1/7 (2006)). In 2009, another amendment to LP Article 6 was adopted to allow the export of CO_2 for transboundary CCS projects (Res LP.3(4)). Due to limited ratification, this amendment has not yet entered into force, which was perceived as an obstacle to the development of CCS by some of its proponents (Bankes, 2019). In 2019, the parties to the LP adopted a Resolution to allow the provisional application of the 2009 amendment between those parties that have ratified it, thus enabling them to proceed with transboundary CCS projects (Res LP.5(14)). Interestingly, the Resolution reiterates that the LP should not 'constitute a barrier to the transboundary movement of CO_2 for disposal as a measure to mitigate climate change and ocean acidification', but also that CCS 'does not remove the obligation under the LP to reduce the need for such disposal and the commitments under the UNFCCC', taking into account the recent IPCC Special Reports.

In respect of ocean fertilisation (OF), a more prohibitive approach has been taken. The idea behind OF is that by adding iron or other nutrients to the water column, the ocean's natural biological pump can be enhanced. Fertilisation leads to an increased growth of marine organisms, which will take up more CO_2 from the atmosphere, and when these eventually sink down, the idea is that this carbon will then be sequestered at great depth (GESAMP Working Group 41, 2019: 42–47). There are, however, considerable potential environmental risks associated with OF, including increased acidification, toxic algal blooms, anoxia, and the formation of even higher concentrations of other GHGs such as methane. Combined with questionable technical and financial feasibility and governance issues, this means that there is currently 'low confidence' that ocean-based manipulations like OF can provide viable mitigation strategies (Pörtner *et al.*, 2019: 525). Following a number of controversial experiments in the early 2000s, the CBD's COP adopted a 'moratorium' resolution in 2008 which stressed the precautionary approach and urged parties to ensure that OF activities (with the exception of small-scale scientific research) do not take place until there is an adequate scientific basis on which to justify such activities' as well as a global regulatory framework in place (CBD COP Decision IX/16C). In the same year, the parties to the LC/LP decided to include OF within the scope of their mandate (IMO Doc. LC30/16 (2008)). This conclusion was not immediately obvious as the definition of 'dumping' mentioned above contains an exception for the 'placement of matter for a purpose other than the mere disposal thereof, provided that such placement is not contrary to the aims' of the LC/LP (LC Article III(1)(b)(ii); and LP Article 1.4.2.2). Hence, the question was whether or not OF qualifies as 'dumping' of the fertilisers or CO_2 contrary to the aims of the LC/LP, or if it could fall under the exception (Scott, 2013). The 2008 Resolution LC-LP.1 removed any doubt, and it prohibited OF for any purposes other than legitimate scientific research. This approach was formalised in an amendment to the LP adopted in 2013 (Res LP.4(8), which has, however, not yet entered into force. Nevertheless, the governance approach in relation to OF with an emphasis on precaution and a strict limitation to legitimate scientific research seems to be consistently reiterated across different regimes, including the LP, the CBD, and in the yearly United Nations General Assembly Resolutions on Oceans, supporting the view that obligations of states under these various regimes and UNCLOS should be interpreted and applied in a consistent manner.

Conclusion: challenges revealed by the climate-oceans nexus

The foregoing discussion has underlined the need and potential for integrated interpretation and implementation of obligations arising at the intersection of the climate and oceans regimes. Yet, it has also revealed a number of challenges arising from this nexus. First, we have seen that due diligence under Part XII of UNCLOS provides a useful mechanism to reinforce the climate change regime by allowing the general obligation of states to protect the marine environment to be interpreted and applied in accordance therewith. However, the concept is no panacea. It will be difficult to hold states to any more ambitious standards of mitigation than what they themselves have committed to as part of their NDCs under the Paris Agreement. Even an emphasis on the precautionary approach does not alter that. Part XII can only be interpreted by reference to what states have 'actually agreed' rather than what they 'should have agreed' (Boyle, 2019: 481). It is furthermore difficult to argue that UNCLOS imposes stricter obligations in relation to climate change impacts on the ocean because, by its very nature, Part XII refers to other (regional) agreements to inform the detailed content of its general obligations. What can be expected of states by virtue of UNCLOS is arguably (at most) that they take measures that result in a 'meaningful lowering' of emissions over time (Boyle, 2016: 219) while 'taking into account' the global targets under the Paris Agreement.

Second, we have seen the limits of the incorporative function of the rules of reference, primarily due to their sectoral nature and the differing degrees of normativity imposed for distinct sources of pollution. For those sources of 'pollution' most central to climate change – land-based sources and pollution through the atmosphere – the relevant articles require merely a 'taking into account' of relevant international standards, thus leaving states considerable discretion in setting their own priorities. The reason for this lies in the nature of the land-based industrial activities from which these types of pollution originate, and their close connection to the economic development of states, which inclines states to preserve as much freedom as possible to balance environmental protection measures with their own developmental needs (Birnie, Boyle and Redgwell, 2009: 454). Yet, with the bulk of marine pollution coming from land-based sources, and more particularly in the light of today's knowledge of the effects of climate change on the oceans, this is now proving to be a much bigger obstacle than at the time UNCLOS was negotiated.

Third, despite the avenues available for integrated interpretation and implementation across regimes, it is clear that systemic integration and coherent implementation of standards originating from separate regimes is not straightforward, especially when some form of purposeful institutional coordination is lacking. States adopting mitigation measures, including the enhancement of the oceanic carbon sink in accordance with their obligations under the Paris Agreement, may choose to implement techniques that successfully sequester CO_2 but which thereby (inadvertently) contribute to or exacerbate ocean acidification or other harm to the marine environment, which in turn they are obliged to prevent under the oceans regime. This phenomenon has also been described as 'problem shifting': when the 'solution' to one problem 'transfers' the problem to somewhere else, or transforms it into another problem (Kim and van Asselt, 2016). Problem shifting poses an additional governance challenge that is almost inevitable under a sectoral approach to (pollution) regulation. Indeed, both CCS and OF could be seen as transferring one source of pollution (CO_2) from the atmosphere into the ocean or seabed, and potentially transforming one problem (atmospheric CO_2 contributing to climate change) into another (enhanced ocean acidification). Problem shifting does not necessarily result in non-compliance with specific obligations; indeed, it may be an (unintended) consequence of complying with one or the other. The point for the present purposes is rather that the governance challenges posed by climate change impacts on the oceans are of such a nature and scope that policy decisions or coordination efforts cannot effectively be made under either the climate change regime or the oceans regime alone. The 'interpretative communities' of distinct regimes typically have diverging views and priorities, even when the objectives of the regimes are in theory mutually supportive (Hey, 2020).

As observed by Redgwell, '[a]lthough the direction of travel in climate and oceans governance is towards processes of integration and conflict avoidance, a degree of mutual myopia persists in the 'twilight zone' of the climate and oceans regimes' (Redgwell, 2019: 456–457). We are starting to see efforts towards improved implementation in a mutually supportive manner from the bottom-up – if nothing else, out of simple necessity, by those states who literally feel the water rising. This should be assisted by enhanced coordination from the top-down at a cross-institutional level: from the UNFCCC to the IMO, and from the CBD to regional treaty bodies, and so on. As always, it boils down to a question of political will rather than law. The preamble of the UNFCCC recognises that 'change in the Earth's climate and its adverse effects are a common concern of humankind'. The preamble of UNCLOS recognises that 'the problems of ocean space are closely interrelated and need to be considered as a whole'. If anything, the climate-oceans nexus underlines that the problems of 'Earth space' are interconnected and need to be considered as a whole. That requires every (international) body to be on board.

References

Bankes, N. (2019) 'Provisional Application of an Amendment to the London Protocol to Facilitate Collaborative CCS Projects', *The NCLOS Blog*, 11 December. Available at: http://site.uit.no/jclos/.

Birnie, P., Boyle, A. and Redgwell, C. (2009) *International Law & the Environment*. 3rd edn. Oxford University Press.

Bodansky, D. (2016) 'The Paris Climate Change Agreement: A New Hope?', *American Journal of International Law*, 110(2): 288–319.

Bodansky, D., Brunnée, J. and Rajamani, L. (2017) *International Climate Change Law*. Oxford University Press.

Boyle, A. (2006) 'Further Development of the Law of the Sea Convention: Mechanisms for Change', in Freestone, D., Barnes, R., and Ong, D. (eds) *The Law of the Sea: Progress and Prospects*. Oxford University Press, pp. 40–62.

Boyle, A. (2016) 'Climate Change, Ocean Governance and UNCLOS', in Barrett, J. and Barnes, R. (eds) *Law of the Sea: UNCLOS as a Living Treaty*. BIICL, pp. 211–230.

Boyle, A. (2019) 'Litigating Climate Change under Part XII of the LOSC', *The International Journal of Marine and Coastal Law*, 34(3): 458–481. https://doi.org/10.1163/15718085-13431097.

Craik, N. and Burns, W. C. G. (2016) *Climate Engineering under the Paris Agreement: A Legal and Policy Primer*. Centre for International Governance Innovation.

Freestone, D. (2009) 'Climate Change and the Oceans', *Carbon & Climate Law Review*, 4: 383.

Galland, G., Harrould-Kolieb, E. and Herr, D. (2012) 'The Ocean and Climate Change Policy', *Climate Policy*, 12(6): 764–771.

Gallo, N. D., Victor, D. G. and Levin, L. A. (2017) 'Ocean Commitments under the Paris Agreement', *Nature Climate Change*, 7(11): 833–838.

Gattuso, J. P. *et al.* (2018) 'Ocean Solutions to Address Climate Change and Its Effects on Marine Ecosystems', *Frontiers in Marine Science*, 5: 337.

GESAMP Working Group 41. (2019) *High Level Review of a Wide Range of Proposed Marine Geoengineering Techniques*. IMO.

Ghaleigh, N. S. (2016) 'Carbon Capture and Storage as a Bridging Technology', in Farber, D. and Peeters, M. (eds) *Climate Change Law*. Edward Elgar Publishing, pp. 189–199.

Hey, E. (2020) 'Regime Interaction and Common Interests in Regulating Human Activities in ABNJ', in Trevisanut, S., Giannopoulos, N., and Roland Holst, R. (eds) *Regime Interaction in Ocean Governance: Problems, Theories and Methods*. Brill, pp. 85–123.

Kaye, S. (2012) 'Climate Change and Maritime Security', in Warner, R. and Schofield, C. (eds) *Climate Change and the Oceans: Gauging the Legal and Policy Currents in the Asia Pacific and Beyond*. Edward Elgar Publishing, pp. 153–166.

Kim, R. E. and van Asselt, H. (2016) 'Global Governance: Problem Shifting in the Anthropocene and the Limits of International Law', in Morgera, E. and Kulovesi, K. (eds) *Research Handbook on international law and natural resources*. Edward Elgar Publishing, pp. 473–495.

Kojima, C. (2018) 'Climate Change and Protection of the Marine Environment', in Craik, N. et al. (eds) *Global Environmental Change and Innovation in International Law*. Cambridge University Press, pp. 138–156.

Masson-Delmotte, V. *et al.* (2018) *IPCC Special Report on the Impacts of Global Warming of 1.5°C above Pre-industrial Levels*. Available at: https://www.ipcc.ch/sr15/.

McGee, J., Brent, K. and Burns, W. (2018) 'Geoengineering the Oceans: An Emerging Frontier in International Climate Change Governance', *Australian Journal of Maritime & Ocean Affairs*, 10(1): 67–80.

Pörtner, H. O. *et al.* (2019) *IPCC Special Report on the Ocean and Cryosphere in a Changing Climate*. Available at: https://www.ipcc.ch/srocc/download-report-2/.

Rajamani, L. (2016) 'The 2015 Paris Agreement: Interplay Between Hard, Soft and Non-Obligations', *Journal of Environmental Law*, 28(2): 337–358.

Rayfuse, R. (2012) 'Climate Change and the Law of the Sea', in Rayfuse, R. and Scott, S. V. (eds) *International Law in the Era of Climate Change*. Edward Elgar Publishing, pp. 147–174.

Redgwell, C. (2016) 'The Never Ending Story: The Role of GAIRS in UNCLOS Implementation in the Offshore Energy Sector', in Barrett, J. and Barnes, R. (eds) *Law of the Sea: UNCLOS as a Living Treaty*. BIICL, pp. 167–186.

Redgwell, C. (2019) 'Treaty Evolution, Adaptation and Change: Is the LOSC "Enough" to Address Climate Change Impacts on the Marine Environment?', *The International Journal of Marine and Coastal Law*, 34(3): 440–457.

Scott, K. N. (2013) 'Regulating Ocean Fertilization under International Law: The Risks', *Carbon & Climate Law Review*, 7(2): 108–116.

Stephens, T. (2015) 'Warming Waters and Souring Seas: Climate Change and Ocean Acidification', in Rothwell, D. et al. (eds) *The Oxford Handbook of the Law of the Sea*. Oxford University Press, pp. 778–798.

Trevisanut, S., Giannopoulos, N. and Roland Holst, R. (eds) (2020) *Regime Interaction in Ocean Governance: Problems, Theories and Methods*. Brill.

Veierud Busch, S. (2018) 'Sea Level Rise and Shifting Maritime Limits: Stable Baselines as a Response to Unstable Coastlines', *Arctic Review on Law and Politics*, 9: 174–194.

4

THE LEGAL IMPLICATIONS OF OCEAN ACIDIFICATION

Beyond the climate change regime

Jennicca Gordon

The global intensification of carbon dioxide (CO_2)-emitting activities that have fundamentally contributed to the climate crisis have also triggered devastating impacts on our oceans and related services. Often referred to as 'climate change's evil twin' (Hull, 2016: 348), acidification occurs as the pH levels of oceans decrease (i.e., become more acidic) as the ocean absorbs excessive CO_2. Oxidation of organic materials also accelerates acidity (Levin and Le Bris, 2015: 766). This presents significant global challenges due to the adverse environmental consequences that the ocean chemistry imbalance has on the marine environment, ecosystem stability, fisheries, food security, and livelihoods. Despite its importance, ocean acidification is not directly addressed by any single treaty regime because its causes and effects span several sectors of international law. This presents significant legal and governance challenges. Ocean acidification is therefore governed by a 'regime complex', which can be defined as 'functionally overlapping parallel regimes and institutions that are non-hierarchical, but which nevertheless affect one another's sphere of operations' (Scott, 2020: 390) (see Chapter 3). In this chapter, multilateral environmental agreements (MEAs) covering oceans, climate change and biodiversity will be critically analysed to assess which treaty regime should take the lead in addressing ocean acidification and what type of new rules will be required to do so effectively.

There are three main international legal structures that involve ocean acidification. The first is the climate change regime, including the 1992 United Nations Framework Convention on Climate Change (UNFCCC), the 1997 Kyoto Protocol and the 2015 Paris Agreement. This UNFCCC regime establishes the universal framework for reducing greenhouse gas (GHG) emissions and adapting to climate change. Second is the 1982 United Nations Convention on the Law of the Sea (UNCLOS), which requires parties to mitigate ocean pollution through its Part XII obligations (see Chapter 2). The third is the 1992 Convention on Biological Diversity (CBD), which establishes comprehensive requirements for parties to conserve biodiversity. Thus far, efforts to create a coherent response to ocean acidification have largely focused on the capacity of these regimes to enable CO_2 emissions reductions, a mitigation element under the climate change regime. However, approaching the issue purely from this basis may negate the ability to focus on ocean acidification as a separate, albeit related, problem to climate change. Addressing ocean acidification solely within the climate change regime could also constrain the range of potential legal responses available, with solutions likely to be more effective when carried out in combination with other fora.

DOI: 10.4324/9781315149745-6

While there is no specific obligation to protect the marine environment from ocean acidification under the climate change regime, there are due diligence obligations that stem from UNCLOS Part XII to prevent, reduce, and control pollution for the protection and preservation of the marine environment. The extent to which these requirements adequately address ocean acidification, and their connection to the climate change regime, will be examined in this chapter. UNCLOS also requires states to take account of ocean acidification as part of their broader climate change mitigation measures, and consequently UNCLOS may play a role in ensuring compliance with the emissions reduction targets set by parties to the Paris Agreement. Although the CBD does not have a specific mandate to proactively address the causes of ocean acidification, exploring the issue under a biodiversity rubric may improve oceanic adaptive capacity and increase the awareness and understanding of the matter within the international community. Key tools for protecting marine biodiversity include cohesive, ecosystem-based approaches to management, which can be in the form of environmental impact assessments (EIAs) and area-based management tools (ABMTs) such as marine protected areas (MPAs). MPAs are a form of long-term, in situ conservation which gives more protection to the area surrounding it, 'due to more stringent regulation of one or more or all human activities, for one or more purposes' (Molenaar, 2014: 3). These are considered as one of the most effective tools in mitigating the effects of ocean acidification by enhancing ocean resilience, safeguarding biodiversity, and maintaining habitat connectivity (Yadav and Gjerde, 2020: 3).

Currently under negotiation at the time of writing is the proposed internationally legally binding instrument for the conservation and sustainable use of marine biological diversity in areas beyond national jurisdiction (BBNJ). This treaty was launched to fill the gaps in the existing management system in areas beyond national jurisdiction (ABNJ). Whether there will be a legal instrument for addressing ocean acidification under this agreement is as yet unclear. However, climate change has been identified by the BBNJ Informal Working Group as relevant in the context of ocean acidification, and a treaty based on safeguarding ocean resilience could help habitats and ecosystems manage the growing threat of acidification through the designation of MPAs and as a result of EIAs. If acidification does fall under the scope of the BBNJ Agreement, it may help to facilitate responses to ocean acidification under UNCLOS in collaboration with other regimes and enhance the effectiveness or the capacity of management activities. Overall, it is clear that mitigation and adaptation to ocean acidification will require a direct, coordinated legal response achieved through the strengthening and harmonisation of existing regimes.

The legal framework for addressing ocean acidification

Ocean acidification is a complex subject, described as being in an 'international legal twilight zone' (Baird et al., 2009: 460). To address the issue under existing frameworks, multifaceted regulation is required, with success likely dependent on collective and harmonised actions being taken. Components of a legal framework for addressing ocean acidification are outlined below.

UN Framework Convention on Climate Change

Ocean acidification comes with a unique and distinct set of challenges compared to climate change. However, climate change and ocean acidification are inextricably linked, in that the most effective means of mitigating ocean acidification is through the reduction of CO_2

emissions. The overall responsibility for reducing GHG emissions falls under the mandate of the UNFCCC. The objectives under Article 2 of the Convention are:

> stabilisation of greenhouse gas concentrations in the atmosphere at a level that would prevent dangerous anthropogenic interference with the climate system. Such a level should be achieved within a time frame sufficient to allow ecosystems to adapt naturally to climate change, to ensure that food production is not threatened and to enable economic development to proceed in a sustainable manner.

There is some debate among academics whether the obligation to consider the impacts of ocean acidification could be interpreted to fall under this provision. This is based on indicators within the UNFCCC that correspond to the atmospheric orientation of the regime, therefore excluding oceanic chemistry (Baird et al., 2009: 463). Furthermore, the UNFCCC has been generally criticised due to insufficient reference to marine issues, although there is evidence that marine issues are becoming more prominent in relation to climate change (Gallo et al., 2017: 833; see Chapter 3).

Moreover, the climate change regime does not limit states to reducing CO_2 emissions exclusively, even though this gas is the primary driver of ocean acidification. GHG emissions are measured by their warming potential, with Article 3(7) of the Kyoto Protocol providing that states may reduce GHG emissions which are 'carbon dioxide equivalent'. The consequence is that states may choose to target certain emissions in their reduction plans, such as methane – a greenhouse gas that has as much as 72 times the global warming potential as the equivalent amount of CO_2 – which may have no overall mitigation effect on ocean acidification. The climate change regime does not take this distinction into consideration, which brings into question the appropriateness of the regime to address ocean acidification. Although some authors suggest that stabilisation of greenhouse gases through the UNFCCC would be sufficient to also cover the requirements to meet ocean acidification reduction (Downing, 2013: 251), it is improbable that this would effectively address the impacts.

The preamble of the Paris Agreement recognises the need to include the oceans when ensuring the 'integrity of all ecosystems'; however, this does not in itself impose any obligations onto states. Nonetheless, Article 4.2 of the Paris Agreement instructs parties to 'prepare, communicate and maintain successive nationally determined contributions'. These serve as a global stocktake to monitor state progress every five years towards the overall goal of '[h]olding the increase in the global average temperature to well below 2°C above pre-industrial levels and pursuing efforts to limit the temperature increase to 1.5°C above pre-industrial levels' (Article 2.1(a)). Nationally determined contributions (NDCs) by states to achieve the Paris Agreement's objectives can provide evidence as to the extent to which the Paris temperature goal can be globally achieved and the possible ways in which commitments could be strengthened. Although there is no direct connection linking global temperature targets with oceanic levels of pH, the IPCC Special Report on Global Warming of 1.5°C identifies that ocean acidification would be less damaging if temperatures were stabilised at 1.5°C through emission reductions (Masson-Delmotte et al., 2018: 282). However, research also indicates that residual harm will occur nonetheless (Meinschausen et al., 2011: 1417). It is notable that more than half of states have included marine components as part of their NDC to the climate change mitigation effort (Gallo et al., 2017: 833). However, research by the Paris Agreement Secretariat analysing the extent of the gap between the current NDCs committed by parties (as of February 2021) and the action levels necessary to meet the Paris

Agreement's long-term goals finds that current pledges 'fall far short' of what is required to keep global temperatures even below 2°C (UNFCCC Synthesis Report by the Secretariat, 2021: 5).

The UN Convention on the Law of the Sea

Beyond the climate change regime, UNCLOS provides a comprehensive framework for the law of the sea, with Part XII addressing the protection of the marine environment. Through Articles 192, 193, and 194(1), there are general obligations on states to protect and preserve the marine environment and to take individually or jointly all necessary measures to 'prevent, reduce and control pollution of the marine environment from any source'. Ocean acidification was not a well-known phenomenon during the drafting of UNCLOS and is not explicitly included in its provisions. However, the definition of 'pollution of the marine environment' has been widely accepted to include anthropogenic CO_2 emissions (Burns, 2007: 36). As per UNCLOS Article 1.1(4), CO_2 is a 'substance' which is introduced into the ocean, causing an increase in acidity likely to 'result in such deleterious effects' as to cause 'harm to living resources and marine life'. States are therefore under an obligation to regulate CO_2 emissions as part of their Article 192 due diligence obligation to protect and preserve the marine environment and through Article 194(5), which emphasises the importance of protection of 'rare and fragile ecosystems as well as the habitats of depleted, threatened or endangered species and other forms of marine life'.

Incorporating rules developed under external agreements is an innovative strategy employed by UNCLOS that limits the amount of detailed regulation and allows it to evolve and 'adapt to new realities' (Barrett and Barnes, 2016: 3). Through Article 212, states must 'take into account internationally agreed rules, standards and recommended practices and procedures'. Article 212(3) outlines that states 'shall endeavour to establish global and regional rules, standards and recommended practices and procedures to prevent, reduce and control pollution'. Finally, through Article 212(2), states are expected to take 'other measures as may be necessary to prevent, reduce and control pollution'. The broad nature of this article means that implementation is expected through non-binding provisions contained within other international agreements. Article 212 has an expansive scope, including air pollution caused by ships and aircraft in any territory. (Exploration of these issues is beyond the scope of this chapter.)

While these provisions were not originally drafted with either ocean acidification or climate change in mind, they can logically be considered as applicable to both in light of the other relevant international rules on the subject. 'Internationally agreed rules and standards' can clearly be interpreted as a reference to standards agreed under the climate change regime when deciding which measures are required under UNCLOS. As a result, the global temperature goals, NDCs and the Paris Agreement global stocktake process are significant in this respect. All parties to UNCLOS are subject to due diligence obligations for the prevention, reduction, and control of pollution of the marine environment. In the absence of external internationally binding rules and standards, does this obligation have any substance in the context of ocean acidification? In other words, can states be required to take additional action beyond their UNFCCC commitments to meet their due diligence obligation to address ocean acidification under UNCLOS?

It could be argued that the climate change regime serves as the *lex specialis* with regard to atmospheric pollution (Scott, 2020: 402). However, expecting states to exceed their climate contributions could be considered unreasonable, particularly as UNCLOS does not provide any additional guidelines or criteria on what would be necessary to comply with due diligence within this context. However, in a number of scenarios, UNCLOS could be very valuable in strengthening UNFCCC mechanisms. As previously stated, the climate change regime does not establish pH stabilisation or specific CO_2 emission targets, and states unilaterally determine their contribution to the global temperature goal. Therefore, it is theoretically possible to meet obligations under the UNFCCC while neglecting to mitigate ocean acidification. In this context, asserting that compliance with such requirements constitutes 'due diligence' appears obsolete. Although there is a lack of precedent over what level of CO_2 reduction should be achieved in order to mitigate ocean acidification, it is arguable that Article 212 requires states to prove that they are taking action to address atmospheric emissions directly affecting the marine environment (Harrison, 2017: 257). Therefore, if parties are negligent in taking account of ocean acidification as part of their mitigation measures against a broader climate change agenda, they could be in violation of UNCLOS.

Furthermore, the role played by UNCLOS in handling disputes could potentially be more significant with regard to emissions reductions under the Paris Agreement for ocean acidification mitigation. Compared to the climate change regime, UNCLOS has strong compulsory dispute settlement procedures under Part XV, which could potentially be used to litigate against polluting states. (The dispute settlement procedures provided by the climate change regime are relatively weak, for example Article 15 of the Paris Agreement sets out a 'facilitative...non-adversarial and non-punitive' compliance mechanism). Therefore, a judicial mechanism is available for disagreements about legal or factual issues and could be used, for example, to resolve fundamental concerns around whether a state has taken sufficient action based on its climate change commitments, either within or outside the climate change regime (Harrison, 2017: 257).

Finally, UNCLOS also imposes a legal obligation on all parties to the Convention to address the causes of marine pollution, regardless of whether they are also signatories to any agreements under the climate change regime. This means that states that are parties to UNCLOS but not to the Paris Agreement could be obliged to fulfil their responsibilities towards preventing excessive CO_2 emissions. As a result, Article 212 offers a legal framework from which ocean acidification mitigation action could be based. However, in practice, litigation against a non-signatory state may be futile as the Paris Agreement has virtually universal participation since the United States re-joined the agreement. There may also be difficulty with jurisdiction if attempting to bring litigation as the UNCLOS case law is not clearly defined, therefore any case based on a dispute within the UNFCCC may not be admissible on grounds of jurisdiction (Boyle, 2012: 836). Additionally, it is also unlikely that the phrasing of UNCLOS Article 212 provisions 'taking into account' and 'endeavour' are strong enough to warrant any specific international commitments to regulate CO_2 emissions through UNCLOS.

Through Articles 204–206 of UNCLOS, parties are also obliged to conduct an EIA to monitor the effects of activities engaged in or permitted. This obligation is derived from the International Court of Justice (ICJ) case on *Legality of the Threat or Use of Nuclear Weapons*, which finds that states have a due diligence requirement to ensure that the environment of other states is respected by the actions taken within a state's jurisdiction and control. Incorporation of ocean acidification into EIAs could thus provide further opportunities for alleviation of the effects of ocean acidification. EIAs can therefore be fundamental to a state's obligation to prevent transboundary harm, to implement the precautionary principle, and to encourage sustainable development.

UNCLOS does not directly require the creation of MPAs in areas within national jurisdiction, rather it includes indirect mechanisms that can be used by states to realise their due diligence commitments with regard to the conservation, preservation, and sustainable use of living resources and biodiversity. With provisions differing for each category of ocean zone, obligations refer to limitations or restrictions on activities such as shipping, fishing, polluting activities, marine research, etc. within specific areas. (For example, Articles 21 and 22 with regard to territorial seas, Articles 58 and 61–68 on exclusive economic zones and Articles 76–85 on the outer continental shelf). Therefore, UNCLOS provides the international framework for legal support for individual state development of MPAs, and other states' approval and compliance with them.

Although there are obvious constrictions on the law of the sea in the direct regulation of CO_2 emissions, it is reasonable to conclude that individual state parties have obligations through UNCLOS as an umbrella framework to adopt regulatory measures to prevent, reduce and control CO_2 emissions from land-based sources, based on a state's due diligence requirements. Therefore, these obligations are not met merely by meeting climate change commitments unless pledges are specifically linked to ocean acidification.

Ocean acidification and biodiversity

The impact of ocean acidification on marine biological diversity is indisputable. The most widely studied effects are on the reduced availability of carbonate ions that are essential for shell formation in a variety of marine organisms. Many of these species are at the very bottom of the marine food chain and the domino effect of their loss has significant negative impacts. Similar effects are also being felt by reef-building corals, which rely on a specific pH balance to extract calcium from the seawater (see Chapter 15). These corals are the ocean's most biodiverse habitats, but acidification affects their ability to develop and their resilience to damage. Their existence is critical for the survival of countless other species, including humans, that are dependent on them for food. Recent studies also suggest that increased ocean acidification affects sensory functions in fish which can impact daily life and survival by influencing homing, settlement, foraging, and predator evasion. This is a particular concern for fisheries already under pressure from poor management and the legacy of overfishing. Therefore, serious socio-economic disruptions are to be expected as a result of the long-term ecological impacts of ocean acidification, including reduced food security and disruptions to livelihoods (Fennel and Vander Zwagg, 2016: 346). The legal frameworks related to these issues are described below.

Convention on Biological Diversity

Unmistakably, the CBD has a role to play in addressing ocean acidification. The three objectives of the agreement are the conservation of biodiversity, the sustainable use of its components, and the fair and equitable sharing of the benefits arising from the utilisation of genetic resources. Specifically, states and international organisations are requested to cooperate with the CBD in order to tackle the causes and effects of ocean acidification through United Nations General Assembly (UNGA) Resolution 63/111. However, the practical effectiveness of targets set through the CBD may be limited due to its ambiguous language and the lack of compliance or enforcement measures. Although there is no mandate for the CBD to be involved directly in ocean acidification mitigation, activities could include awareness-raising, improving collaboration between MEAs, and increasing adaptive capacity of the oceans.

In 2008, the Conference of the Parties (COP) to the CBD recognised ocean acidification as a threat to achieving the Convention objectives through its Decision IX/16. The COP also recognised ocean acidification as a significant issue requiring continuing consideration under the Programme of Work on marine and coastal biodiversity in Decision X/29. The COP encouraged incorporation of the most up-to-date understanding of ocean acidification into the strategies and action plans on natural biodiversity on both a national and local scale, with integration into marine and coastal management in consideration of their continued and improved protection. In 2012, a Joint Expert Review Meeting initiated successive expert review processes, through which the impacts of ocean acidification on marine biodiversity could be assessed collaboratively with relevant organisations. Through this, opportunities for information sharing, avoiding duplications, developing joint work programmes and gap-filling were identified. This is in line with the scientific cooperation provision of Sustainable Development Goal 14.3 (see Chapter 25), which specifically targets ocean acidification as a threat that must be addressed and minimised. The proactivity of the CBD to establish formal partnership arrangements between other MEAs such as UNCLOS and the UNFCCC shows that its membership recognises the importance of enhancing cooperation between MEAs due to their distinct yet related mandates.

As stated previously, tools for building ecosystem resilience include undertaking EIAs and establishing MPAs. While these will not stop ocean acidification, they can help to alleviate its effects by eliminating other stressors. Although direct CO_2 reduction measures within the CBD are limited, Article 14.1(a) obliges states to carry out EIAs should a project be 'likely to have significant adverse effects on biological diversity, with a view to avoiding or minimising such effects'. In Article 14.1(c), the importance of collaboration between states to minimise the negative impacts that activities have on biodiversity is also emphasised. Furthermore, the CBD contains mechanisms to establish a system of MPAs through Article 8, Aichi Biodiversity Target 11, and Articles 6 and 10 which oblige state parties to establish MPAs as part of their national strategy for marine biodiversity. As per Article 4, the scope of the CBD covers activities in all areas that could have a detrimental effect on biodiversity, including ABNJ. Although there is some debate as to whether the jurisdictional scope is applicable to ABNJ (Gjerde and Rulska-Domino, 2012: 360), the CBD does nevertheless contain some designation mechanisms in ABNJ, such as ecologically or biologically significant areas, as per COP 11 Decision XI/17. However, the CBD, which advocates MPAs as *in situ* protective mechanisms, is limited both by its own scope of application and by UNCLOS (as per CBD Articles 3, 4(b), and 8). In practice, it seems that the CBD recognises that it plays a role in the development of scientific criteria for the selection of MPAs but not in the actual designation of MPAs (Drankier, 2012: 298).

Biodiversity beyond national jurisdiction

Ocean acidification in ABNJ presents unique management challenges under the current law of the sea framework. While UNCLOS establishes an expansive framework for the protection and preservation of the marine environment, the principle of the freedoms of the high seas has led to a multiplicity of overlapping regimes for ABNJ conservation due to the different areas of jurisdiction and the systems of various treaties and institutions which have mandates over different sectors. Ocean acidification impacts ABNJ due to the combined effect of multiple stressors, including overfishing, destructive fishery practices, intensification of global shipping and marine pollution (Levin and Cheung, 2016: 3). Deep ocean characteristics in ABNJ create further vulnerabilities of species and ecosystems to acidification. Specifically, the slow growth rate of many marine organisms, which rely on the long-term stability of ecosystems to sustain their growth environment and therefore guarantee their long-term survival (Levin and Cheung, 2016: 3). Although the majority of habitats and species that are being directly affected by acidification are located within the coastal water boundaries of national jurisdictions (e.g., coral reefs, molluscs, crustaceans, etc.), the opportunity to enhance ocean resilience and adaptive capacity of marine biodiversity and ecosystems in ABNJ (as in areas within national jurisdiction) can be achieved through the mechanisms of EIAs and ABMTs such as MPAs.

These measures must consider the full range of impacts that global environmental change, including ocean acidification, ocean warming, and deoxygenation, is having on biodiversity in the marine environment and contribute to a meaningful development of best practice. In terms of law-making, this emphasises the need for multi-sectoral, collaborative integration of policies and management of conservation in ABNJ. This is the scope and content of the new international legally binding BBNJ Agreement currently under negotiation at the United Nations. Under the auspices of UNCLOS, the instrument will address the conservation and sustainable use of biodiversity in ABNJ. Four elements of the BBNJ package are currently under negotiation: (1) marine genetic resources, (2) ABMTs including MPAs, (3) EIAs, and (4) capacity building and the transfer of marine technology. Although the BBNJ Agreement does not have a mandate to address the causes of ocean acidification, it does provide an opportunity to enhance ocean resilience and fill gaps in the existing ABNJ management system. As per UNGA Resolution 69/292, this process *must not undermine* existing relevant instruments and frameworks, and thus the issue of institutional arrangements underpins all these aspects, and indeed the BBNJ Agreement as a whole. The vision for BBNJ is to build upon UNCLOS to strengthen the existing governance structure, safeguard marine biodiversity, and prioritise collaboration and science in the ABNJ management regime.

Integration of ocean acidification considerations into BBNJ could be achieved through ABMTs, EIAs, and capacity building and technology transfer. ABMTs could include prioritisation of fully or highly protected MPAs and MPA networks created with a special focus on locations with a range of biodiversity and potential for adaptation, adoption of management procedures that integrate with other forms of area-based conservation measures, and accounting for future oceanic changes as a result of climate change and ocean acidification. EIAs could include taking account of collective climate change impacts including acidification, deoxygenation, and warming; evaluating alternatives to planned activities and strengthening criteria for particularly vulnerable locations. Capacity building and technology transfer could include developing global scientific capacity to monitor pH levels of the ocean and increasing institutional capability to respond to impacts in a timely way.

However, there are a number of challenges. First, increasing the number of MPAs may not necessarily result in a higher degree of marine environmental protection. Due to the unavoidable tension between high seas freedoms and conservation measures, MPAs that are not effectively managed or enforced may only exist as so-called 'paper parks', with no evidence that they are actually contributing to marine protection (Matz- Lück and Fuchs, 2014: 155; see Chapter 21). This tension cannot be resolved simply by giving priority to marine environmental protection; measures may only protect the ocean effectively if accompanied by enforcement regulations. Currently, there are limited mechanisms available for restricting access or usage rights in ABNJ, with regulation of MPAs and associated habitat protection tools such as Particularly Sensitive Sea Areas falling under the purview of regional fisheries management organisations; certain MEAs with ABNJ authority, such as the OSPAR Convention; and global bodies such as the International Maritime Organisation and the International Seabed Authority, operating throughout the high seas as permitted by the MARPOL Convention. Additionally, there are practical challenges with regulation of high seas MPAs, including the policing and surveillance of remote locations, and challenges arising from poorly planned and poorly supported MPAs.

Next, there are also challenges associated with undertaking EIAs in ABNJ. Existing EIA obligations in ABNJ are uncomprehensive and poorly enforced by states. Provisions only refer to general obligations and do not detail any precise procedural, methodological, or implementing provisions to carry out an EIA (Warner, 2018: 39). There is also no direct requirement to take account of the impacts of ocean acidification in marine environments in an EIA (Warner, 2018: 39). The EIA obligation in ABNJ is also fragmented between different sectors and regions. Some activities in ABNJ such as fishing, deep sea mining, and shipping have limited EIA requirements in their processes, and others, such as regions like Antarctica, offer more comprehensive coverage for their activities in ABNJ (for example Article 3 of the Madrid Protocol; see Chapter 19). Therefore, while EIAs may play a role in addressing ocean acidification, in the absence of other mitigation measures to minimise the impacts of ocean acidification, these are likely to be inadequate.

The opinions expressed by states during BBNJ negotiations to date show a fragmentation of views between ensuring a comprehensive agreement is reached and 'not undermining' existing relevant legal instruments, frameworks and bodies. For example, the revised draft text of the BBNJ Agreement contains a draft Article 4 that replicates the UNGA's not-undermining clause, and as such has stirred some controversy. One of the key concerns in negotiations is whether primary authority over the designation, management, and enforcement of MPAs in ABNJ should be assigned to regional institutions or whether authority should be assigned to a global institution that would operate in all parts of the world. This fundamental issue is indicative of the dichotomy between freedoms of the high seas and protection of the marine environment and is crucially important for regulation of future ocean acidification adaptation measures under BBNJ. Broadly speaking, three different governance models have been proposed by states regarding this institutional architecture. These can be generally grouped into global, regional and hybrid approaches, with proposals spanning along a spectrum of options (Gjerde et al., 2019: 37).

Although a compromise between these governance models has not yet resulted in a solution in the BBNJ negotiations, the hybrid approach reflects a balance of competing interests between the global and the regional approaches and is the model most likely to be adopted in the end. This institutional arrangement would progress ocean governance further than the regional 'status quo' model, but would not extend powers to a global, centralised decision-making body as per the global approach. The hybrid approach takes the view that existing regional and sectoral bodies are the most efficient mechanisms for regulating activities in ABNJ since they have the expertise, experience and knowledge required for the adoption of measures (DOALOS, 2017). However, the main challenge with the hybrid approach is akin to that of the regional approach. If it relies too heavily on the existing governance organisations, it is uncertain how much, if at all, this approach will strengthen the current state of ocean governance.

Although negotiations are driven by the fact that the new BBNJ treaty should not undermine current regional and sectoral systems, it could be more helpful for negotiators to think in terms of complementarity and continuity, rather than attempting to delineate individual and overlapping mandates geographically and substantially. If a flexible and pragmatic approach towards the role of current and prospective institutions can be adopted, a substantive and effective agreement may be reached that will serve to allow a shift in focus towards an ecosystem-based governance approach to MPAs and EIAs in ABNJ. The BBNJ negotiations therefore provide an opportunity for ocean acidification mitigation if the final approach results in a robust institutional setup. A modern legal and regulatory structure could allow for greater cross-sectoral and cross-regional cooperation around the globe.

Conclusion

Ocean acidification is a complex issue with the potential to have extensive global and multi-sectoral impacts. It is a unique problem from a governance perspective since it shares common roots with climate change. However, addressing the mitigation of climate change causes may not necessarily assist in the equivalent reduction of ocean acidification or the mitigation of its impacts. Therefore, the current regulatory landscape is confused and has impeded effective responses under the climate, biodiversity and ocean regimes. Although the framework structure of UNCLOS is innovative and novel in incorporating external rules and standards, in the context of ocean acidification, this may be considered a barrier to successful mitigation. Assuming the Agreements under the climate change regime represent the generally accepted 'international standards' through UNCLOS Article 212, it could be argued that compliance with their obligations equates to compliance with UNCLOS due-diligence requirements. While there has been ambition recently to limit global temperature rise through the Paris Agreement, no comparable target has been set for addressing pH variation in the ocean. However, UNCLOS is a dynamic instrument, that has been characterised as a 'living treaty' due to its capacity to evolve in the face of new realities (Barrett and Barnes, 2016: 3). Consequently, states could be required by their UNCLOS due diligence obligation to prevent, reduce, and control ocean acidification – to go further than to merely comply with UNFCCC requirements, unless these deliberately address acidification.

To enhance ocean resilience and adaptive capacity in ABNJ, it is crucial that the new BBNJ Agreement establishes effective processes for the designation and management of MPAs, as well as comprehensive, precise, and joined-up EIAs. A flexible, holistic, hybrid approach to developing robust institutional arrangements, providing an effective mechanism for facilitating cooperation and filling existing governance gaps, is potentially where the BBNJ treaty could make the most difference. This may go a long way towards transitioning away from the traditional, fragmented, sector-based approach to ocean management and towards comprehensive protection through an ecosystem-based governance model. However, whether these measures will make a meaningful contribution to protection against ocean acidification will remain to be seen. Although it cannot be said that the law of the marine environment as it currently stands offers comprehensive protection against ocean acidification, the issues related to it are such that they must be taken into consideration in conjunction with a coordinated, holistic and harmonised approach with other MEAs, and not treated in isolation. This offers a unique opportunity for collaboration between MEAs with different mandates and purposes but with one common objective: to protect the planet, which includes saving our oceans through law.

References

Baird, R., Simons, M. and Stephens, T. (2009) *'Ocean Acidification: A Litmus Test for International Law'*, Carbon and Climate Law Review vol 3, no 4, p13.

Barrett, J. and Barnes, R. (2016) *'Law of the Sea – UNCLOS as a Living Treaty'* (eds) The British Institute of International and Comparative Law, Cambridge University Press.

Boyle, A. (2012) *'Law of the Sea Perspectives on Climate Change'*, International Journal of Marine and Coastal Law, vol 27, no 4, pp831–838.

Burns, W. (2007) *'Potential Causes of Action for Climate Change and Impacts under the United Nations Fish Stocks Agreement'*, Sustainable Development Law and Policy, vol 7, no 2, pp34–38.

Division of Oceans Affairs and the Law of the Sea (DOALOS). (2017), in Preparatory Committee established by General Assembly resolution 69/292, http://www.un.org/depts/los/biodiversity/prepcom_files/Chairs_streamlined_non-paper_to_delegations.pdf

Downing, Y. (2013) *'Ocean Acidification and Protection Under International Law from Negative Effects: A Burning Issue Amongst a Sea Of Regimes?'*, Cambridge Journal of International and Comparative Law, vol 2, no 2, pp242–273.

Drankier, P. (2012) 'Marine Protected Areas in Areas Beyond National Jurisdiction', Report on Research Question 2 of the Study on 'Biological Diversity and Governance of the High Seas' commissioned by the Netherlands Ministry of Affairs, Agriculture and Innovation.

Fennel, K. and Vander Zwagg, D. (2016) *'Ocean Acidification: Scientific Surges, Lagging Law and Policy Responses'*, in R. Warner and S. Kaye (eds) Routledge Handbook of Maritime Regulation and Enforcement, Routledge.

Gallo, N., Victor, D. and Levin, L. (2017) *'Ocean Commitments under the Paris Agreement'*, Nature Climate Change, vol 7, pp833–838.

Gjerde, K., Clark, N. and Harden-Davies, H. (2019) *'Building a Platform for the Future: The Relationship of the Expected New Agreement for Marine Biodiversity in Areas beyond National Jurisdiction and the UN Convention on the Law of the Sea'*, Ocean Yearbook, vol 33, no 1, pp1–44.

Gjerde, K. and Rulska-Domino, A. (2012) *'Marine Protected Areas beyond National Jurisdiction: Some Practical Perspectives for Moving Ahead'*, The International Journal of Marine and Coastal Law, vol 27, no 2, pp351–373.

Harrison, J. (2017) *Saving the Oceans Through Law*, The International Legal Framework for the Protection of the Marine Environment. Oxford University Press, Oxford.

Hull, E. (2016) *'Ocean Acidification: Legal and Policy Responses to Address Climate Change's Evil Twin'*, Washington Journal of Environmental Law and Policy, vol 6, no 2, pp348–386.

Levin, L. and Cheung, W. (2016) 'Policy Brief: Climate Change in Oceans beyond National Jurisdictions', https://nereusprogram.org/reports/policy-brief-climate-change-in-oceans-beyond-national-jurisdictions/

Levin, L. and Le Bris, N. (2015) *'The Deep Ocean under Climate Change'*, Science, vol 350, no 6262, pp766–768.

Masson-Delmotte, V. et al. (2018) 'Global warming of 1.5°C. An IPCC Special Report on the impacts of global warming of 1.5°C above pre-industrial levels and related global greenhouse gas emission pathways, in the context of strengthening the global response to the threat of climate change, sustainable development, and efforts to eradicate poverty' Intergovernmental Panel on Climate Change, http://www.ipcc.ch/report/sr15

Matz-Lück, N. and Fuchs, J. (2014) *'The Impact of OSPAR on Protected Area Management beyond National Jurisdiction: Effective Regional Cooperation or a Network of Paper Parks?'*, Marine Policy, vol 49, pp155–166.

Meinschausen, M., Raper, S.C. and Wigley, T.M. (2011) *'Emulating Coupled Atmosphere-Ocean and Carbon Cycle Models with a Simpler Model, MAGICC6–Part 1: Model Description and Calibration'*, Atmospheric Chemistry and Physics, vol 11, no 4, pp1417–1456.

Molenaar, E.J. (2014) 'The International Legal Framework Relating to Area-Based Measures for Merchant Shipping at the Global Level', Presentation at the A-LEX Interdisciplinary Seminar, https://tinyurl.com/4x9z4phj

Scott, K. (2020) *'Ocean Acidification: A Due Diligence Obligation under the LOSC?'*, The International Journal of Marine and Coastal Law, vol 35, no 2, pp382–408.

Synthesis Report by the Secretariat. (2021) 'Nationally Determined Contributions under the Paris Agreement', FCCC/PA/CMA/2021/2

Warner, R. (2018) '*Oceans in Transition: Incorporating Climate-Change Impacts into Environmental Impact Assessment for Marine Areas beyond National Jurisdiction*', Ecology Law Quarterly, vol 45, pp31–51.

Yadav, S. and Gjerde, K. (2020) '*The Ocean, Climate Change and Resilience: Making Ocean Areas Beyond National Jurisdiction More Resilient to Climate Change and Other Anthropogenic Activities*', Marine Policy, vol 122, p104184.

5

REGIMES FOR OCEAN MANAGEMENT

Regional seas programmes and blue carbon ecosystems

Luciana Fernandes Coelho and Nata Tavonvunchai

This chapter aims to contribute to the debate about the Anthropocene by assessing the interplay between the ocean governance and climate change regimes at the regional level. In the Anthropocene, humankind is acting as a geological force (Chakrabarty 2018). Climate change is defined as the 'change of climate which is attributed directly or indirectly to human activity that alters the composition of the global atmosphere and which is in addition to natural climate variability observed over comparable time periods' (e.g., increased greenhouse gas (GHG) levels) (UNFCCC, Article 1(2)). It has been flagged as an acute example of humanity as a geological agent (Pörtner et al. 2021). Impacts of climate change are evidenced in the ocean by the changing of food-web dynamics, risk of species extinction (e.g., coral reefs and associated species), ocean acidification, and sea-level rise (Aswani et al. 2018, Pörtner *et al.* 2019) (see Chapters 4 and 14). Regimes can be broadly defined as 'sets of norms, decision-making procedures and organisations coalescing around functional issue-areas' (Young 2012: 9). Treaties have a predominant role in regime creation (Trevisanut, Giannopoulos, and Holst 2020). Taking this into account, the United Nations Convention on the Law of the Sea (hereafter LOSC or Convention) is considered the backbone of the ocean governance regime. Meanwhile, the climate change regime encompasses the norms, institutions, and processes coalescing around the United Nations Framework Convention on Climate Change (UNFCCC) and related instruments, such as the Paris Agreement (see Chapter 2).

There is an increasing body of literature investigating the interplay between regimes. In international law, regime interaction is conceived as 'where rules, institutions and the operation of one legal regime is affected by another legal regime' (Trevisanut, Giannopoulos and Holst 2020: 4). Although this definition is limited to the legal field, by using a broad definition of regimes, it could be expanded to mean the influence of rules, institutions, and the operation of the ocean governance regime in the climate change regime and vice versa. While climate change was seldom on the international agenda at the time the LOSC was negotiated (Stephens 2015), there is currently significant attention on the role of the ocean in carbon storage (i.e., acting as a carbon sink), which is when atmospheric carbon is absorbed and stored (McDonald, McGee and Barnes 2020). Indeed, there is much convergence between the two regimes (see Chapter 3). For instance, from a normative point of view, safeguarding marine ecosystems falls under the obligation of states to take measures to protect and preserve

DOI: 10.4324/9781315149745-7

ecosystems (Article 194.5, LOSC) and under the obligation of Parties to conserve and enhance carbon sinks (Article 4(d), UNFCCC and Article 5(1), Paris Agreement). Therefore, cooperation between institutions whose mandate can serve both regimes is desirable.

This chapter focuses on carbon storage through natural means. Coastal marine ecosystems, specifically seagrass, salt marshes, and mangroves, have a significant role to play in ocean-based climate change governance as they absorb atmospheric carbon and store it in sediment or biomass (Macreadie et al., 2019). 'Blue carbon' is the organic carbon stored by coastal ecosystems, such as seagrass, salt marshes, and mangroves (Silver et al. 2015, Martin et al. 2016, Macreadie et al. 2019). Considering that these ecosystems are responsible for 50–71% of the carbon stored in ocean sediment (Nellemann and Corcoran 2009), their degradation through climate change (Hoegh-Guldberg et al. 2019) and direct human activity (e.g., coastal development and increased pollution) (Moraes 2019, Pendleton et al. 2012), contribute to the release of GHG emissions (UNESCO 2020).

Against this background, debates on policy and economic incentives for sustainably managing blue carbon ecosystems have gained traction in ocean governance, promoting opportunities to verify how the ocean and climate regimes have been interacting. At a global level, communication between these regimes has been substantially investigated in the literature (Gattuso et al. 2018). Furthermore, increasing acknowledgment of the crucial role of regional ocean governance in delivering sustainability (Wright et al. 2017) has inspired studies on the climate/ocean nexus focused on particular regions (Warner and Schofield 2012). However, there is a paucity of publications assessing how this interplay takes place (or not) in Regional Seas Programmes (RSPs). These programmes are under the auspices of the United Nations Environmental Programme (UNEP) and described as the 'principal platform[s] for implementing global conventions, multilateral agreements, global programmes and initiatives [to conserve marine and coastal environments] at the regional level' (Oral 2015: 361). Therefore, the programmes have the mandate to serve both the ocean governance and the climate change regimes, promoting cooperation between them in the protection of blue carbon ecosystems.

At the RSPs' 14th Global Meeting in 2012, it was recognised that, although RSPs already protected blue carbon ecosystems indirectly through action plans and protocols, more effort was needed to develop specific actions targeting these ecosystems. It was recommended to align the work of the programmes with UNEP's Blue Carbon Initiative towards this goal (UNEP 2012). In addition, RSPs would be responsible for evaluating and informing blue carbon methodologies with the idea of making them practical and relevant to each region (UNEP 2012). However, no reference to a follow-up discussion on blue carbon ecosystems at RSPs could be found at UNEP's website or in the literature. Hence, it is unclear whether additional steps have been taken. Considering the mandate of RSPs in bridging regional and global ocean initiatives, as well as the ocean governance and climate change regimes, this chapter examines whether and how RSPs have been promoting the conservation of blue carbon ecosystems through ecosystem-centred approaches (i.e., ecosystem-based adaptation (EbA)). EbA is defined as 'a strategy for the integrated management of land, water, and living resources that promotes conservation and sustainable use in an equitable way' (CBD 2010) and is considered one of the best strategies to promote the conservation and sustainable use of marine ecosystems (Agardy et al. 2011). We give an overview of the RSPs' structure and assess those that have adopted instruments promoting EbA. The challenges and good practices of RSPs in protecting blue carbon ecosystems are analysed by comparing and contrasting Progress Reports on the implementation of the Regional Seas Strategic Directions 2017–2020 (UNEP 2019). In concluding remarks, we propose some possible ways forward.

Regionalism in the ocean governance and climate change regimes: the cases of regional seas programmes

Overview of the multiple formats of regional seas programmes

If 'the law of the sea is inherently global' (Boyle 2000: 20), the same can be said about climate change law, as both regulate environmental components not limited to States' boundaries. Nonetheless, both international legal frameworks acknowledge the benefits of complementary regional approaches to manage natural resources, engage local stakeholders, and address regional realities such as geographical features, ecosystems, and social-policy-economic situations (approaches loosely referred to in this chapter as norms, action plans, processes, and institutions). For instance, the LOSC champions cooperation and the adoption of regional standards concerning the protection and preservation of the marine environment (Articles 123, 197, 212(3), 276, and 277). Similarly, regional responses to mitigate and adapt to climate change are fostered in the UNFCCC (Article 4(1)) and Paris Agreement (Article 7(2)). Indeed, the regional level is said to be primed to facilitate ambitious and purposeful goals, coordination, and accountability (Billé et al. 2017).

Despite the multitude of arrangements within the umbrella of regionalism (Betsill 2005) and the absence of a prescriptive definition for 'marine region' in the LOSC, the UNFCCC and the Paris Agreement (Giannopoulos 2021), the Regional Fisheries Bodies (RFBs), RSPs and the Large Marine Ecosystems (LMEs) are depicted as significant mechanisms for promoting marine science-policy interface at regional level (Billé et al. 2017). Among those, RSPs have been very active (Oral 2015), and in some cases are the sole entry point for environmental conservation (Billé et al. 2017). RSPs are action-oriented mechanisms dedicated to conserving the marine and coastal environment. They have a twofold mandate to bridge the implementation of Multilateral Environmental Agreements (MEAs) at the regional and local levels, and to adopt agreements and action plans responding to regional priorities through regular meetings of governing and technical bodies (UNEP n.d.). The first RSP, inaugurated in the Mediterranean region in 1974, aimed at coordinating action for preventing, mitigating, and responding to pollution by States sharing a common marine space (Haas 1991, Oral 2015). Over the years, the number of RSPs and their mandates have expanded to incorporate other local environmental emergencies. For instance, although not a global concern, the rapid influx of *sargassum* seaweed in the Caribbean since 2011 has been monitored by the Caribbean Region RSP, which seeks to propose policy actions (UNEP 2021).

The 18 RSPs cover an extensive oceanscape across 146 countries (UNEP 2016). UNEP provides technical support and acts as the policy coordination body, stimulating cooperation between countries and with other mechanisms (Vallega 2002). Table 5.1 lists the various regions covered by RSPs and their relationships to UNEP.

The legal and policy structure of RSPs are generally based on action plans underpinned with a framework convention followed by subject-based protocols (Haas 1991, Akiwumi and Melvasalo 1998). However, no treaty has been adopted by the East Asia Seas, the Arctic Region, NOWPAP, or SAS RSPs (see Chapter 19). Institutional support to RSPs is usually provided by a centralised body, a Secretariat or Regional Coordinating Unit (RCU), and/or by decentralised Regional Activity Centres (RACs). The RCU provides administrative and diplomatic functions, whereas the RAC provides legal and technical assistance for implementing action plans. Additional bodies and individuals (e.g., academia and nongovernmental organisations (NGOs)) may provide technical inputs as part of a Regional Activity Network (RAN) (UNEP n.d.: 2). Indeed, the participation of external contributors, notably

Table 5.1 Regional seas programmes

UNEP-administered RSPs	*RSPs established by UNEP but not administered by it*	*RSPs cooperating with UNEP*
Caribbean Region	Black Sea Region	Arctic Region
East Asian Seas (COBSEA)	North-East Pacific Region	Antarctic Region
Eastern Africa Region (Western Indian Ocean) (WIO)	Red Sea and Gulf of Aden (PERSGA)	Baltic Sea (HELCOM)
Mediterranean Region	Regional Organisation for Protection of the Marine Environment (ROPME) Sea Area	North-East Atlantic Region (OSPAR)
North-West Pacific Region (NOWPAP)	South Asian Seas (SAS)	
Western Africa Region	South-East Pacific Region	
Caspian Sea (administered by the UNEP's Regional Office for Europe)	Pacific Region	

Source: Based on UNEP (n.d.).

NGOs, has been acknowledged for the relative success of the Mediterranean Region, OSPAR, and Caribbean Region programmes by providing an entry point for scientific evidence and community engagement (Van Dyke 2013).

While initial financial support comes from UNEP (Haas 1991, Akiwumi and Melvasalo 1998), the aim is that RSPs will eventually be sustained by member States, polluter-/user-pay schemes, and international contributions (Oral 2015: 351). Additional sources of financial support may come from bilateral agreements, international and regional schemes, and the private sector (e.g., the Global Environmental Facility and the World Bank) (Rochette and Billé 2012). Intermittent funding and limited resource capacity have been underscored as major barriers to implementing the instruments adopted and achieving sustainability (Billé and Rochette 2013, Van Dyke 2013). Additional setbacks to achieve the objectives of some RSPs include the absence of monitoring mechanisms to evaluate policies' implementation, institutional arrangements tailored towards a sole environmental agenda with limited incorporation of EbA, and limited cooperation and coordination within RSPs and between RSPs and other global, regional, and local initiatives (Chung 2010).

From an environmental agenda to ecosystem-based management

Ecosystem-based adaptation specifically and climate change generally are topics gaining prominence among RSPs. The emergence of environmental instruments, particularly LOSC and the Convention on Biological Diversity (CBD), was a first step in this direction, inspiring amendments in constitutive treaties of some RSPs to incorporate legal mechanisms, such as the precautionary principle (Oral 2015). Debates on adopting management approaches utilising ecosystem-based approaches (EbA), or ecosystem-based management (EbM), have gained momentum. More recently, RSPs are increasingly mentioning blue carbon and blue carbon resources. Exploring how RSPs support natural blue carbon resources and other EbAs would be a useful next step to take in assessing how the interplay between ocean governance and climate change regimes can be more effective. By acknowledging anthropogenic impacts and taking a holistic approach to drivers affecting marine resources and ecosystems rather than focusing on their individual components (Curtin and Prellezo 2010, Churchill 2015), EbA is an appropriate governance approach for managing the interconnected nature of marine ecosystems (Agardy et al. 2011). EbM tools include marine protected areas (MPAs), marine spatial planning (MSP), and integrated coastal zone management (ICZM).

MPAs are the most commonly used EbM tool for marine conservation (Moraes 2019). Indeed, the literature concurs about its effectiveness not only in preventing ecosystem degradation and restoring ecosystems (Howard et al. 2017) but also in contributing to other ecological and socio-economic benefits, such as increasing biodiversity and improving fisheries (Bennett and Dearden 2014). MPAs are also said to be useful in safeguarding blue carbon through building ecosystem resilience (Moraes 2019). When incorporated into well-managed and efficient MPAs, a focus on blue carbon can complement the delivery of other positive benefits (Moraes 2019), such as ecosystem restoration, additional funding, and policy support for MPAs (Howard et al. 2017). Conversely, it is estimated that only 24% of protected areas globally are 'soundly' managed (Bennett and Dearden 2014). Even though marine spaces are dynamic and interconnected to a much larger spatial area, MPAs are usually designed in isolation and without the consideration of the wider ecosystem to which they are connected (Agardy et al. 2011, Rees et al. 2018). MPAs are often too small relative to the conservation targets they set out to achieve, and they fall short in addressing degradation in their unprotected surrounding areas (Agardy et al. 2011). Therefore, MPA mismanagement can limit the achievement of positive outcomes (Bennett and Dearden 2014), eventually leading to ecosystem degradation rather than protection, which can result in blue carbon ecosystems becoming 'blue carbon emitters' (Moraes 2019).

To reach their potential and be resilient, MPAs would benefit from integration into wider management strategies aligning with EbA (Rees et al. 2018), such as ICZM and MSP, and/or a wider network of MPAs, rather than as 'islands of protection' (Bennett and Dearden 2014). This is in accordance with the Aichi Biodiversity Targets' goal of connecting protected areas of interest with other protected areas and conservation measures that include protection and restoration of ecosystems in their mandates (Moraes 2019). MSP and ICZM have a broader conceptual and spatial approach, seeking to incorporate 'social, economic, cultural, political, and environmental considerations' across the broader 'land and seascape' (Bennett and Dearden 2014: 103). Both also acknowledge the dynamics of ecosystem processes and are aligned with a regional approach (Agardy et al. 2011). In addition, MSP considers the use and regulations of its adjacent areas (i.e., zoning) creating 'buffer zones' (e.g., blue corridors), providing bases to create MPA networks (Rees et al. 2018). The broader perspective of MSPs can also better represent a region's ecological state and protect other important areas that may not have been selected as MPA sites (e.g., areas 'too threatened' as a result of 'high human impact') (Rees et al. 2018: 248). Similar to MSP, ICZM aims to foster integrated, holistic planning that involves a multi-perspective approach (Visbeck et al. 2014).

As previously alluded to, regional approaches must be balanced with local planning and be inclusive to local stakeholders in order to effectively deliver intended outcomes (Agardy et al. 2011). Therefore, the adoption of each EbM tool requires the involvement of local communities, indigenous peoples, and vulnerable populations to ensure more cohesion and equity in conservation planning (Rees et al. 2018). The adoption of legal instruments incorporating EbM tools differs among RSPs (UNEP n.d.: 3). Several RSPs have adopted instruments supporting the establishment of MPAs, and PERSGA went further and adopted a protocol creating a network of MPAs (Oral 2015, UNEP n.d. 8). The Mediterranean Sea and Western Africa Region enacted Protocols on ICZM in 2008 and 2017, respectively (UNEP n.d. 5, UNEP n.d. 6), and the Eastern Africa Region is developing an instrument on the topic (UNEP n.d.: 7). In 2016, the Baltic Sea programme adopted a Guideline for implementing MSP (Baltic Marine Environment Protection Commission 2016). More generally, OSPAR, HELCOM, and the Mediterranean programmes are studying how to incorporate EbM in their action plans and strategies, in line with the European Union Marine Strategy Framework Directive (MSFD) (van Leeuwen et al. 2014). In 2016, the Western African Region adopted a Protocol on Sustainable Mangrove Management, which is specific to blue carbon (UNEP n.d. 5).

UNEP is adopting medium-term strategies seeking to align its work with emerging environmental topics, such as the Sustainable Development Goals (SDGs) (UNEP 2016, UNEP n.d. 3; see Chapter 25). The 2018–2021 Strategy focuses on climate change, resilience, biodiversity, environmental governance, waste and chemical product management, resource efficiency, and monitoring (UNEP 2016). There has also been evidence of RSPs collaborating with other regional mechanisms, such as RFBs and LMEs (Billé et al. 2017), and regional platforms, such as in the case where the WIO RSP received funding from the European Union for one of its projects (WIO Region 2019). Since 2014, Regional Seas Strategic Directions have been elaborated aiming at improving the governance of RSPs and aligning their work with initiatives developed by other bodies in the same region (UNEP n.d.: 4). Four key shared priorities were identified at the 2017–2020 Strategic Directions: pollution, climate change, and ocean acidification, extraction of living and non-living resources, and governance. Both the 2017–2020 RSPs Strategic Direction and the 2018–2021 UNEP Medium-Term Strategy foster the implementation of policies and actions to address climate change and biodiversity loss and strengthen governance. This goes hand-in-hand with the realisation that some RSPs already have adopted protocols and action plans aiming to implement EbA and protect blue carbon ecosystems. Moving ahead from the adoption of instruments to the implementation stage, the following section looks at the 2019 RSP's reports informing the Strategic Directions' implementation.

Regional seas programmes and climate change: a synthesis of blue carbon ecosystems

RSP protection of blue carbon ecosystems

RSPs are asked to periodically submit reports to UNEP informing it of actions adopted and challenges faced by them in implementing the Strategic Directions. As the reports are structured around the same topics, they provide a valuable opportunity for comparing the programmes. The latest implementation reports were submitted in 2019 and addressed actions implemented since 2016. (While RSPs not mentioned in this section may also be taking action towards implementing EbA and protecting blue carbon ecosystems, our assessment considers only the eight RSPs that submitted reports.) Our brief analysis seeks to identify common challenges and best practices by examining the extent of actions in place to incorporate EbM tools and protect blue carbon ecosystems. Though the scope of EbA activities can be considered broadly, with species conservation and water quality/pollution addressed by UNEP (UNEP, 2012), we look only at direct action for ecosystems. Moreover, while coral reef ecosystems have been mentioned and support blue carbon ecosystems, they are not within the definition of blue carbon ecosystems considered here (Lovelock and Duarte 2019, Guerra-Vargas, Gillis, and Mancera-Pineda 2020). For this reason, coral reefs will also not be the focus of the assessment but may be used as examples of EbA usage. (For more on coral reefs, see Chapter 15)

A heat map facilitates understanding the level of commitment (see Table 5.2). The key EbA activities in regard to conservation are examined by considering the main themes from the reports, which are (1) establishing legal and policy framework and organisational setup (standards); (2) monitoring and evaluation (M&E); (3) capacity building (training); and (4) external funding. Cells in dark grey in Table 5.2 indicate that sufficient action has been implemented in that the activity has become commonplace for the RSP. Medium grey cells in the table indicate initiatives currently being piloted. Light grey cells refer to activities acknowledged as important, but where substantial action has not yet taken place. The column on the right indicates which programmes have mentioned actions to protect blue carbon ecosystems.

Mitigating the impacts of marine litter seems to have been on the core of the RSPs agenda, whereas ecosystem protection has been addressed by several, but not all programmes. The Caribbean Region and PERSGA have progressed the most in terms of ecosystem conservation activities, followed by WIO, COBSEA, and the Pacific Region. Neither the SAS nor the Caspian Sea makes specific mention of blue carbon or EbA in their reports. Generally, where blue carbon ecosystem conservation was mentioned, this has only been as stand-alone initiatives, which does not follow the ecosystem approach. In addition, most initiatives towards blue carbon are only at the ideation or pilot phase, and they are focused solely on mangroves and seagrass. As the definition of blue carbon also includes salt marshes, it may be worthwhile for RSPs to also champion actions to protect these ecosystems, where applicable.

The Caribbean Region, WIO, and the PERSGA appear to be the 'leading' programmes with respect to blue carbon and EbA actions as these topics are alluded to most in their reports. This could be a result of their strong institutional, legal, and policy frameworks as the literature emphasises these to be relevant factors of progress in shifting from a single-focus environmental-harm perspective focused on pollution to a more holistic ecosystem approach (Billé et al. 2017). Strong frameworks are also said to facilitate external funding, of which the Caribbean Region and WIO are beneficiaries (Cartagena Convention Secretariat 2019, WIO Region 2019). While both of these RSPs are under UNEP's administration, this is not a necessary factor for a purposeful framework or effectiveness (e.g., see the Red Sea RSP) (PERSGA 2019). Similarly, even though COBSEA works without a framework convention, it was able to achieve some progress on ecosystem protection according to its report (East Asian Seas n.d.).

Table 5.2 High-level summary of RSP analysis for marine ecosystem protection

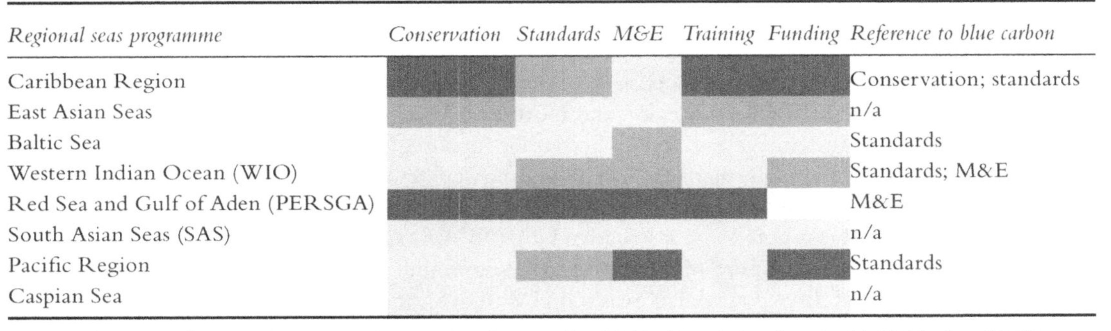

Regional seas programme	*Conservation*	*Standards*	*M&E*	*Training*	*Funding*	*Reference to blue carbon*
Caribbean Region						Conservation; standards
East Asian Seas						n/a
Baltic Sea						Standards
Western Indian Ocean (WIO)						Standards; M&E
Red Sea and Gulf of Aden (PERSGA)						M&E
South Asian Seas (SAS)						n/a
Pacific Region						Standards
Caspian Sea						n/a

Sources (from top to bottom): Cartagena Convention Secretariat (2019), East Asian Seas (n.d.), Baltic Sea (2019), Western Indian Ocean Region (2019), PERSGA (2019), South Asian Seas (SAS) Region (2019), South Pacific (2019), Tehran Convention (2019).

Some reports convey a fair amount of emphasis on MPAs as a tool to strategically manage marine areas (Agardy et al. 2011). For example, the PERSGA identifies MPAs as suitable locations for EbM activities (PERSGA 2019). While MSP has been mentioned by RSPs, actions and legal instruments about it are mostly in the pilot stages. As identified from the WIO's MSP training workshop, the region still needs high-level policy decisions, clear goals, and objectives for MSP implementation (WIO Region 2019). The conclusion was that this requires, at the institutional level, a legally responsible entity as well as sector-neutral MSP, and the establishment of intersectoral MSP steering committees to share data, support implementation, and monitor progress (WIO Region 2019). Within the WIO, it is recommended to embed ecosystem-based MSPs in national legislation and policies, and in the ratification of the ICZM Protocol (WIO Region 2019).

Opportunities for effective conservation of blue carbon ecosystems within RSPs

The implementing reports and existing literature provide a basis for saying that, besides political will and legal and policy framework, RSPs must address three main issues to effectively incorporate EbM measures: (1) finance and human resource capacity, (2) collaboration and community engagement, and (3) M&E. Capacity limitations were noted in almost all the reports (Cartagena Convention Secretariat 2019, PERSGA 2019, SAS Region 2019, South Pacific 2019, WIO Region 2019, East Asian Seas n.d.). This is partially due to few RSPs having adapted their institutional structure and mandates from a solely environmental perspective to EbM (Billé et al. 2017). Financial shortfall has also posed challenges in improving capacity remarkably for COBSEA, Caribbean Region, WIO, and the Mediterranean Sea (Billé et al. 2017). However, from the progress report, the Caribbean Region (Cartagena Convention Secretariat 2019) and South Pacific (2019) received comparatively considerable external funding, including from the private sector. The significance of collaboration with the local community, within RSPs, and with other inter-regional institutions was echoed across various reports (Cartagena Convention Secretariat 2019, PERSGA 2019, SAS Region 2019, South Pacific 2019, WIO Region 2019, East Asian Seas n.d.). At the community level, increasing public awareness and engaging local communities in policymaking and marine environment protection complemented private investments and proved fruitful for some RSPs (Cartagena Convention Secretariat 2019, WIO Region 2019). For example, community involvement in EbM and MPA management has been used to increase monitoring capacity in MPAs in PERGSA (PERSGA 2019).

Coordination and cooperation between States within the same regional area and RSPs are also key to providing substantial output (Hegland, Raakjær and Tatenhove 2015). This can be done with a focus on thematic issues through exchanging information and/or enhanced integration of new mechanisms with existing ones (Billé et al. 2017). In this regard, the collaboration between the Caribbean Region and Northeast Atlantic RSPs to sustainably manage marine and coastal resources exemplifies action for inter-regional collaboration and knowledge sharing to sustainably manage marine and coastal resources (Cartagena Convention Secretariat 2019). Receiving input from and exposure to the work of other RSPs, especially those nearby, has been noted as constructive by COBSEA. Collaboration has been pointed out as one of the most important aspects of the Programme in many reports (SAS Region 2019, South Pacific 2019, East Asian Seas n.d.). Moreover, increased synergy and awareness of RSPs' work, both among RSPs and by UNEP's offices, was highlighted as relevant to promote effective implementation of the commitments and avoid duplicated efforts (Cartagena Convention Secretariat 2019). While inter-regional collaboration is important, it was noted that setting up these governance structures can be resource-intensive in terms of time taken and finance needed (Hegland, Raakjær and Tatenhove 2015, Cartagena Convention Secretariat 2019). This feeds back into the previously mentioned challenge of resource limitations.

Limited capacity in M&E was also pointed out as a barrier to achieving conservation goals (Cartagena Convention Secretariat 2019, SAS Region 2019, South Pacific 2019, WIO Region 2019, East Asian Seas n.d.), and perhaps the most significant one, as baseline data is essential for effective ecosystem conservation policy (Moraes 2019, Fanning et al. 2021), enforcement and compliance (Agardy et al. 2011). Indeed, restricted development in M&E can lead to uncertainty in the performance of conservation initiatives as there would be no 'quantifiable goals' set to have been achieved and no baseline to assess progress against (WIO). Conversely, improved enforcement capacity has been shown to improve ecological, social, and economic outcomes of MPAs (Moraes 2019). Using the example of blue carbon, monitoring is crucial to understand carbon storage because limited baseline data on the coverage and condition of blue carbon ecosystems poses a challenge when tracking progress (Moraes 2019). This is particularly the case in Asia, Africa, South America, and the South Pacific, where datasets on seagrass and salt marshes are often underestimated, missing, or incomplete (Moraes 2019). The institutional frameworks of RSPs complement the corresponding countries' national strategies on ecosystem protection, with the latter ultimately responsible for implementation (Billé et al. 2017). Adopting EbM has also been recommended for national strategies (Moraes 2019). Conservation enforcement and building baseline knowledge rely on and is directly related to, training (Moraes 2019), linking back to capacity limitations.

Conclusion

Addressing the worldwide challenges of marine environmental governance requires cooperation and coordination among different norms, institutions, and operations at global, regional, and local levels. The protection of blue carbon ecosystems is mutually supported under the legal framework of the ocean governance and climate change regimes, making a comparison of them an insightful case to visualise how RSPs can promote synergistic interactions between regimes. Regional ocean governance approaches are considered most effective at addressing local challenges as well as setting more ambitious targets. RSPs are relevant mechanisms for enforcing international obligations at the local level and promoting cooperation to address regional challenges. Nonetheless, limited consistency among legal and policy frameworks, institutional settings, human resources and technical capacity, collaboration and community engagement, M&E, and funding, have been barriers to achieving RSPs' objectives (Billé et al. 2017).

UNEP and RSP Medium-term Strategies and Strategic Directions seek to align RSP's work with global discussions, in particular climate change and biodiversity loss. However, the achievement of this objective is uncertain. From analysing the progress reports submitted in the review of the Regional Seas Strategic Directions (2017–2020), some RSPs have incorporated EbM measures, including MPA, MSP, and/or ICZM. Extending RSP mandates and institutional capacity would be a next step to localising marine ecosystem conservation. This would be followed by diversifying funding mechanisms, including cooperation with the private sector (see Chapter 10), and enhancing support for training and capacity development. Fostering collaboration and cooperation with local communities, between RSPs and with other mechanisms (e.g., RFB, LME, NGOs, universities, and citizen science initiatives), could improve community engagement and knowledge sharing, avoid duplicated efforts, and enabling M&E. Moreover, protecting blue carbon ecosystems, notably mangroves, seagrass, and salt marshes, which have relevance to addressing GHG emissions and mitigating climate change (e.g., by protecting coastlines from storm impact and rising sea levels), can be an impactful avenue in which RSPs can enhance interrelated ecological and socio-economic benefits. Taking that into account, the progress of the Caribbean Region, WIO, and the PERSGA could be used as examples to inspire other programmes for effective marine governance amidst global environmental change.

Acknowledgements

Luciana Fernandes Coelho would like to acknowledge the funding of the World Maritime University-Sasakawa Global Ocean Institute by The Nippon Foundation and the financial support of the Land-to-Ocean Leadership Programme provided by the Swedish Agency for Marine and Water Management and the German Federal Ministry of Transport and Digital Infrastructure. Nata Tavonvunchai would like to thank Siân Rees for her valuable feedback in the development of this paper and Eric Zusman for sharing this publication opportunity.

References

Agardy, T., di Sciara, G. N. and Christie, P. (2011) 'Mind the Gap: Addressing the Shortcomings of Marine Protected Areas through Large Scale Marine Spatial Planning', *Marine Policy*, 35(2), pp. 226–232.

Akiwumi, P. and Melvasalo, T. (1998) 'UNEP's Regional Seas Programme: Approach, Experience and Future Plans', *Marine Policy*, 22(3), pp. 229–234.

Aswani, S., *et al.* (2018) 'Marine Resource Management and Conservation in the Anthropocene', *Environmental Conservation*, 45(2), pp. 192–202.

Baltic Marine Environment Protection Commission. (2016) 'Guideline for the Implementation of Ecosystem-based Approach in Maritime Spatial Planning (MSP) in the Baltic Sea Area.' Available at: <https://helcom.fi/media/documents/Guideline-for-the-implementation-of-ecosystem-based-approach-in-MSP-in-the-Baltic-Sea-area_June-2016.pdf>.

Baltic Sea. (2019) 'Progress Report on the Implementation of the Regional Seas Strategic Directions 2017–2020.' Available at: <https://www.unep.org/events/un-environment-event/regional-seas-programme-annual-meeting>.

Bennett, N. J. and Dearden, P. (2014) 'From Measuring Outcomes to Providing Inputs: Governance, Management, and Local Development for More Effective Marine Protected Areas', *Marine Policy*, 50, pp. 96–110.

Betsill, M. M. (2007). 'Regional Governance of Global Climate Change: The North American Commission for Environmental Cooperation', *Global Environmental Politics*, 7(2), pp. 11–27.

Billé, R. and Rochette, J. (2013) 'Bridging the Gap between Legal and Institutional Developments within Regional Seas Frameworks', *The International Journal of Marine and Coastal Law*, 28(3), pp. 433–463.

Billé, R., *et al.* (2017) 'Regional Oceans Governance: Making Regional Seas Programmes, Regional Fishery Bodies and Large Marine Ecosystem Mechanisms Work Better Together', Nunes, P.A.L.D. Svensson, L. E. and Markandya, A. (eds.) *Handbook on the Economics and Management of Sustainable Oceans*. Edward Elgar Publishing, pp. 493–518.

Boyle, A. (2000) 'Globalism and Regionalism in the Protection of the Marine Environment', Vidas, D. (ed.), *Protecting the Polar Marine Environment: Law and Policy for Pollution Prevention*. Cambridge University Press, pp. 19–33.

Cartagena Convention Secretariat. (2019) Progress Report on the Implementation of the Regional Seas Strategic Directions 2017–2020. Available at: <https://www.unep.org/events/un-environment-event/regional-seas-programme-annual-meeting>.

Chakrabarty, D. (2018) 'The Seventh History and Theory Lecture: Anthropocene Time', *History and Theory*, 57(1), pp. 5–32.

Chung, S. Y. (2010) 'Strengthening Regional Governance to Protect the Marine Environment in Northeast Asia: From a Fragmented to An Integrated Approach', *Marine Policy*, 34(3), pp. 549–556.

Churchill, R. (2015) 'The LOSC Regime for Protection of the Marine Environment – Fit for the Twenty-First Century?', Rayfuse, R. (ed.), *Research Handbook on International Marine Environmental Law*. Edward Elgar Publishing, pp. 3–30.
Convention on Biological Diversity (CBD). (2010) 'Ecosystem Approach.' Available at: <https://www.cbd.int/ecosystem/description.shtml>. Accessed: June, 2021.
Curtin, R. and Prellezo, R. (2010) 'Understanding Marine Ecosystem Based Management: A Literature Review', *Marine Policy*, 34(5), pp. 821–830.
East Asian Seas. (n.d.) Progress Report on the Implementation of the Regional Seas Strategic Directions 2017–2020. Available at: <https://www.unep.org/events/un-environment-event/regional-seas-programme-annual-meeting>.
Fanning, L., *et al.* (2021) 'Challenges to Implementing Regional Ocean Governance in the Wider Caribbean Region', *Frontiers in Marine Science*, 8, p. 667273.
Gattuso, J. P., *et al.* (2018) 'Ocean Solutions to Address Climate Change and Its Effects on Marine Ecosystems', *Frontiers in Marine Science*, 5 (Oct), p. 337.
Giannopoulos, N. (2021) 'Regionalism and Marine Environmental Protection: The Case of Offshore Energy Production', *SSRN*, pp. 1–20. Available at: <https://ssrn.com/abstract=3770726>.
Guerra-Vargas, L. A., Gillis, L. G. and Mancera-Pineda, J. E. (2020) 'Stronger Together: Do Coral Reefs Enhance Seagrass Meadows 'Blue Carbon' Potential?', *Frontiers in Marine Science*, 7 (July), pp. 1–15.
Haas, P. M. (1991) 'Save the Seas: UNEP's Regional Seas Programme and the Coordination of Regional Pollution Control Efforts', *Ocean Yearbook*, 9(1), pp. 188–212.
Hegland, T. J., Raakjær, J. and van Tatenhove, J. (2015) 'Implementing Ecosystem-Based Marine Management as a Process of Regionalisation: Some Lessons from the Baltic Sea', *Ocean and Coastal Management*, 117, pp. 14–22.
Hoegh-Guldberg, O., *et al.* (2019) 'The Ocean as a Solution to Climate Change: Five Opportunities for Action.' Available at: <http://www.oceanpanel.org/climate>.
Howard, J., *et al.* (2017) 'The Potential to Integrate Blue Carbon into MPA Design and Management', *Aquatic Conservation: Marine and Freshwater Ecosystems*, 27, pp. 100–115.
Lovelock, C. E. and Duarte, C. M. (2019) 'Dimensions of Blue Carbon and Emerging Perspectives', *Biology Letters*, 15(3), pp. 1–5.
Macreadie, P. I., *et al.* (2019) 'The Future of Blue Carbon Science', *Nature Communications*, 10(1), pp. 1–13.
McDonald, J., McGee, J. and Barnes, R. (2020) 'Oceans and Coasts in the Era of Anthropogenic Climate Change', *Research Handbook on Climate Change, Oceans and Coasts*. Research Handbooks in Climate Law series. Edward Elgar Publishing, pp. 2–26.
Moraes, O. (2019) 'Blue Carbon in Area-Based Coastal and Marine Management Schemes – A Review', *Journal of the Indian Ocean Region*, 15(2), pp. 193–212.
Nellemann, C. and Corcoran, E. (eds.) (2009) 'Blue Carbon. A Rapid Response Assessment.' Available at: <https://www.grida.no/publications/145>.
Oral, N. (2015) 'Forty Years of the UNEP Regional Seas Programme: From Past to Future', Rayfuse, R. (ed.), *Research Handbook on International Marine Environmental Law*. Edward Elgar Publishing, pp. 339–362.
PERSGA. (2019) Progress Report on the Implementation of the Regional Seas Strategic Directions 2017–2020. Available at: <https://www.unep.org/events/un-environment-event/regional-seas-programme-annual-meeting>.
Pörtner, H. O., *et al.* (2019). *IPCC Special Report on the Ocean and Cryosphere in a Changing Climate*. IPCC Intergovernmental Panel on Climate Change: Geneva, Switzerland.
Pörtner, H. O., *et al.* (2021). *Scientific Outcome of the IPBES-IPCC Co-sponsored Workshop on Biodiversity and Climate Change*. IPBES Secretariat.
Rees, S. E., *et al.* (2018) 'Defining the Qualitative Elements of Aichi Biodiversity Target 11 with Regard to the Marine and Coastal Environment in Order to Strengthen Global Efforts for Marine Biodiversity Conservation Outlined in the United Nations Sustainable Development Goal 14', *Marine Policy*, 93, pp. 241–250.
Rochette, J. and Billé, R. (2012) 'ICZM Protocols to Regional Seas Conventions: What? Why? How?', *Marine Policy*, 36(5), pp. 977–984.
Silver, J. J., *et al.* (2015) 'Blue Economy and Competing Discourses in International Oceans Governance', *Journal of Environment and Development*, 24(2), pp. 135–160.

South Asian Seas (SAS) Region. (2019) Progress Report on the Implementation of the Regional Seas Strategic Directions 2017–2020. Available at: <https://www.unep.org/events/un-environment-event/regional-seas-programme-annual-meeting>.

South Pacific. (2019) Progress Report on the Implementation of the Regional Seas Strategic Directions 2017–2020. Available at: <https://www.unep.org/events/un-environment-event/regional-seas-programme-annual-meeting>.

Stephens, T. (2015) 'Ocean Acidification', Rayfuse, R. (ed.), *Research Handbook on International Marine Environmental Law*. Edward Elgar Publishing, pp. 431–450.

Tehran Convention. (2019) 'Progress Report on the Implementation of the Regional Seas Strategic Directions 2017–2020.' Available at: <https://www.unep.org/events/un-environment-event/regional-seas-programme-annual-meeting>.

Trevisanut, S., Giannopoulos, N. and Holst, R. R. (2020) 'Introduction: Regime Interaction in Ocean Governance', Trevisanut, S., Giannopoulos, N. and Holst, R. R. (eds.), *Regime Interaction in Ocean Governance: Problems, Theories and Methods*. Brill Nijhoff, pp. 1–21.

United Nations Educational, Scientific and Cultural Organization (UNESCO). (2020) *UNESCO Marine World Heritage: Custodians of the globe's blue carbon assets*. Paris.

United Nations Environment Programme (UNEP). (2012) 'Blue Carbon – Opportunities for the Regional Seas Conventions and Action Plans,' 14th Global Meeting of the Regional Seas Conventions and Action Plans. Nairobi, Kenya, pp. 229–232.

United Nations Environment Programme (UNEP). (2016) 'Regional Seas Strategic Directions 2017–2020.' Available at: <https://wedocs.unep.org/bitstream/handle/20.500.11822/7621/-UNEP_medium-term_strategy_2018-2021-2016MTS_2018-2021.pdf.pdf?sequence=3&isAllowed=y>.

United Nations Environment Programme (UNEP). (2019) 'Regional Seas Programme Annual Meeting.' Available at: <https://www.unep.org/events/un-environment-event/regional-seas-programme-annual-meeting>.

United Nations Environment Programme - Caribbean Environment Programme. (2021). 'Sargassum White Paper – Turning the Crisis into an Opportunity. Ninth Meeting of the Scientific and Technical Advisory Committee (STAC) to the Protocol Concerning Specially Protected Areas and Wildlife (SPAW) in the Wider Caribbean Region.' Kingston, Jamaica.

United Nations Environment Programme (UNEP). (n.d. 1), 'Regional Seas Programmes.' Available at: <https://www.unep.org/explore-topics/oceans-seas/what-we-do/regional-seas-programme>.

United Nations Environment Programme (UNEP). (n.d. 2), 'Caribbean Environment Programme.' Available at: <https://www.unep.org/explore-topics/oceans-seas/what-we-do/working-regional-seas/regional-seas-programmes/wider?_ga=2.89564700.62420896.1621852230-1214584557.1615322790>.

United Nations Environment Programme (UNEP). (n.d. 3), 'Why does Working with Regional Seas Matter?' Available at: <https://www.unep.org/explore-topics/oceans-seas/what-we-do/working-regional-seas/why-does-working-regional-seas-matter>.

United Nations Environment Programme (UNEP). (n.d. 4), 'Strategy,' Available at: <https://www.unep.org/explore-topics/oceans-seas/what-we-do/working-regional-seas/strategy>.

United Nations Environment Programme (UNEP). (n.d. 5), 'West and Central Africa,' Available at: <https://www.unep.org/explore-topics/oceans-seas/what-we-do/working-regional-seas/regional-seas-programmes/west-and?_ga=2.161988154.761600029.1627498520-1397653992.1626861442>.

United Nations Environment Programme (UNEP). (n.d. 6), 'ICZM Protocol,' Available at: <https://www.unep.org/unepmap/who-we-are/contracting-parties/iczm-protocol>.

United Nations Environment Programme (UNEP). (n.d. 7), 'Eastern Africa Region,' Available at: <https://www.unep.org/explore-topics/oceans-seas/what-we-do/working-regional-seas/regional-seas-programmes/eastern-africa?_ga=2.199613352.761600029.1627498520-1397653992.1626861442>.

United Nations Environment Programme (UNEP). (n.d. 8), 'Red Sea and Gulf of Aden,' Available at: <https://www.unep.org/explore-topics/oceans-seas/what-we-do/working-regional-seas/regional-seas-programmes/red-sea-and?_ga=2.195016619.761600029.1627498520-1397653992.1626861442>.

Vallega, A. (2002) 'The Regional Seas in the 21st Century: An Overview,' *Ocean & Coastal Management*, 45(11–12), pp. 925–934.

Van Dyke, J. M. (2013) 'Whither the UNEP Regional Seas Programmes?', *Regions, Institutions, and Law of the Sea*. Brill Nijhoff, pp. 87–110.

van Leeuwen, J., *et al.* (2014). 'Implementing the Marine Strategy Framework Directive: A Policy Perspective on Regulatory, Institutional and Stakeholder Impediments to Effective Implementation', *Marine Policy*, 50(part B), pp. 325–330.

Visbeck, M., *et al.* (2014) 'Securing Blue Wealth: The Need for a Special Sustainable Development Goal for the Ocean and Coasts', *Marine Policy*, 48, pp. 184–191.

Warner, R. and Schofield, C. (eds) (2012) *Climate Change and the Oceans: Gauging the Legal and Policy Currents in the Asia Pacific and Beyond*. Edward Elgar Publishing.

Western Indian Ocean Region. (2019) 'Progress Report on the Implementation of the Regional Seas Strategic Directions 2017–2020.' Available at: <https://www.unep.org/events/un-environment-event/regional-seas-programme-annual-meeting>.

Wright, G., *et al.* (2017) 'Partnering for a Sustainable Ocean: The Role of Regional Ocean Governance in Implementing Sustainable Development Goal 14.' Available at: <https://publications.iass-potsdam.de/pubman/item/item_2353384>.

Young, M. (2012) 'Introduction: The Productive Friction between Regimes', Young, M. (ed.) *Regime Interaction in International Law: Facing Fragmentation*. Cambridge University Press, pp. 1–19.

6

BLUE DIMENSIONS OF THE EUROPEAN GREEN DEAL

Climate action at sea

Gabriela A. Oanta

The European Green Deal (EGD) proposed by the European Commission on 11 December 2019 (COM(2019) 640 final) seeks to strengthen the commitment of the European Union (EU) to respond to the challenges of global environment change. With it, the European Commission is launching a new growth strategy based on achieving a modern, resource-efficient, and competitive economy that is climate-neutral by 2050 (Krämer 2020: 268). In other words, the aim is to balance the quantity of carbon dioxide emitted into the EU's atmosphere with the quantity removed by various means in order to achieve net-zero emissions by 2050 at the latest. The new EU strategy laid out in the EGD is aligned with the European Council's main priorities for the 2019–2024 period, which were published in June 2019, including 'building a climate-neutral, green, fair and social Europe' (European Council 2019: 2 and 5). It is likewise aligned with the European Parliament's January 2020 call for the necessary transition to a climate-neutral society by 2050 and for that transition to become a European success story (European Parliament 2020: 2).

To effectively address the global environment change arising from global warming and biodiversity loss, the EGD proposes mainstreaming the effort to make the EU climate-neutral across all European policies, including its Integrated Maritime Policy (IMP) and maritime sectoral policies, including the Common Fisheries Policy and policies concerning maritime transport, maritime energy, blue growth and the blue economy, maritime spatial planning, international ocean governance, and seafarers. Blue growth is understood as the European Commission's initiative to further harness the potential of Europe's oceans, seas, and coasts for sustainable jobs and growth (COM(2012) 494 final: 2). The blue economy includes all sectoral and cross-sectoral economic activities based on or related to the oceans, seas, and coasts. It is considered that blue economy traditional sectors contribute to about 1.5% of the EU-27 GDP and provide about 4.5 million direct jobs, which represents 2.3% of the EU-27 total employment (European Commission 2021: ii and 2). In this context, the sustainable blue economy (COM(2021) 240 final) is at the heart of the EGD (Wolf *et al.* 2021).

Climate change and the marine environment are closely interconnected: climate change impacts the health of seas and oceans, while seas and oceans – the planet's 'blue lungs' – play a fundamental role in regulating the climate and lie at the centre of the climate system (Bodansky 2021: 317; Sobrino 2021: 22–23) (see Chapter 3). In recent years,

 DOI: 10.4324/9781315149745-8

oceans have gone from being considered agents of climate change to being recognised as an integral part of climate challenge solutions (Dundas *et al.* 2020: 1). Thus, a commitment to a greener EU requires a commitment to a bluer EU. In this regard, it is worth noting that the EU depends on the seas and oceans economically, environmentally, and socially. Europe's interaction with the sea is intense, varied, and vital to its well-being and prosperity. The surface area of EU-wide territory – a true maritime peninsula – is more sea than land; the maritime areas under its sovereignty or jurisdiction, including the outermost regions, are the most extensive in the world (spanning almost 20 million km^2), and its Exclusive Economic Zone (EEZ) is the largest on the planet. Despite the fact that Greenland is under Danish sovereignty, its waters have to be excluded from the EU's EEZ as it comes under an autonomous regime and does not belong to the EU. It is also worth mentioning that Greenland is one of the overseas countries and territories associated with the EU. Additionally, the EU's coastline (nearly 70,000 km long) is three times longer than that of the United States and twice as long as the Russian Federation's, meaning that more than two-thirds of EU Member States' external borders are maritime. All of this is compounded by another reality: no European resident lives more than 700 km from the coast, and almost half of the population lives less than 50 km from the sea, concentrated in urban areas along the coast. Furthermore, 3.5 billion tonnes of cargo and 350 million passengers pass through Europe's 1,200 seaports each year. The EU's merchant fleet accounts for 40% of the global fleet, it has the fifth largest fishery sector in the world, 70% of its external trade and 40% of its internal trade is shipped by sea, and 90% of its oil is imported by sea (European Commission 2020: 1–48).

With the adoption of the EGD and its implementation through its Roadmap (COM(2019) 640 final. Annex) and the EU's multiyear financial framework for the 2021–2027 period, including the Next Generation EU fund, the Union is applying a holistic approach in which maritime sectoral policies and the IMP play a prominent role. The EU has been creating and implementing these policies mainly in the last few decades, albeit without recognising the central role of climate implications for them. Likewise, the Union has been developing an environmental policy since 1972 to address issues related to its response to the challenges posed by climate change, environmental degradation, the need to preserve and restore biodiversity, the protection and restoration of vulnerable ecosystems and the environmental status of its surrounding seas and oceans, as well as participating in international conferences and conventions on the subject (see, e.g., the 2015 United Nations Paris Agreement on climate change).

However, it was not until more recently that the EU, through the EGD, sought to 'blue' its green policy or 'green' its IMP, with the aim of becoming climate-neutral by 2050. Since its adoption in December 2019, this novel approach has undergone significant developments as a result of the publication of various European Commission initiatives, although the COVID-19 pandemic caused some delays in the progress on its roadmap. These initiatives aim to facilitate a transition to a climate-neutral economy and sustainable society. Such a transition would increase the EU's influence in the ever-growing number of international forums devoted to the climate and biodiversity preservation. A sustainable society goes beyond the scope of just an environmental agenda and affects the EU's policies and actions as a whole. As we will see, this includes Europe's maritime sectoral policies and the IMP. Hence, this European green strategy (including its climate-related actions) should continue to be developed and implemented in harmony with the EU's policies impacting seas and oceans. In other words, the transition should also have an oceanic dimension and be extended to the European common maritime area (Sobrino and Oanta 2020: 29–31).

From the Integrated Maritime Policy to the EGD

On 10 October 2007, the European Commission published the IMP Green Paper (COM(2007) 575 final). In this soft-law text, the European Commission provides for the coordination of various maritime policies, including those related to maritime spatial planning, maritime surveillance, transport, fisheries, research, and the environment. One of the aims of the IMP Green Paper was to enhance Europe's capacity to face the challenges of climate change. Consequently, one of the European Commission's concerns was to coordinate a series of maritime activities, including those related to the marine environment, in which climate change was considered an important issue. The EU's holistic approach to maritime issues, which drove the IMP, also highlighted the environmental dimension of various maritime sectoral policies included in the IMP (Oanta 2019; Sobrino 2018).

Through the IMP, the European authorities sought to develop and implement integrated, coordinated, consistent, transparent, and sustainable decision-making in various maritime sectors in relation to the oceans, seas and coastal, island, and outermost regions (namely: French Guiana, Guadeloupe, Martinique, Mayotte, Reunion Island, and Saint-Martin (France), Azores and Madeira (Portugal), and the Canary Islands (Spain)). To this end, based on the IMP Green Paper, the Commission launched an action plan to make better use of the EU's maritime resources, especially in the areas of maritime transport, employment, scientific research, fisheries, and protection of the marine environment. This time, for the IMP's implementation, the Commission proposed adopting a strategy to mitigate the effects of climate change on coastal areas, launching pilot actions to reduce the impact of, and adapt to, climate change in such areas, and supporting research to predict, mitigate, and adapt to the effects of climate change on maritime activities, the marine environment, coastal zones, and islands.

Two years after the IMP Green Paper's publication, the Commission issued a Communication on the development of the IMP's international dimension (COM(2009) 536 final), in which it stressed that marine areas were being affected by climate change, while at the same time noting that these areas could also be the setting for effective mitigation strategies, such as the development of new energy sources (alternative renewables) or techniques for storing CO_2 emissions. It also recognised the need to give greater importance to the impact of oceans and coasts on climate. These lines of action were reinforced with the publication, on 29 August 2012, of the Green Paper 'Marine Knowledge 2020: from seabed mapping to ocean forecasting' (COM(2012) 473 final), which identified several efforts the EU had to make to adapt to climate change, including launching the European Climate Adaptation Platform (Climate-Adapt), which would feature a section on EU environmental and fishery policies with a view to identifying the impacts of climate change and Europe's vulnerabilities. The goal in this field was to help policymakers develop climate change adaptation measures, especially in coastal areas. Five years after the IMP Green Paper was adopted, the Commission published a report on the policy's progress (COM(2012) 491 final). The report emphasises the need for the EU to adapt to and mitigate the consequences of climate change. For example, it stated that climate change could have disastrous consequences for coastal regions – in particular, threats to coastal defences and the risk of erosion, flooding, and rising sea levels – and that it could also have heightened impacts in combination with other pressures on the marine environment.

Also, in the framework of the IMP, since 2007, the EU adopted a series of communications, strategies, and statements covering a wide range of issues affecting the environmental status of the seas and oceans. Among them, special attention should be called, first, to the 17

June 2008 Marine Strategy Framework Directive (Directive 2008/56/EC), which became the first environmental pillar of the IMP laying out a strategy in this area for the 2008–2018 period; and, second, the Communication adopted in September 2012 on 'Blue Growth opportunities for marine and maritime sustainable growth' (COM(2012) 494 final), which aimed to launch a joint initiative with the Member States, regions and all interested parties to harness the potential of the blue economy and, in particular, maritime and coastal tourism, offshore renewable energy, marine mineral resources, aquaculture, and blue biotechnology. It highlights the role of the blue economy in the fight against climate change and notes that it must be truly sustainable, as the use of marine resources depends, directly or indirectly, on the quality and long-term resilience of the oceans. All these aspects have an obvious climate and marine environmental impact, which today are mentioned in the documents being adopted to implement the EGD, which consider it necessary for the oceans to occupy a prominent place within it. In other words, they argue that it is necessary to endow the EGD with a blue dimension and for the oceanic dimension to be mainstreamed as an essential element of the new EU strategy.

The EU considered it necessary to deepen its international maritime policy through the IMP. It did this through the Communication 'International ocean governance: an agenda for the future of our oceans' (JOIN(2016) 49 final), jointly adopted, in November 2016, by the European Commission and the High Representative of the Union for Foreign Affairs and Security Policy. This Communication marks a shift from a sectoral approach to an integrated one, in which the EU commits to healthy oceans in order to effectively address various global challenges, including climate change and its impact. The Union's interest in these issues is very significant, as its Member States have jurisdiction over more than 10% of the planet's oceans. This all but translates to an obligation for them to continue assuming a leadership role as global actors to strengthen international governance related to the oceans, achieve Goal 14 of the United Nations 2030 Agenda for Sustainable Development and contribute to sustainable blue growth and maritime security. The Report, published on the first two years of the implementation of the Communication JOIN(2016) 49 final, recalls that oceans are both climate regulators and extremely vulnerable to the impacts of climate change (JOIN(2019) 4 final: 3). The EU's efforts have been aimed at ensuring adaptation to the effects of climate change on oceans and their uses, culminating in the November 2018 publication of the new European strategic vision to cut emissions by at least 55% (compared to 1990 values) by 2030 and to make the Union climate-neutral by 2050 (COM(2018) 773 final: 5). This strategy emphasises the conservation, use, and management of marine resources and ecosystems as one of the priorities for climate change mitigation and adaptation. This entire approach is on display in the strategy launched in the EGD.

The EGD and the climate dimension at sea: the European Climate Law

For the EU, blue sectors (marine living resources, marine non-living resources, marine renewable energy, port activities, shipbuilding and repair, maritime transport, and coastal tourism) are currently key to restoring the state of health of the environment. They are present in the EGD and they are an integral part of its roadmap to achieve resource efficiency by transitioning to a clean and circular economy, halt climate change, reverse biodiversity loss, and reduce pollution (Sobrino 2021: 33). The Union has adopted several initiatives to date to implement the EGD, including the Biodiversity Strategy for 2030 (COM(2020) 380 final), which provides, among other things, that 30% of the European Maritime and Fisheries Fund should contribute to climate action; the 'Farm to Fork' Strategy for a fair, healthy and environmentally friendly food system (COM(2020) 381 final); the EGD Investment Plan (COM(2020) 21 final), which provides that the EU will allocate 25% of its budget to climate and environmental spending and announces the intention to stimulate 'green investment' with the support of the European Investment Bank; the Proposal for a Regulation establishing the Just Transition Fund (COM(2020) 22 final), which aims to mobilise at least 150 billion euros between 2021 and 2027 in the most-affected regions by the transition towards a climate-neutral economy; a new Circular Economy Action Plan for a cleaner and more competitive Europe (COM(2020) 98 final); the Communication on a new approach for a sustainable blue economy in the EU (COM(2021) 240 final); and the proposal for a European Climate Law (COM(2020) 80 final). During 2022, new legislative initiatives were to be published by the European Commission with the aim to deepen its climate ambitions in different fields such as the release of in relation to the emissions trading system, energy efficiency, renewable energy, and governance, and protecting the ozone layer, amongst other things, as well as a new climate change adaptation strategy. These initiatives are reflected in the institutional organisation of the European Commission itself, which has made the EGD its primary objective for the 2019–2024 period. Indeed, the European Commission has organised a College of Commissioners around the EGD, which is made up by Frans Timmermans (Executive Vice-President of the European Commission and responsible for the EGD) and the commissioners responsible for health and food safety; energy; environment, oceans, and fisheries; agriculture; cohesion and reforms; and transport.

The March 2020 proposal for a European Climate Law was the farthest-reaching measure put forward under the umbrella of the EGD. It is the culmination of various initiatives proposed by the European Commission in recent years. The 2030 Climate Target Plan (COM(2020) 562 final) proposes reducing net greenhouse gas emissions by at least 55% by 2030 and considers offshore renewable energy to be a key element of the future energy system the EU needs in order to enable its transition to climate neutrality. The European Climate Pact, published on 9 December 2020 (COM(2020) 788 final) and expected to be expanded soon to include ocean, rural and coastal areas as well, reinforces the Union's intention to review the political instruments deemed pertinent to achieving additional reductions in greenhouse gas emissions. The EU Strategy on Adaptation to Climate Change, published by the Commission in February 2021 (COM(2021) 82 final), considers that protecting and restoring wetlands, peatlands, and coastal and marine ecosystems would be very beneficial for climate change resilience and would likewise contribute to multiple EGD objectives. All these initiatives accompanying the proposed European Climate Law will not only allow the EU to act internally, but also to be a powerful actor on the international stage (Krämer 2020: 303–304), such as the Commission for the Conservation of Antarctic Marine Living Resources (in relation to the creation and promotion of new marine protected areas) or the United Nations (by means of its intervention in the international negotiations for the future agreement on the conservation and sustainable use of the marine biological diversity of areas beyond national jurisdiction).

As for the legal course to be followed by the proposed European Climate Law, on 9 January 2020, the Commission published the roadmap for this regulatory initiative, which was open for public consultation for several weeks so that anyone who might be affected by the future proposal (government agencies, companies, civil society organisations, private individuals, etc.) could make comments and contributions. The result of this public consultation was the presentation, by the Commission, on 4 March 2020, of a proposal for a Regulation on a 'European Climate Law' (COM(2020) 80 final), with the aim of setting in legislation the EU's objective of achieving climate neutrality by 2050. The Commission is also tasked with reviewing the policies (including the IMP and maritime sectoral policies) and EU legislation to determine whether they are consistent with the climate neutrality objective and the pathway defined for the next three decades. On 5 May 2021, the Council and the European Parliament reached a provisional agreement on the proposal for the European Climate Law, and the EU is expected to develop a legally binding regulation – as opposed to a mere strategy – for its objective of achieving climate neutrality by 2050. The proposed subject matter and scope of the European Climate Law is to establish 'a framework for the irreversible and gradual reduction of anthropogenic greenhouse gas emissions by sources and enhancement of removals by sinks regulated in Union law' (Article 1) by 2050 at the latest and, thus, to meet the objective assumed under Article 2.1(a) of the Paris Agreement on climate change (Skjaerseth 2021: 26) of

> [h]olding the increase in the global average temperature to well below 2°C above pre-industrial levels and pursuing efforts to limit the temperature increase to 1.5°C above pre-industrial levels, recognising that this would significantly reduce the risks and impacts of climate change.

There is no doubt that several EU maritime activities, such as maritime traffic, among others, will be directly affected by the measures to be adopted to reduce greenhouse gas emissions. This is compounded by the fact that the seas and oceans are natural sinks that enable the absorption of much of those gases. These realities are likely to be reflected in the proposals for the revision and updating of EU climate and energy legislation, which the Commission is expected to release in 2021 and 2022, with the aim of meeting the agreed climate targets for 2030 and 2050, respectively.

The sustainable blue economy

The 'blue economy' is a potential field that the EGD can mine to achieve the transition to a climate-neutral economy by 2050 (Sobrino 2021: 24). A sustainable blue economy is not a new approach for the EU. It has already developed a sustainable approach to ocean management through, among other things, its IMP, the Marine Strategy Framework Directive, its maritime transport policy, the Common Fisheries Policy, and its system for fighting illegal, unreported and unregulated (IUU) fishing. The blue economy, which, according to the EGD, should be sustainable in the Union, will have a key role to play in reducing pressure on the earth's resources and halting climate change. Given the increasing recognition of the role of oceans in adapting to and mitigating climate change, the Commission aims to use the EGD to achieve more sustainable management of the maritime space, especially to facilitate access to the growing potential of marine renewable energy, while at the same time raising awareness of its zero-tolerance approach to IUU fishing (COM(2019) 640 final: 14).

It is precisely these reasons that determined the Commission to publish, on 17 May 2021, the 'Communication on a new approach for a sustainable blue economy in the EU: Transforming the EU's Blue Economy for a Sustainable Future' (COM(2021) 240 final). After hypothetically comparing the global blue economy to the seventh largest economy in the world and the ocean to a member of the G7, the European Commission notes that its approach to the European blue economy also includes an ocean policy. Especially relevant seem to be the contributions of oceans to energy production, to the greening of transport, and to sustainable food production. The European Commission underlines that 'contribution of a healthy ocean is essential for a sustainable economy. We need to better connect the green and the blue policies, while extending our approach beyond EU borders and leading the way on international ocean governance' (COM(2021) 240 final: 2). Therefore, oceans and this blue economy are considered *sine qua nons* of the transformation proposed in the EGD, without which neither the climate crisis nor the biodiversity crisis can be successfully addressed. The EU's new sustainable blue economy approach complements the other Commission initiatives mentioned above. It focuses on decarbonisation, conservation of natural capital, the circular economy, and responsible food production. It emphasises climate neutrality and zero pollution, the circular economy and preventing waste, biodiversity and investing in nature, coastal resilience, and responsible food systems (COM(2021) 240 final: 3–10).

Regarding the EU's aim to reduce greenhouse gas emissions by at least 55% (compared to 1990 levels) by 2030 and to become climate-neutral by 2050, the European Commission considers that offshore renewable energy could be a very useful tool, together with decarbonisation of maritime transport and fishing operations. In support of these initiatives, the Commission proposes creating a 'Blue Forum' for users of the sea to coordinate a dialogue between the fisheries, aquaculture, shipping, tourism, and renewable energy sectors; promoting the use of EU funds to green maritime transport by using the European Maritime Fisheries and Aquaculture Fund to support fishing fleets in adopting cleaner engines and techniques, while endeavouring to ensure that these upgrades do not result in overcapacity or overfishing; pursuing the goal of zero-emission ports; and helping Member States prepare for and respond to maritime pollution accidents through the EU Civil Protection Mechanism and the various actions of the European Maritime Safety Agency (COM(2021) 240 final: 4). The European Commission likewise considers that the blue economy could reduce the impact of human activities on the sea by playing a vital role in combating marine pollution, which, in turn, can also be regarded as an opportunity to create jobs (COM(2021) 240 final: 5–6). To this end, the European Commission highlights the pollution of the various marine spaces caused by human activity through nutrients (leading to eutrophication, contaminants, litter (largely made up of plastics) and underwater noise. In order to meet the EGD targets in this area, the European Commission proposes taking action to halve plastic litter at sea, nutrient loss into the sea and the use and risk of chemical pesticides by 2030; taking action to restrict intentionally added micro-plastics; ensuring that litter caught in fishing operations is reported at port and that plastic fishing gear is collected and recycled after its use; and revising the Ship Recycling Regulation (Regulation (EU) No 1257/2013; OJ L 330/1, 10.12.2013) and the EU requirements for decommissioning offshore platforms (COM(2021) 240 final: 5). Biodiversity conservation, protection, and restoration can contribute considerably to achieving the EGD's decarbonisation targets. The European Commission will thus table a proposal for legally binding EU targets to restore degraded ecosystems; propose a new action plan by the end of 2021 to conserve fisheries resources and protect marine ecosystems; work to identify and designate additional marine protected areas; and promote and support local participatory initiatives that combine the regeneration of marine resources with the preservation of local livelihoods.

Adaptation to climate change has to be grounded in natural solutions, based on wetlands, coastal ecosystems, and landscapes, including marine ones, rather than the construction of new 'grey infrastructure'. Such solutions would create a new sector of the blue economy. To this end, the new EU Strategy on Adaptation to Climate Change focuses on the need to work in order to close the knowledge gap and stimulate innovation for increased climate resilience, as well as to ensure the necessary capacity to better anticipate the effects of extreme weather events and regional sea-level rise (COM(2021) 82 final: 3–17; COM(2021) 240 final: 8).

Better use of marine resources and choosing alternative sources of food and feed could also help alleviate the pressure on the climate. The EGD and some of the initiatives adopted for its implementation have emphasised this need, while at the same time highlighting the need for the EU to have responsible and sustainable food systems, which could be achieved in the blue economy. In this regard, the European Commission is expected to table, by 2023, a legislative proposal for a framework that will include fishery and aquaculture products that will enable a real transition to a sustainable food system; table, in 2022, a legislative proposal for modern, sustainable marketing standards for seafood and, the same year, adopt a dedicated initiative on algae to support the development of a specific algae industry in the EU; support the digital transition of fisheries control; assess the potential and the research and investment needs related to cell-based seafood; and strengthen fisheries management in the Mediterranean and the Black Sea in accordance with the Common Fisheries Policy (COM(2021) 240 final: 10).

In addition to the blue economy agenda, the Communication on the new approach for a sustainable blue economy announces the publication, in 2021 and 2022, of other new initiatives aimed at supporting the development of a sustainable blue economy with a clear climate dimension in line with the EGD targets. These initiatives will seek better knowledge of the oceans through the preparation, by 2022, of an Ocean Observation Initiative. So far, the Commission has already adopted the roadmap for the upcoming adoption of this initiative, which is expected for the third quarter of 2021. In addition, the European Commission proposed the creation, in 2021, of a Blue Economic Observatory; and the release of a stable methodology that integrates the concept of 'natural capital' in economic decisions. They will pursue, through the development of a pan-European innovation ecosystem for a blue economy, marine and maritime research to make the EU climate-neutral by 2050, protect and restore marine ecosystems and make the blue economy a font of ideas and action to generate sustainable innovation (COM(2021) 240 final: 10–15).

Such initiatives will require a strong commitment by the EU to both public and private investment. In addition to the Union funding approved for the 2021–2027 period to be allocated to the EGD, the European Commission will work with the European Investment Bank and the European Investment Fund in this area (COM(2021) 240 final: 13). The financial aspect takes on even greater importance in the context of the Covid-19 pandemic (Fleming and Mauger 2021: 174–175), which caused, among other things, a widespread slowdown in the job market. In this context, the transition to a sustainable blue economy is seen as an opportunity for the economy in general and for generating employment in particular, enabling the creation of blue careers and jobs, while at the same time improving working conditions and public perception of careers in the marine and maritime world (COM(2021) 240 final: 14–15).

All these EU initiatives and approaches related to the EGD must be based on a set of conditions for sustainable governance to enable the successful transition to climate neutrality by 2050. In this regard, the Communication highlights the need for a regulatory framework on spatial planning (namely 'a process by which the relevant Member State's authorities analyse and organise human activities in marine areas to achieve ecological, economic and social objectives'; Article 3 of Directive 2014/89/EU; OJ L 257/135, 28.8.2014) that has broad support, for citizen engagement and ocean literacy, for regional cooperation in sea basins, for maritime security, and to promote a sustainable blue economy abroad in the context of international maritime governance, in which the EU has come to play a very prominent role in recent decades (COM(2021) 240 final: 15–20).

Conclusion

This chapter has sought to show how sea and ocean sustainability poses a great challenge to the EU and is closely linked to climate change. The EU is aware of this reality and, for more than a decade, has been taking a comprehensive approach to addressing it. This approach is quite clear in the inclusion of the demands of the sustainable use of the seas and oceans in the decision-making process regarding the EU's environmental considerations, including those affecting the climate. Global environment change affects both marine areas and the climate, and a solution to global warming and the loss of biodiversity, including marine biodiversity – both of which the EU has emphasised – requires the coordinated, consistent, and transparent action of all European policies, including the IMP and the EU's sectoral maritime policies, at the heart of which the future European Climate Law and the sustainable blue economy lie. All of these feed into the EGD roadmap, which is marked by the blue dimension of the environment. They will position the EU, with the EGD's goal of making the Union climate-neutral by 2050, in the vanguard in its approach to the intrinsic relationship between climate and the sea.

Acknowledgements

This research was supported by the 'Grupo con Potencial de Crecimiento' ['Group with Growth Potential'] project funded by the Galician government (Spain) for the period January 2020-December 2022.

References

Bodansky, D. (2021) 'The Ocean and Climate Change Law: Exploring the Relationship', in R. Barnes and R. Long (eds.) *Frontiers in International Environmental Law: Oceans and Climate Changes*. Leiden: Brill/Nijhoff.

COM(2007) 575 final, Communication from the Commission to the European Parliament, the Council, the European Economic and Social Committee and the Committee of the Regions: An Integrated Maritime Policy for the European Union, Brussels, 10.10.2007.

COM(2009) 536 final, Communication from the Commission to the European Parliament, the Council, the European Economic and Social Committee and the Committee of the Regions: Developing the international dimension of the Integrated Maritime Policy of the European Union, Brussels, 15.10.2009.

COM(2012) 473 final, Green Paper: Marine Knowledge 2020 from seabed mapping to ocean forecasting, Brussels, 29.8.2012.

COM(2012) 491 final, Report from the Commission to the European Parliament, the Council, the European Economic and Social Committee and the Committee of the Regions: Progress of the EU's Integrated Maritime Policy, Brussels, 11.9.2012.

COM(2012) 494 final, Communication from the Commission to the European Parliament, the European Council, the Council, the European Economic and Social Committee and the Committee of the Regions: Blue Growth opportunities for marine and maritime sustainable growth, Brussels, 13.9.2012.

COM(2018) 773 final, Communication from the Commission to the European Parliament, the European Council, the Council, the European Economic and Social Committee, the Committee of the Regions and the European Investment Bank: A Clean Planet for all. A European strategic long-term vision for a prosperous, modern, competitive and climate neutral economy, Brussels, 28.11.2018.

COM(2019) 640 final, Communication from the Commission to the European Parliament, the European Council, the Council, the European Economic and Social Committee and the Committee of the Regions: The European Green Deal, Brussels, 11.12.2019.

COM(2019) 640 final Annex, Annex to the Communication from the Commission to the European Parliament, the European Council, the Council, the European Economic and Social Committee and the Committee of the Regions: The European Green Deal, Brussels, 11.12.2019.
COM(2020) 21 final, Communication from the Commission to the European Parliament, the Council, the European Economic and Social Committee and the Committee of the Regions: Sustainable Europe Investment Plan European Green Deal Investment Plan, Brussels, 14.1.2020.
COM(2020) 22 final, Proposal for a Regulation of the European Parliament and of the Council establishing the Just Transition Fund, Brussels, 14.1.2020.
COM(2020) 80 final, Proposal for a Regulation of the European Parliament and of the Council establishing the framework for achieving climate neutrality and amending Regulation (EU) 2018/1999 (European Climate Law) Brussels, 4.3.2020.
COM(2020) 98 final, Communication from the Commission to the European Parliament, the Council, the European Economic and Social Committee and the Committee of the Regions: A new Circular Economy Action Plan For a cleaner and more competitive Europe, Brussels, 11.3.2020.
COM(2020) 380 final, Communication from the Commission to the European Parliament, the Council, the European Economic and Social Committee and the Committee of the Regions: EU Biodiversity Strategy for 2030. Bringing nature back into our lives, Brussels, 20.5.2020.
COM(2020) 381 final, Communication from the Commission to the European Parliament, the Council, the European Economic and Social Committee and the Committee of the Regions: A Farm to Fork Strategy for a fair, healthy and environmentally-friendly food system, Brussels, 20.5.2020.
COM(2020) 562 final, Communication from the Commission to the European Parliament, the Council, the European Economic and Social Committee and the Committee of the Regions: Stepping up Europe's 2030 climate ambition Investing in a climate-neutral future for the benefit of our people, Brussels, 17.9.2020.
COM(2020) 788 final, Communication from the Commission to the European Parliament, the Council, the European Economic and Social Committee and the Committee of the Regions: European Climate Pact, Brussels, 9.12.2020.
COM(2021) 82 final, Communication from the Commission to the European Parliament, the Council, the European Economic and Social Committee and the Committee of the Regions: Forging a climate-resilient Europe - the new EU Strategy on Adaptation to Climate Change, Brussels, 24.2.2021.
COM(2021) 240 final, Communication from the Commission to the European Parliament, the European Council, the Council, the European Economic and Social Committee and the Committee of the Regions on a new approach for a sustainable blue economy in the EU: Transforming the EU's Blue Economy for a Sustainable Future, Brussels, 17.5.2021.
Directive 2008/56/EC of the European Parliament and the Council of 17 June 2008 establishing a framework for community action in the field of marine environmental policy, OJ L 164/19, 25.6.2008.
Directive 2014/89/EU of the European Parliament and of the Council of 23 July 2014 establishing a framework for maritime spatial planning, OJ L 257/135, 28.8.2014.
Dundas, S.J. *et al.* (2020) 'Integrating Oceans into Climate Policy: Any Green New Deal Needs a Splash of Blue', *Conservation Letters*, 13: 1–12.
European Commission. (2020) *Facts and Figures on the Common Fisheries Policy: Basic Statistical Data*, Luxembourg: Publications Office of the European Union.
European Commission. (2021) *The EU Blue Economy Report 2021.*
European Council. (2019) *A New Strategic Agenda 2019–2024*, Brussels.
European Parliament. (2020) *European Parliament Resolution of 15 January 2020 on the European Green Deal*, P9_TA (2020) 0005.
Fleming, R.C. and Mauger, R. (2021) 'Green and Just? An Update on the "European Green Deal"', *Journal for European Environmental Planning Law*, 18: 164–180.
JOIN(2016) 49 final, Joint Communication of the European Commission and the High Representative of the Union for Foreign Affairs and Security Policy to the European Parliament, the Council, the European Economic and Social Committee and the Committee of the Regions: International ocean governance: an agenda for the future of our oceans, Brussels, 10.11.2016.
JOIN(2019) 4 final, Joint Report of the European Commission and the High Representative of the Union for Foreign Affairs and Security Policy to the European Parliament and the Council: Improving International Ocean Governance – Two years of progress, Brussels, 15.3.2019.

Krämer, L. (2020) 'Planning for Climate and the Environment: the EU Green Deal', *Journal for European Environmental & Planning Law*, 17: 267–306.

Oanta, G.A. (2019) *Textos jurídicos para el estudio de la Política Marítima Integrada de la Unión Europea*, Barcelona: Bosch Editor.

Regulation (EU) No 1257/2013 of the European Parliament and of the Council of 20 November 2013 on ship recycling and amending Regulation (EC) No 1013/2006 and Directive 2009/16/EC, OJ L 330/1, 10.12.2013.

Skjaerseth, J.B. (2021) 'Towards a European Green Deal: The evolution of EU Climate and Energy Policy Mixes', *International Environmental Agreements*, 21: 25–41.

Sobrino Heredia, J.M. (2018) 'El mar y la UE: la necesidad de una política marítima integrada', *Revista de Derecho Comunitario Europeo*, 61: 835–849.

Sobrino Heredia, J.M. (2021) 'La Política Marítima Integrada se tiñe de verde: la dimensión oceánica del Pacto Verde Europeo', in F. Fernández Prol (coord) *Pesca marítima y crecimiento sostenible: Análisis en clave jurídica*, Barcelona: Bosch Editor.

Sobrino Heredia, J.M. and Oanta, G.A. (coords) (2020) *La construcción jurídica de un espacio marítimo común europeo*, Barcelona: Bosch Editor.

Wolf, S. *et al.* (2021) 'The European Green Deal – More than Climate Neutrality', *Intereconomics*, 56: 99–107.

7

LEADERSHIP

Actors and their strategies in marine environmental governance

Małgorzata Zachara-Szymańska

Research examining international cooperation and governance provides compelling evidence that leadership is a crucial determinant of success or failure in efforts to address complex challenges (Collinge and Gibney, 2010; Modelski, 1999; Rhodes and Hart, 2014). The realities of marine governance are among the most far-reaching of the world's contemporary wicked problems (Grint, 2005), rooted in such complex systemic risks as climate change, resource exploitation, and biodiversity loss, which outpace the existing approaches to management and protection. The impact of human activities on marine biodiversity is highly variegated, so governance of these areas requires an appropriately synergistic approach, ranging from the level of an individual, through group efforts, to decision-making processes at the level of international organisations. Leadership is an indispensable component for assuring the proper identification of the problems, communication, and trust building between the diverse partners present in the governance spectrum, and effective collective action. While there are a variety of definitions and concepts covering the notion of leadership, in the context of the transition to sustainability, it should be understood as a multilevel activity oriented towards creating the capacity to act (Horlings, 2010).

The global marine governance system is based on a framework of international regulations enshrined in the UN Convention on the Law of the Sea (UNCLOS) and sustainable development goals (SDGs), but it suffers from fragmentation and non-compliance, resulting in drifts between policy and practice. The major challenge of global marine governance is to balance demands for development with the need to protect the environment and to achieve social and economic objectives in a sustainable way. While the health of the oceans has long been recognised as a common good that should be a shared aspiration of societies, the integration between stakeholder interests necessary for the conservation and sustainable use of the oceans is yet to be achieved. Marine ecosystems are characterised by geomorphological and ecological interdependencies, but this is not matched by the integration of the framework of their governance; therefore, responsibilities are dispersed across a number of bodies, actors, and institutions. This raises the question of effectiveness and divergences between the normative framework expressed in the marine governance statutes, documents, declared political strategies, and actual imperfect practices, often leading to over-exploitation, habitat destruction, and species loss (Haas et al., (2021); Pinto et al., 2014).

DOI: 10.4324/9781315149745-9

Leadership and marine governance

Given the complexity of this issue, leadership processes are a key factor in shaping the dynamics, and indicating the directions of, marine governance. Leadership can be seen as a mechanism for translating global policy concepts into an operational management practice adapted to local circumstances. All forms of governance in this sphere are concentrated around leadership processes viewed as the management of stakeholder relationships with regard to resource use and distribution. Leaders, by initiating and developing relations, create governance systems in which the complementarities, and the contradictions of the environmental policies and interests of the actors involved, can be negotiated. Problems embedded in the management of marine resources are being addressed on the basis of a variety of polyarchic forms, ranging from privatisation and institutionalisation to consensual or interactive governance (Kooiman et al., 2005; Ostrom, 1990; Valdés, 2017). Marine governance is being shaped by individual actors characterised by different worldviews and goals as well as groups and institutions responding to different drivers of change. Therefore, leadership processes within this area cover relations within and between states, formal and informal institutions, civil society and local communities, and the market, as well as individual citizens. Leaders establish institutions in which they interact by recognising needs, addressing change, and creating scenarios for the future. The parties involved are engaged in processes of constant negotiations, advancing further their agendas, and transforming the existing institutional rules of the game.

Marine governance is framed by the issues of the transition to sustainability and the protection of global public goods. These meta-topics stand in the background of the natural, human, and governance process-drivers. Natural marine resources are fluid, characterised by intensive material movement across the borders drawn by administrators and governors. The oceans can only be managed on the basis of universal cooperation, without which a bad equilibrium of selfish strategies on the part of states and businesses leads to further disruption. Decision-makers are divided between the various orientations shaping their perspectives: the principles of the 'Blue Economy', which seek to promote economic growth alongside the preservation or improvement of livelihoods, are different from strategies that see maritime governance as a primary instrument of environmental protection, and those that are driven mostly by profit and short-term goals.

Leadership within this framework has to be viewed not exclusively as concerning those in formal positions of authority, but rather as a relational and discursively constructed process. It covers the mechanisms responsible for institutional, cultural, and political relations taking place on many levels – from presidential and prime-ministerial offices, through the public to opinion leaders, expert bodies, corporate actors, and nongovernmental organisations (NGOs) to individual citizens. Looking at things through the leadership lens helps us see who can enter into cooperation, under what conditions and for what reasons, what perspectives this cooperation will develop and how transformative it will be for the global picture. Marine leadership processes are therefore more comprehensive than marine governance, as they encompass the variety of agents creating the dynamics of the mutual relations and are oriented towards obtaining the most effective possible sustainability outcome. Leadership should be seen as a driving force behind the behaviours of the individual agents as well as a framework for the further development of institutional settings.

Given the multidimensional nature of marine governance, the leadership processes within this field can be divided into three types: the structural, which is driven primarily by power relations and hierarchies within the domain of the international politics; the cognitive, which concerns knowledge and awareness; and the relational, joining the other components to enable collective action.

Structural leadership

Structural leadership mainly occurs between states and/or international organisations and is organised around hierarchies of interest and the bargaining leverage provided by material power. Leadership at the structural level covers the development of institutional approaches to marine problems as well as designing the approach through which this problem will be discussed. Structural leaders, as state agents or international organisations, are oriented towards positioning the ocean high on the global political agenda and generating interest and the political will necessary for the implementation of protective standards. One of the major objectives of this level of leadership is to create institutional and legal framework for marine governance – the rulebook that will guide the behaviours of the stakeholders. The primary tier of structural leaders is composed of the United Nations and a number of international organizations that address governance in specific sectors: the International Maritime Organization (IMO), the Food and Agriculture Organization (FAO), Regional Fisheries Management Organizations and ad-hoc coalitions (e.g., the High-Level Panel on Sustainable Ocean Economy). These marine power bodies and networks are responsible for expanding the foundations for comprehensive marine governance, ranging from the United Nations Law of the Sea Convention (UNCLOS), international customary law, and international treaties to national, state, and local level laws derived from tradition, legislation, and the courts. Their efforts resulted in defining the marine protection priorities in Agenda 2030 and its Sustainable Development Goal for the ocean (SDG 14).

The basis for marine governance was established during the construction of the post-World War II international order, rooted in institutionalised structures of governance and responsibility for the provision and protection of public common goods. This system has long dealt with the consequences of environmental changes across the globe implementing, policies and instruments that can have broad, if not near-universal applicability. As structural leaders, global institutions negotiate around and initiate activities in the common ground a plethora of stakeholders, as well as coordinating networks of leadership processes, expanding the results of the trade-offs and deals across national and regional political boundaries that lead to compromise in multilateral negotiations. The complex process of agreeing on the 'constitution of the sea' provides an insight into the leadership dynamics that resulted in the establishment of today's regulatory framework.

The impulse for change has been provided by the desire to constrain the attempts of the nation states to assert sovereign rights over increasingly larger areas of the ocean. The actions of the United States, which in 1945 extended such control to all the natural resources on its continental shelf, provoked other maritime nations around the globe to issue similar claims. Chile, Peru, and Ecuador extended their claim to 200 nautical miles to include their fishing grounds. It has been widely recognised that the changing environment of sea exploration requires a broader codification of marine resource use and protection, but different orientations of the parties have provided the major obstacle in the negotiation process. As seas and coastlines are interconnected and shared between states, the United Nations – a body composed of nation states oriented towards protecting the common public good – represented the optimal normative stance that the resources of the seabed are the 'common heritage' of humankind. However, the boundaries and practical design of this concept had to be decided.

The nature of the structural decision-making at this historical moment also influenced the outcomes, as no international organization can be stronger than the most powerful of its member states. Though institutions gain autonomy and independence over the course of their development, there are limits to their authority as they rely structurally on the dynamics of power within the parts they are composed of. The leadership component expresses the persuasive powers of the actors involved, and their final positions on the individual level (state delegates others in formal positions of power), group level (member states or coalitions, e.g., the Group 77 – composed of developing states that, due to cooperative strategies, become an effective bargaining force), or strategic alliances (Nordquist et al., 1985). Although not all of the claims could be moderated due to the contradictions between some of them, the institutional logic of the process relied heavily on the co-production of leadership. The disputes focused on major tensions that underlie the general development of the law of the sea. This can be illustrated by the evolution of the United States' position during the UNCLOS negotiation process, revealing power dynamics and agency patterns in the process. The U.S., being the most prominent actor in the international system, took a leadership role in the negotiations in the 1960s and 1970s focused on preserving the principles of freedom of navigation and other vital security concerns as well as protecting the right of the U.S. to engage in deep sea mining.

The regime of the seabed and ocean floor beyond the limits of national jurisdiction proposed in the Convention favoured developing nations in terms of deep sea mining technology transfer, which the United States opposed, arguing that this solution would deprive the multinational firms of competitive advantage, even though they had made substantial investments in research and development and expanded their innovative capabilities. This negotiating position was expressed in a sequence of moves substantially influencing the course of the process. The U.S. Congress debated unilateral action in deep sea mining, which triggered a legal response from the Group of 77 that proved the unlawfulness of the planned legislation, but also increased the bargaining leverage of the U.S. Its position influenced the negotiation strategies of the other actors, some of which unilaterally established fisheries zones or exclusive economic zones up to a maximum 200 nautical miles. This negatively impacted the receptiveness of the high protective standards promoted by the UN, and the ability to balance interests of coastal states and other users of the maritime realm. The United States pursued its commercial agenda and encouraged followers to take a similar path, thereby creating a pattern for an emergent custom, which had to be recognised by the conference. This impeded effective movement towards quick and/or widespread acceptance of the Convention in the 1980s. Despite the fact that the desired goals were eventually reached, the U.S. resisted ratification.

In this deep sea mining issue, the United States played the role of a group leader; however, in the evolution of processes such as this the impact of individual leaders is crucial for the creation of authority and the legitimacy of international process. The history of all institutions is crafted by the efforts of the individuals who provided the conceptual basis and ethical standards upon which organic systems are developed. In the case of UNCLOS, the mechanism of leadership relations manifested itself in the formation of the informal clubs (the Castaneda Group), long-term personal rapport between individuals, efforts at persuasion, and the ability to steer discussion (making or breaking the consensus). The power of authority manifested itself when the president of the III UNCLOS Conference, H.S. Amerasinghe, having been excluded from his Sri Lanka's delegation after a domestic power shift, remained in position because in the eyes of the parties involved, his presence was indispensable to the success of the process (Nordquist et al., 1985).

The UNCLOS negotiation saga reveals the mechanisms of the structural leadership, providing insight into power hierarchies, interests, and the mechanics of the development of global rules. The articles of the Convention were negotiated with 157 nation states over 11 sessions, for nearly two decades, resulting in one of the most comprehensive political and legislative projects ever undertaken by the United Nations. The negotiation process reveals the role of structural leaders, interacting to provide globally agreed international rules and procedures, influencing regional actions, and generating national legal frameworks and integrated policies. The universal rules, constructed at the UN level, indicate a general code of conduct but do not necessarily respond to competing spatial claims and conflicts between maritime economic activities and biodiversity in maritime regions. Leadership processes, played out within specific areas, construct the dynamics of decision-making and ensure progress in the building of structures of governance. Structural leaders manage expectation and create a vision, defining what is possible and what is desired in marine governance.

Cognitive leadership

Cognitive leadership concerns knowledge – how it is produced, shared, and transformed, providing answers for the basic conceptual frames of marine governance based on the questions of 'what', 'who' and 'how'. Cognitive leadership processes ensure social connectivity among actors that influence how societies make and implement decisions. They create the background for the permanent negotiation of the social contract of the human use of the oceans. Almost half of the global population live in coastal areas and therefore are directly related to marine ecosystems, but all humans rely on the marine realm in their survival and well-being strategies. The cause-effect relationship in ocean protection begins with conscious transformations of individual behaviour, as the ocean plays a critical role in supporting human well-being, from providing food, livelihoods, and recreational opportunities to regulating the global climate (Atkins et al., 2011). The cultural and social ties that link individual experiences with the idea of marine protection vary, so recognition of this diversity and plurality of values, as well as collective responsibility, provide a foundation for building a sustainable relation with the marine environment.

The idea of 'ocean citizenship' – expressing a relationship between the lives of the people and the health of the coastal and marine areas – conveys a desire for a greater public understanding of marine environmental relations (Fletcher and Potts, 2007). The distance from the sea, whether physical or psychological, is thereby reduced, bringing an understanding that healthy ocean is essential to all life on Earth. The citizenship agenda has typically focused on governance issues, such as political literacy, involvement in local decisions, and social and moral responsibility. In the case of the global ocean framework, efforts are directed at building recognition of interdependence and generating a sense of duty and stewardship, reflected in enhanced participation in environmental networks or activities around consumer choices. Cognitive leadership processes within this sphere refer to different dimensions of marine protection awareness-building, covered by the label of 'ocean literacy', defined as the 'understanding of the ocean's influence on you and your influence on the ocean' (Cava et al., 2005: 5). They include education and marine activism, designed to transform popular views, and actions that are less harmful to and more protective of the oceans, their ecosystems, and related populations.

Ocean environmental awareness and sustainable education have long been seen as key for sustainable ocean and coastal development, as they are directly related to increased public support for protective measures and policies. Participants in the attempt to spread ocean literacy have included artists, water sport athletes, surfers, scientists, and celebrities. The gallery of public leaders opens with the popular French explorer Jacques-Yves Cousteau in the 1950s and many subsequent public figures and celebrities who are able to spread awareness about the deteriorating condition of the marine environment. International nongovernmental organizations, like Oceana, Greenpeace, and the WWF conduct outreach to the global public using a variety of means, from media presence, through art, documentaries, learning-at-sea expeditions, public cleaning events on beaches and coasts, and constant monitoring efforts. These activities are oriented towards engaging individuals in a relationship with the global ocean, generating deep learning, and stimulating changes in behaviour. They range from the most widespread ideas for the environment – such as support for limitations on the use of plastic and making informed lifestyle choices – to the shaping of standards directly related to oceans' health, such as clearing of litter from beaches or furthering the understanding of rights and responsibilities in the use of marine areas. As a result of enhanced popular understanding of environmental problems and potential solutions, individuals modify their behaviour and are better equipped to take part in the change-oriented dialogue on the environment.

Cognitive leadership also covers creative and diverse picture of the global ocean society, providing a new dimension to the traditional views associated with the social structure, and the various roles played by individuals in coastal communities and other users of maritime resources. This may be illustrated by the increased impact of the Gender and Water Alliance (GWA), which focuses on illuminating the issue of gender in marine governance. Research and public awareness activities underline the fact that the contribution of women in fish supply and marine processes has long been neglected not only by politicians but also by scholars and policymakers. Activities begun in the 1980s were based on research revealing the role of women in the maritime industries of the Western and Southern countries. This process of recognition of the gender factor has been expressed in exposing the crucial role of women in the implementation of Sustainable Development Goal 14. This information base, combined with the development of international networks, promoted women's participation in sea-related sectors and revealed their role as economic actors.

The scientific grounds of this process are part of another project: to frame the purposes and possibilities of the existing epistemic community of marine scholars engaged with the understanding of people's relation with the coastal and marine environment. In this area, scientific and technological leaders work on projects that help assess the condition of the marine ecosystems and develop innovative solutions to address the key sustainability threats. Technologies are used to reduce the negative impacts of plastic pollution, improve maritime security and disaster prediction, as well as emergency response systems, marine spatial planning, and ocean stewardship. Scientific understanding of new environmental concerns, e.g., ocean acidification (OA), is emerging. The potential for power associated with tides and ocean currents is being researched in the context of alternative renewable energy resources. Due to technological progress, green shipping is being increasingly influential as an industry trend oriented at decreasing emissions, consuming less energy, and generally remedying environmental impacts. On the basis of cognitive leadership not only does new research develop and cutting-edge technological solutions become implemented, but attention is redirected also towards existing scientific, indigenous, and traditional knowledge, accumulated through many generations of close interaction between people and the natural world (Weiss et al., 2012). The discovery and documentation of cultural values and customary practices of coastal communities provides broadening exposure to activities which have long been implemented in marine governance in Micronesia, Kiribati, Belize, Malaysia, Australia, and number of other places (Gee et al., 2017; Poe, Norman, and Levin 2014).

Leaders are needed to provide and use various kinds of evidence on the human impact on the ocean, which result in knowledge-based decision-making. With a constantly increasing range of interactions with the global ocean occurring, cognitive leadership processes play an important role in providing a necessary counterbalance for the marine governance models rooted in international power games and the protection of national interests (Cash et al., 2003).

Relational leadership

Understanding global marine governance is not possible without decoding the dynamics in play within multi-actor governance arrangements and social networks, created as a result of the multiplicity of linkages taking place at different spatial, temporal, and organisational scales. These can only be funded and brought into existence with the engagement of leaders caring for incentives, framing the processes, ensuring effective communication, and advocating for change. Relational leadership in marine governance includes all the interactive components needed to produce effective collective action and to construct an operating framework of management and control. Through the emotional and functional ties of social and organisational networks, power is mediated, and an inclusive environmental agenda can be built around an agreed set of principles. For sustainable management to be achieved, people and governance systems from the various realms have to interact. Relations between actors help to overcome inconsistencies in sectoral approaches and conflicts between stakeholders and enable communication with regard to all levels of human presence in the marine environment. Relational leaders not only bring the parties involved to the negotiating table, but they also increase the resilience of the structure, assuring its constant development. Thus, does relational leadership cover all situations in which joint actions and cooperation between actors involved in structuring marine governance occur? Such relations have partially been institutionalised; however, in real-life situations, they also work beneath the surface of the formal processual level.

While structural leadership involves states, as primary actors operating on the international scene, relational leaders often address conflicts around the adoption of the environmental measures necessary for the enhancement of protection standards, but not in a way that is directly connected to the interests of states. There is a leadership continuum within marine governance, in which rules established at the structural level are disseminated and implemented through the efforts of intermediaries providing resources and training. As a result, the net of leadership responsibilities and functions is being extended, bringing tangible results in terms of the design and implementation of environmental standards. The system of marine standards is based on international relationships, ranging from decision-makers to technical experts, NGO agents, and local communities' representatives. Through systematic connection the network of knowledge expands, spreading rules, know-how, solutions, and tools. The net of influences is filling the governance gap created by the states which are either unable to implement high standards of protective measures or are unlikely to intervene in issues not directly related to their immediate benefit. Furthermore, many states lack the economic, technological and leadership potential with which to build protective measures on their territorial waters and in the high seas, where most of the crimes against the marine environment are committed. As a consequence, globally accepted regulations – protective conventions and guidelines – cannot be fully implemented, as local governing bodies, natural parks or other vulnerable marine environments lack the tools and resources to influence situations. Here, a co-management of arrangements for the conservation of natural resources is required to improve local financial and institutional capacity. Such solutions enable a transfer of power and authority from national government agencies to local communities and sub-national governments. This mechanism can be illustrated by the scale of activities of the WildAid Marine Program, a publicly sponsored NGO working on enhancing the effectiveness of marine protected areas (MPAs) around the world.

The cooperation model involves the active participation of local governments and communities in providing resources and enforcement capacity to protect marine species, as in the case of illegal shark fishing in the Galapagos. The WWF identified the Galapagos Islands as an international conservation priority. Cooperative strategies initiated in the Wildlife Marine Program led to the implementation of surveillance systems and resulted in staff training and certification in the areas of marine sanctuaries. When the technical framework was ready and the base for the protective culture built, the local governor, the Galapagos National Park, was able to embrace its leadership role, promoting best practice and mentorship in Ecuador and beyond. As a result of the establishment of this project, the population of sharks in the Galapagos became the densest in the world, while a decade ago an estimated 12,000 sharks were poached annually. Such cooperative strategies respond to the 'imperative of integration', identified as a basic condition of effective marine governance by empirical experience around the world. This imperative highlights the necessity of introducing global standards to the practice of marine governors at all of the levels: from village councils or Indian panchayats through sectorial fishery organizations to Food and Agriculture Organization (FAO) (Cicin-Sain, 1993). The leadership task of relating management practices, based on ecological understanding of the social mechanisms existing in different geographical settings, cultures, and ecosystems is essential for building the global framework.

Relational leadership not only solves day-to-day inconsistencies, contributing to system's coherence but helps overcome the problem of fit – lack of congruence between the attributes of a system of governance (i.e., rules, rule-making systems, and actor-networks) and the nature of a social-ecological system. Mancur Olson' s book on collective action describes the 'roving bandits' problem, which in the case of marine resources takes the form of the exploiters able to operate on large geographical scales in their search for target animal species or other assets (Olson, 2000). The scale of the possible damage that can be done by the roving bandits has been expanded by globalisation processes that rapidly increase international demand for high-value marine products and disseminated superb communication options and technological tools to facilitate them (Berkes et al., 2006). A lack of governance systems, information, and trust precludes reciprocal exchanges, feedback loops, synergy effects, and the creation of other mechanisms for enhancing and amplifying progress in governance. Relational leadership thus not only enables cooperation but also empowers and creates a driving force for direct participation in protection-oriented networks.

Conclusion

Global environmental change in marine ecosystems undisputedly influences a leadership vacuum in various aspects of governance, as strengthening resilience not only requires effective institutional solutions but also trust and a sense of agency. The major challenge is posed by the number of actors involved in governance processes, represented across a wide spectrum: the privileged and the underprivileged, the organised and the unorganised, those with much political and bargaining power and those with very little. For an effective system to be built they all have to be involved in negotiating their rights of access to common resources and rules of engagement. The process involves not only establishing a common understanding of needs and rights, but defining the meaning of social justice, food security, power and human-environment relations that vary across cultures. These components are based in the cognitive domains addressing what is required for an effective response to the challenges outlined above. They have also resulted from transformed hierarchies of interests that enable the redirection of attention and resources towards the protection rather than the exploitation of the marine base.

Managing environmental change in marine ecosystems has to be based on global political leadership, as the problem is a model illustration of the influence of global realities on the conditions and perspectives of particular societies, and the fact that it cannot be managed unilaterally. The structural, cognitive, and relational manifestations of leadership are oriented towards formatting and stabilising a systemic response to global environmental change, as the scale of the challenges involved is likely to grow in the decades to come. Given the specifics of the marine governance arena, leadership processes are oriented mainly towards responding to fragmentation and steering relations in order to influence two major processes: interaction and competition management.

Interaction. The whole concept of marine governance is concentrated around regulating human interaction with the global ocean. Various interactions are embedded in its nature, as even geographically and culturally distant stakeholders are connected in social, economic, and technological interdependencies (Jouffray et al., 2020). Interaction per se does not provide solutions for wicked problems of sustainable change in marine policies, but it serves as an entry condition for any attempt to do so. Neither organizations operating globally, nor governments or researchers are in possession of sufficient knowledge and experience to base an overall policy system on: 'Academics, policymakers and users have to interact "to get the picture right"' (Kooiman et al., 2005).

Competition management. The depletion of marine resources has been widely indicated as the primary reason for stricter management and more efficient governance. From the 'fish wars' cycles through pollution and the introduction of invasive species to the illegal exploitation of high-value natural resources by criminal networks the scale of global ocean degradation reflects the intensity of global competition. Leadership is becoming increasingly important in this area because the new operational system for marine governance will be shaped not only by creating a balance of power within the group of the most important international players, but more than ever before by securing the sustainability of the accepted solutions. Marine governance cannot be limited to laws and regulations protecting marine areas and resources, as its strategic importance creates strong incentives for free-riding or other opportunistic behaviours. Therefore, the actors involved have to build a dynamic governance structure through which they will be able to govern themselves with a view to receiving joint benefits (Ostrom, 1990: 29).

Leadership is a crucial factor in ensuring agency, empowering stakeholders, and enhancing their capacity to make decisions and to act upon them, and thus to influence the behaviours of government and institutions (Hass, et al. 2021). Given the strategic significance of the world's seaways, frameworks for global marine governance have been slow to develop. Their contribution to the global environmental change relies on leadership, which not only initiates governance frameworks but also recognises that good governance is a dynamic process requiring new ways of bringing international actors together to address the accumulation of new problems and transformations.

References

Atkins, J. P., Burdon, D., Elliott, M., Gregory, A. J. (2011) 'Management of the marine environment: Integrating ecosystem services and societal benefits with the DPSIR framework in a systems approach', *Marine Pollution Bulletin*, vol 62, pp215–226.

Berkes, F., Hughes, T. P., Steneck, R. S., Wilson, J. A., Bellwood, D. R., Crona, B., Folke, C., Gunderson, L. H., Leslie, H. M., Norberg, J., et al. (2006) 'Globalization, roving bandits, and marine resources', *Science*, vol 311, pp1557–1558.

Cash, D. W., Clark, W. C., Alcock, F., Dickson, N. M., Eckley, N., Guston, D. H., Jäger, J., Mitchell, R. B. (2003) 'Knowledge systems for sustainable development', *Protocols of the National Academy of Science USA*, vol 100, no 14, pp8086–8091.

Cava, F., Schoedinger, S., Strang, C., Tuddenham, P. (2005) 'Science content and standards for ocean literacy: A report on ocean literacy', http://coexploration.org/oceanliteracy/documents/OLit2004-05_Final_Report.pdf, accessed March 01, 2021.

Cicin-Sain, B. (1993) 'Sustainable development and integrated coastal management', *Ocean and Coastal Management*, vol 21, pp11–43.

Collinge, C. and Gibney, J. (2010) 'Connecting place, policy and leadership', *Policy Studies*, vol 31, no 4, pp379–391.

Fletcher, S. and Potts, J. (2007) 'Ocean citizenship: An emergent geographical concept', *Coastal Management*, vol 35, pp511–524.

Gee, K., Kannen, A., Adlam, R., Brooks, C., Chapman, M., Cormier, R., et al. (2017) 'Identifying culturally significant areas for marine spatial planning', *Ocean and Coastal Management*, vol 136, pp139–147.

Grint, K. (2005) 'Problems, problems, problems: The social construction of 'leadership', *Human Relations*, vol 58, pp1467–1494.

Haas, B., Mackay, M., Novaglio, C., Fullbrook, L., Murunga, M., Sbrocchi, C., McDonald, J., McCormack, P. C., Alexander, K., Fudge, M., Goldsworthy, L., Boschetti, F., Dutton, I., Dutra, L., McGee, J., Rousseau, Y., Spain, E., Stephenson, R., Vince, J., Wilcox, C., and Haward, M. (2021) 'The future of ocean governance', *Haward Reviews in Fish Biology and Fishery*, pp1–18.

Horlings, I. (ed) (2010) *Vital coalitions, vitalregions; cooperation, in sustainable, regional development*. Wageningen Academic Publishers, Wageningen.

Jouffray, J. B., Blasiak, R., Norström, A. V., Österblom, H., and Nyström, M. (2020) 'The blue acceleration: The trajectory of human expansion into the ocean', *One Earth*, vol 2, no 1, pp43–54.

Kooiman, J., Bavinck, M., Jentoft, S., Pullin, R. (2005) *Fish for life*. Amsterdam University Press, Amsterdam.

Modelski, G. (1999) From leadership to organization: The evolution of global politics. In Bornschier, V., Chase-Dunn, C. (eds.) *The Future of Global Conflict*. London, Sage: 11–39.

Nordquist, M. H. (ed) (1985) *United Nations Convention on the Law of the Sea 1982, A Commentary*, vol. 1, LaicMartinus-Nijhoff Publishers, Dordrecht, Boston.

Olson, M. (2000) *Power and prosperity*. Basic Books, New York.

Ostrom, E. (1990) *Governing the commons. The evolution of institutions for collective action*. Cambridge University Press, Cambridge.

Pinto, R., De Jonge, V. N., Marques, J. C. (2014) 'Linking biodiversity indicators, ecosystem functioning, provision of services and human well-being in estuarine systems: Application of a conceptual framework', *Ecological Indicators*, vol 36, pp 644–655.

Rhodes, R. A. W., and Hart, P. (2014) Puzzles of political leadership. In Rhodes, R.A.W. and T Hart, P. (eds.) *The handbook of political leadership*. Oxford: Oxford University Press: 1–21.

Valdés, L. (ed) (2017) *IOC-UNESCO Global Ocean Science Report. The current status of ocean science around the world*. UNESCO Publishing, Paris.

PART 3

Non-state actors in marine environmental governance

8

EXPERTS

Scientific knowledge for ocean protection

Alice B.M. Vadrot

Ocean ecosystems are among the most difficult places on earth to monitor, govern, and protect. Although single states are unlikely to shoulder ocean protection alone in the light of global environmental change, shared efforts are hampered by a lack of know-how needed to manage oceans and share marine resources in a sustainable and equitable way. Science and expert knowledge play a pivotal role in this regard: they inform policy and international treaty-making by providing scientific advice ideally found on the 'best available science'. At the same time, scientists and other experts have their own stakes and interests in ocean governance and need to respond to emerging challenges, such as global environmental change, growing demands by policymakers and industry, and the changing context in which marine scientific research (MSR) is conducted and used. The expertise we already possess in monitoring, governing, and protecting ocean ecosystems has been shaped by the political, legal, technological, and economic conditions under which knowledge about the ocean has been constructed and applied.

For a long time, a lack of scientific knowledge about the ocean was perceived as the main obstacle to advancing efforts to protect ocean ecosystems and respond to global environmental changes. This gap was steadily closed thanks to global initiatives and collective efforts to develop global monitoring programmes, data banks, and assessment reports, such as the Census of Marine Life, the World Ocean Assessments, and the United Nations Decade of Ocean Science, which was launched in 2021. These initiatives have greatly enhanced our knowledge of the impacts of climate change, ocean acidification, overfishing, and pollution on marine ecosystems. Together with significant innovations in MSR — from data collection to data analysis — they have been particularly relevant for strengthening the environmental focus of the Law of the Sea (Stephens and Rothwell, 2015: 560).

Correspondingly, governments and other stakeholders involved in agreement making are increasingly requesting scientific expertise to underpin the development and implementation of ocean protection measures. Such measures include the identification, designation, and management of marine protected areas (MPAs) and other area-based management tools (ABMTs). To ensure that the implications of political decisions for the environment are considered beforehand, both state and non-state actors may commission strategic environmental assessments (SEA) and environmental impact assessments (EIAs). Scientific evidence promises the 'objective' resolution of trade-offs between different forms of use of marine

DOI: 10.4324/9781315149745-11

ecosystems, and across different sectors, such as fishing, shipping, energy, conservation, tourism, and security. The more human activities expand into the ocean, the more experts will be needed to assess the environmental risks of extracting marine resources, especially if these are in remote or less well-known areas, such as the seabed beyond national waters.

The role of science and experts in ocean governance is an emerging research field that has benefitted from studies in other environmental fields, such as climate change and biodiversity, and from insights gained across a number of scientific disciplines, including global environmental politics, the history of science, Science and Technology Studies (STS), sociology, anthropology, and law. The history of science, for instance, has greatly contributed to elucidating the influence of politics, the economy, and the military on the emergence of oceanography, while scholars from the fields of political science, sociology, and STS have cast light on scientific advice practices and the manifold dynamics at the interface between science and policy, or knowledge and power.

This chapter introduces the reader to prominent themes related to the role of science and expertise in ocean governance, with a specific emphasis on emerging challenges. It starts from the assumption that we cannot apprehend it in isolation. Legal, (geo) political, economic, and technological conditions affecting access to, and the assessment and measuring of ocean space, its ecosystems, and its resources have significantly shaped the development of MSR, policy advice practices, and the role attributed to experts in ocean governance. The first part of this chapter introduces readers to the context in which ocean expertise has been produced. The second part provides insights into theoretical reflections and research addressing the role of experts. I conclude by outlining the relevance of this emerging field of study and discussing issues that may become pressing in the near future.

Changed conditions for producing ocean expertise

It is often said that we know more about the moon than we know about the ocean. As Hannigan half ironically noted, today 'the ocean is on the verge of being transformed from a 'half-known life' and 'submarine aliens to an emerging focus of global attention and concern' (Hannigan, 2016: 2). In this section, I illuminate the historical, legal, political, and technological conditions for the production and use of marine expertise in response to global environmental change. Although knowledge about the ocean is not restricted to scientific expertise, marine scientific research, especially ocean science, has significantly influenced how we monitor, govern, and protect it. What does this research entail? What are the conditions for its production? And in what ways has the focus on environmental protection manifested itself as a key pillar of MSR in the twenty-first century?

Marine scientific research: an expensive and extensive discipline

For decades, diplomats and legal scholars have sought to define marine scientific research without much success. Legal scholars tend to use the term to designate 'any form of scientific investigation, fundamental or applied, concerned with the marine environment, i.e., that has the marine environment, as its object' (Birnie, 1995: 242). The United Nations Convention on the Law of the Sea (UNCLOS; see Chapter 2) includes provisions on MSR, but it does not provide a legal definition; neither does it list activities that fall under this term (Stephens and Rothwell, 2015: 561). According to the *Global Ocean Science Report*, ocean science includes 'physical, biological, chemical, geological, hydrographic, health and social sciences, as well as engineering, the humanities and multidisciplinary research on the relationship between humans and the ocean'; besides, it 'seeks to understand complex, multiscale socio-ecological systems and services, which requires observations and multidisciplinary and collaborative research.' (IOC-UNESCO, 2017: 19).

MSR topics are often interconnected: research on one phenomenon, such as deep-sea ecosystems, may reveal important facts about other phenomena, such as the effects of pollution or deep-sea mining. As with the geomorphology of continental margins, knowledge is needed for several reasons. States need physical oceanographic data in order to establish the outer limits of their continental shelves beyond 200 nautical miles (UNCLOS, 1982). Data may also be required to analyse seismic activities responsible for tsunamis. Operations to search for lost airplanes or vessels rely on it. Last but not least, predictions of ocean change as a result of climate change rest on this data, too. These examples demonstrate the difficulty of distinguishing between pure science and applied research; the boundary is often blurred by overlaps between the interests of scientists, state actors, industry, and the military in exploring marine phenomena.

From a legal and political point of view, internationally agreed definitions of what MSR is and what activities it encompasses are highly political. Moreover, the unequal distribution of marine science and technology poses a challenge for international collaboration between regions (Global Ocean Science Report, 2020). Bibliometric studies revealed imbalances in ocean science, pointing to 'path dependencies that shape the asymmetric relationships within transregional collaborations until today' (Partelow et al. 2020: 20) and the unequal distribution of 'collaboration capital' and research topics between the Global North and the Global South (Tolochko and Vadrot, 2021a, 2021b). Global imbalances in research capacities and technological infrastructures, indeed, constitute a dividing line between developed countries that have the economic means to explore the oceans and developing countries that lack the means to monitor and commodify ocean resources. For this reason, closing the gap between the Global North and Global South through capacity building and technology transfer has been at the core of recent debates on how to strengthen science-policy interrelations in ocean governance (Visbeck, 2018).

Insights from the history of science and oceanography

Historians of science and oceanography have significantly increased our understanding of the connections between scientific development and its social context (Mills, 1993). The history of ocean science has meticulously studied the dependence of early oceanography on a combination of ships, expeditions, and instruments and has explored the role of science in fisheries and post-World War II oceanography. Studies on the intersection between ocean science and global geopolitics reaching back to the nineteenth century have revealed close ties between knowledge and power, and how our ocean knowledge relates 'to empire, the extension of national power, and naval warfare, especially involving submarines' (Rozwadowski, 2014: 336). Oceanography 'has taken shape within particular conceptions of place, interests, and sovereignty—not least because the ocean sciences are involved with the practical and epistemological control of the earth's resources and are therefore intimately connected with the realities of the nation-state and its institutions, on the one hand, and a globalised economy, on the other hand.' (Rozwadowski, 2014: 336). The co-production of ocean science, where scientific, economic, military, and political interests converged, was also described by the sociologist Vanderpool (1983), who explored the impact of the Law of the Sea on knowledge generation in marine scientific research, arguing that UNCLOS negotiations in the 1970ies disclosed the political-economic context of science. At the same time, the extension of national spheres of influence to 200-nautical-mile Exclusive Economic Zones has also shaped scientific inquiry by extending obligations to monitor fish stocks in these areas.

Historians of science, sociologists and STS scholars have thus sensitised us to the fact that oceanography is 'fundamentally social' (Mills, 1993: 9) and that '"societal" factors may at time affect the content of science and its evaluation of scientists' (Mills, 1993: 14). Ocean science is 'situated knowledge', produced in laboratories or on research vessels through collective efforts and technological advances that have changed human ability to explore the ocean. The ocean is not only a space for transport, a battlefield of geopolitical powers, and a trove of resources such as fish, salt, minerals and, today, genetic material to be used and protected from overexploitation (Steinberg, 2001), but also a setting for scientific inquiry. Through technological and scientific research, particularly from the beginning of the nineteenth century, the ocean emerged as a space to be systematically mapped, monitored, and represented. For their part, scholars from the field of historical geography have studied ocean knowledge production in terms of a variety of scientific, cultural, and technical practices (Laloë, 2016). Different periods 'have perceived, been able to perceive, and produced knowledge about the ocean space in diverging but characteristic ways' (Laloë, 2016: 7). While in the early nineteenth century individual natural philosophers could only study isolated elements of the sea, technology and systematic approaches have introduced a rapid shift to 'big science' and 'big ecology', which emerged as a response to the need for information to address major environmental problems (Coleman, 2010).

During the Cold War, oceanography was both a model for international cooperation and a major outlet for military research funds (Hamblin, 2005: 260). Indeed, after 1945 funding for ocean science increased significantly, most notably because armed services and the Navy needed to develop a 'superior understanding of the ocean environment' (Hamblin, 2005: 261). Military concerns shaped the form and content of oceanography and led to significant advances in science (Oreskes, 2003; Reidy, 2008). One example is the discovery of hydrothermal vents on the seafloor thanks to a novel technology that enabled humans to work in the ocean environment. These vents were first described in 1977, and since then only about 10% of their habitats have been explored. While the early discovery of hydrothermal vents was linked to military interest, in the past decade they have attracted increased attention, not only of environmental researchers but also of companies that are intent on exploiting minerals for commercial purposes. Yet another example is underwater acoustics, which the US Navy developed to detect and track Soviet submarines; later, marine scientists employed this technology to take the ocean's temperature through sound in the Acoustic Thermometry of Ocean Climate project (Hannigan, 2016: 86). Thus, these technological developments were not only significant from a security point of view; they played a prominent role in understanding global environmental change, too.

Mounting environmental concerns and accelerated knowledge production

Since the 1950s, dynamic, model-based methods have played a crucial role across environmental sciences, including oceanography, and have consistently informed views about global environmental change affecting the ocean (Hamblin, 2014). While for a long time overfishing was considered the most problematic resource exploitation practice, in the 1970s concern grew over increased extraction of both non-living (sand, petroleum, and gravel) and living resources (previously untapped marine genetic resources) from marine environments (Hannigan, 2016, 105). Technologies for deep-sea mining down to depths of 4,000 metres are rapidly maturing, with several companies investing into scientific expeditions to assess the prospects of extracting mineral resources from the deep seabed. In the 1970s, environmentalists focused on human impacts in shallow coastal waters, and occasionally on the effects of offshore oil spills, drift-net fishing, toxic dumping, and the decimation of whale and dolphin species. The effects of expanding human activity on open-ocean and deep-sea environments — including deep-water fisheries, deep-sea mining, and deep-water oil and gas drilling — were poorly understood and out of sight (Baker et al., 2001: 5). It is only in the past two or three decades that scientists have turned their attention to the cumulative results of human activity and how these have led to a 'serial decline in the ocean's health and resilience' (IPSO, 2013: 3). As a result, data collection to assess ocean change has intensified and research on global environmental change has accelerated.

Knowledge and data on marine ecosystems and resources are emerging as valuable assets. Also, monitoring practices and sampling technologies contribute to the scientific representation of the ocean and its manifestation as a site where various technological innovations compete for scientific legitimacy and marketability. For instance, new technologies now enable the collection and storage of vast amounts of different types of data, as well as their integration. This development predated new paradigms in deep-sea ecology and the constitution of new scientific disciplines, such as 'biodiversity informatics' or 'bioinformatics' informing modelling and scenario building efforts. It is expected that by linking field observations to remote sensing and by increasing efforts to streamline, integrate, and disseminate environmental data, it will become easier to tackle uncertainty and, more specifically, to overcome the taxonomic impediment.

What has been outlined here is a rapid shift to 'big science' and 'big ecology', which are complex scientific endeavours in which large investments, often in the form of government funding, are made. 'Big science' often involves international, collaborative efforts amongst many scientists and institutions and is often characterised by expensive shared equipment and infrastructure for data production, management, curation, and circulation (Vermeulen, 2013). Efforts by so-called 'data and scenarios communities' to further develop and harmonise methodologies, data sets, and indicators created new concepts such as 'Ocean Essential Variables' (OBV); the latter is to bolster an integrated framework for sustained ocean observation. These concepts emerged in relation to, and are legitimised by, state parties to Multilateral Environmental Agreements (MEAs) and regional management organisations, as well as by assessment exercises, scientific committees and advisory bodies, and other forms of scientific and expert advice established to underpin ocean protection with the 'best available science'.

Science-policy interrelations: a complex web

Scientific expertise has become indispensable in policymaking processes, given its increasingly important role in SEAs and EIAs, especially as regards complex global environmental issues. In the past decade, several international efforts emerged to strengthen the knowledge base available to policymakers tackling ocean change. A number of new institutions and practices appeared whose aim was to improve science-policy interaction and facilitate the uptake of knowledge by policymakers, including scientific committees, intergovernmental scientific and technical bodies, global assessment bodies and procedures, and ad-hoc advisory groups and panels. Although these differ in size, purpose, scale, and degree of institutionalisation, they all share the general objective of developing recommendations to policymakers without being policy prescriptive.

The question of how to strengthen science-policy interfaces for ocean management and protection cannot be addressed without considering the context and level of policymaking, the stage in the policymaking process, the specific institutional setting, the degree of formalisation, the availability of expertise, and the specific policy advice culture. Moreover, science and expertise can be used in several ways: to legitimise political action or non-action and as policy instruments. Policymakers may also selectively refer to expert knowledge related to their own priorities and strategically select experts who resonate with their own interests. Environmental knowledge is often contested, either because of a lack of consensus among science or a 'lay-expert' divide (Peterson, 2019). According to Collins and Evans (2006), there is a spectrum between 'experts' and 'laypeople' contributing to policymaking with 'referred expertise', 'interactional expertise', or 'contributory expertise'; each indicating a different degree of expertise that someone has in a technical or scientific field relevant to policy. As Cash et al. (2003) argue, there is a need to negotiate 'boundaries' between science and policy (Star and Griesemer, 1989), between disciplines, across nations, and across levels. In this endeavour, concepts such as 'usable knowledge' and 'epistemic communities' may help us explore the multi-faceted roles of experts.

Science for policy: usable knowledge as an imperative

Usable knowledge is defined as a relevant body of scientific knowledge or as 'accurate information that is of use to politicians and policymakers' (Haas, 2005: 387). Natural scientists often emphasise that they have a societal responsibility to present data objectively and advocate for conservation. Without any scientific evidence, it 'may be difficult to convince the international community to suspend fishing in an area' (Mossop, 2018: 6) or to participate in policy by developing indicators on the state of the marine environment. Scientists are supposed to translate data for society and to share them, along with infrastructure and resources. By collaborating and sharing their data amongst each other, marine scientists are expected to provide a consistent evidence-based message and thus add credence to their advice to political leaders. The research community is meant to describe natural environments, and coordinate observation and monitoring activities, as well as study the effectiveness of conservation measures, by collecting baseline data (Wright et al., 2018). Its role is to orchestrate common interests needed to manage areas beyond national waters and — being funded by public sources — it should have 'an ethical obligation to involve in research and monitoring the different users of the maritime space, such as governments, civil society, NGOs, and economic actors (see Chapters 9–11). This would improve study designs and execution by incorporating historic, local, and/or traditional knowledge from different knowledge systems and make new resources available' (Barbier et al., 2018: 532).

In fact, many of these insights and claims are developed within ocean science communities themselves; they are testimony to the emerging normative dimension of MSR and the imperative to produce 'usable knowledge' for policymakers and other stakeholders. The need for information and the fact that the data are changing rapidly call for dynamic interaction between science and policy (Lubchenco, 1995: 8). While scientists constantly face the challenge of communicating the state of the art of their research, they must also per se address the right issues in an appropriate manner. Relevant scientific information for supporting natural resource management and assist environmental decision-making 'includes basic information about the system and people's choices for altering future system states' (Lubchenco et al., 2019: 107). Furthermore, science for ocean policy shall fulfil certain criteria (Sullivan et al., 2017). It needs to be clear and succinct, meaning that communication of scientific knowledge should be organised in ways that are understandable to non-scientists while preserving the meaning and relative certainty of findings. By addressing issues faced 'now', by recognising constituents, mandates, and interests, and by being non-prescriptive, science for policy should ideally be timely and relevant. It should also be legitimate and credible; that is, it should represent consensus findings across relevant scientific communities and disciplines (Cash et al., 2003) by integrating multiple perspectives based on transparent processes and inclusiveness. Science for policy should also be actionable and go beyond diagnosing a problem: it must provide options (Palmer, 2012).

As described in the previous section, the scientific substantiation of global environmental change has increased the demands made by policymakers on science and created the imperative of usable knowledge and inter- and transdisciplinarity (Till et al., 2018); scientists are expected to actively engage in policy advice at diverse policy levels and to interact with knowledge holders and stakeholder groups from different sectors. This emerging imperative is mirrored by several global initiatives and collective efforts to develop global monitoring programmes, data banks, and assessment reports, such as the Census of Marine Life, the first World Ocean Assessment (UN, 2017) in 2010, the second World Ocean Assessment (UN, 2021), the Ocean Biogeographic Information System (OBIS), the Global Ocean Observing System (GOOS), and the UN Decade of Ocean Science (IOC-UNESCO, 2020).

Epistemic communities

Social science scholars from diverse scientific disciplines have become interested in the institutional and political dimensions of links between environmental science, policy, and politics, particularly since the Global Environmental Assessment (GEA) project, launched in 1995 (see, e.g., Jasanoff, 2004). Many studies contest the idea of science 'speaking truth to power', which had assumed that a clear distinction between 'objective knowledge' and 'subjective values' was not only possible, but desirable (Wildavsky, 1979; Haas, 2005). The 'speaking truth to power' paradigm implies that science per se is not associated with power, its core principle is credibility, and its findings are value-free. Hence, a clear distinction was made between science and policy, and between science and power. Even though such an understanding might seem outdated, it has been a major issue in political science for a long time.

One very prominent example is the concept of 'epistemic community' informing the constructivist turn in International Relations theory. Constructivist approaches challenge the dominant realist and liberal paradigms in IR by assuming that reality is socially constructed and shaped by norms, identity, and ideas, implying that material objects do not merely exist but become constructed by the meaning that different actors assign. This concept of an epistemic community assumes that a network of experts who share the same values and epistemologies will emerge and enable the political recognition of a problem previously identified by science. Haas defined it thus: 'a network of professionals with recognised expertise and competence in a particular domain and an authoritative claim to policy-relevant knowledge within that domain or issue area' (Haas, 1992: 46). This made sense of the observation that a group of scientists might succeed in transforming knowledge claims into a political agenda. Such a group might be described as an organised advocacy coalition formed around shared norms and beliefs that would include scientific methods, theories, and paradigms regarding well-defined issues. As such, epistemic communities may be involved in agenda-setting, problem-understanding, and the development of policy fields. They may influence political processes both as experts and advocates, especially if they have reached a specific degree of institutionalisation.

According to Haas and Stevens, one condition is the availability of usable knowledge: if it is successfully constructed and transmitted to policymakers, policymakers may, through social learning, reflect scientific consensus about environmental sustainability, which leads to more effective regimes (Haas and Stevens, 2011). Lodge et al. (2007) applied this thinking to fisheries and found a lack of political will amongst fishery managers and the users of marine resources to follow scientific advice and enforce related management measures, which led to ineffective agreements. In order to demonstrate the role played by epistemic communities to enact social learning, Gonçalves (2021) applied the epistemic community model to the case of regional fisheries management organisations. She used process tracing, elite interviews, and a systematic study of meeting reports concerning three such organisations: the Commission for the Conservation of Antarctic Marine Living Resources (CCAMLR); the International Commission for the Conservation of Atlantic Tunas (ICCAT); and the Commission for the Conservation of Southern Bluefin Tuna (CCSBT). She found that the ways in which epistemic communities mattered in practice differed (Goncalves, 2021).

In the case of ICCAT, for instance, the Standing Committee on Research and Statistics (SCRS) was successful in addressing the Bluefin Tuna crisis and anticipated social learning regarding the need to reduce the fish catch amongst the more than 50 state parties that needed to reach agreement. Similarly, the CCSBT, where initially countries such as Japan, Australia, and New Zealand struggled to define their Total Allowable Catch, saw a turning point when states established a Scientific Committee (SC) composed of independent scientists as an advisory body to the Commission. Amongst other things, the SC oversees the assessment of the status and trends regarding the Southern Bluefin Tuna population, coordinates research, and makes recommendations to the Commission on the basis of consensus. It took five years to fully establish the SC; this, according to Gonçalves, anticipated social learning and confirms Haas's epistemic community model: it presupposes a process that clearly separates the reaching of consensus amongst experts from the consideration of scientific information and specific recommendations by policymakers.

Although the epistemic community model implies a separation between science and policy, it does not preclude the possibility that science may have political consequences. Scientific information may support specific policy options that are beneficial for some but detrimental to others. This points to an inherent democratic deficit: if those who will suffer from the effects of the application of scientific knowledge have not taken part in its development, they will tend to view such processes as an illegitimate and exploitative set of discursive practices (Lidskog and Sundqvist, 2002). This also points to the instrumentalisation of science and knowledge to legitimise political decisions or non-decisions, either by arguing that there is a lack of certainty regarding the scope of the problem, or a lack of evidence for the appropriateness of specific policy measures.

Beyond epistemic communities

Although the 'epistemic community' model does hold an explanatory power, describing how specific policy fields and institutional arrangements (in other words, 'regimes') have emerged internationally as a response to environmental degradation, most notably in the wake of the 1992 Earth Summit in Rio, it has been criticised for assuming that knowledge is inherently objective and for neglecting the societal and cultural context within which scientific findings are reflected (Litfin, 1984). In particular, STS scholars have pointed out the need to consider the dynamics involved in the production of (scientific) knowledge. The practical work of scientific investigation, for example in the laboratory or at other micro-levels (Latour, 1987), or the many ways in which science and society are 'co-produced' and enable each other (Jasanoff, 1990). From this perspective, solutions for addressing global environmental change require solutions to problems of knowledge (generation) and vice-versa. Science itself incorporates political and ethical principles into the production of knowledge insofar as it constantly separates facts from non-facts and science from non-science (Gieryn, 1995).

This raises several issues, especially with regard to legitimisation mechanisms, democratic accountability, power, and influence. 'Epistemic selectivities' may emerge within institutions or international negotiation settings that favour some knowledge forms and epistemologies over others (Vadrot, 2014). Studying expertise by focusing on international negotiations has indeed become an important avenue for research into science-policy interrelation practices (Vadrot, 2020). Negotiations are 'windows into networks of global environmental governance' (Campbell, et al., 2014: 2) and allow non-state actors (e.g., scientific networks, associations, and alliances) — such as the Deep-Ocean Stewardship Initiative, the International Union for Conservation of Nature, the World Wildlife Fund, and the High Seas Alliance — to participate as observers.

Such field sites include Conferences of the Parties of MEAs, treaty negotiations, and meetings of global assessment bodies such as the IPBES and the IPCC. Many MEAs have established subsidiary bodies under the conventions, for instance Scientific and Technical Bodies or Committees that emerged as important objects of research to understand the role of science in environmental agreement making (Chasek, 2019; Kohler, 2020). Recently, participant observation and ethnographic studies have been systematically developed for these specific 'field sites' (Campbell et al., 2014) and in order to explore struggles over environmental knowledge in practice (Hughes and Vadrot, 2019; Vadrot, 2020). Gray et al. (2014), for instance, analysed the role of the concept of 'marine protected area' thanks to data collected through Collaborative Event Ethnography during the 10th Conference of the Parties of the CBD. The MPA concept has accommodated the distinct and sometimes conflicting perceptions of various states and non-actors. Gray et al.'s analysis of the diverse narratives circulating in the negotiation room revealed tensions between a science-driven top-down approach and a more bottom-up attitude that acknowledges the importance of local and indigenous knowledge. These conflicting views converged into a decision to adopt a regional approach to the designation of Ecologically or Biologically Significant Marine Areas. Thus, while the MPA concept was broad enough to accommodate different interests, the authors found a lack of debate on whether MPAs were actually a suitable marine conservation tool, which led to an 'emphasis on counting MPAs as a means of measuring progress on marine conservation [which] is misleading' (Gray et al., 2014: 80).

Since 2018, some scholars have started to study ongoing negotiations towards a new, international, legally binding instrument under the United Nations Convention on the Law of the Sea that addresses the conservation and sustainable use of marine biological diversity of areas beyond national jurisdiction (BBNJ Treaty) (Tessnow-von Wysocki and Vadrot, 2020). The BBNJ Treaty seeks to close legal gaps related to marine genetic resources; area-based management tools, including marine protected areas; environmental impact assessments; and capacity building and the transfer of marine technology. Although these four 'package elements' make high demands on technical expertise, there is no formalised process that would ensure the systematic integration of (scientific) knowledge during the negotiation. Nonetheless, experts do play a role by being part of national delegations, providing information on diverse topics during side events, publishing on relevant topics, and disseminating relevant literature to national delegates. As scholars interested in the complex web constituting science-policy interrelations, we need both conceptual and methodological innovations and fresh ideas on how to strengthen the role of expertise in ocean governance across levels.

Conclusion

The comprehensive study of the role of expertise is an emerging field within ocean governance scholarship and deserves to receive increased attention for three main reasons. First, rising pressures on the marine environment, especially due to global environmental change, call for rapid policy responses that require expertise and thus increase the demands placed on MSR to respond to global environmental change. Second, these new demands have triggered a debate amongst scholars and practitioners about the type of expertise that is needed, partly revealing the weak spots of ocean science, namely, data scarcity, lack of inter- and transdisciplinarity, and the North-South Divide. Thirdly, the rapid proliferation of institutions, processes, and initiatives designed to increase the uptake of scientific knowledge about the ocean calls for a scholarly debate and for research into science-policy interrelations, especially as regards formalised processes such as scientific and technical bodies or assessments.

The role of experts and the institutional conditions for the uptake of science may vary significantly depending on the scale of intervention, the level of policymaking, and the culture of policy advice. This poses a challenge for the systematic study of the role of science and expertise in ocean governance — impeding the development of universal explanations or models that go beyond single case studies. The specific political, legal, technological, and economic conditions under which ocean knowledge has been constructed and used need to be carefully assessed when developing models for the uptake of (scientific) knowledge and the design of new advisory processes and institutions. Moreover, the role of experts and expertise cannot be disconnected from persisting, global research capacity imbalances between the Global North and Global South. These have come to the attention of both policymakers and scientists seeking to strengthen international scientific collaboration, as intended by the UN Decade of Ocean Science. Debates over deeply rooted inequalities between traditional maritime powers and, for instance, Small Island Developing States have created a new sensitivity to the need for capacity building and marine technology transfer (see Chapter 14). This is at the heart of many discussions on strengthening expertise for ocean protection, as in the BBNJ Treaty case.

There is a need to diversify the available knowledge base to respond to global environmental change, as some (especially STS-inspired approaches) have argued. This would include recognising the contribution of knowledge originating from all relevant sectors and knowledge holders as well as from local and indigenous communities. The parallel rise of scientific, political, economic, and military interests in the ocean as a space to be known and governed — on the basis of complex scientific endeavours into which vast sums are invested — has not only shaped MSR, but also determined the conditions for interaction between experts, on the one hand, and policy and practice, on the other hand. As scholars, it is our duty to question our assumptions about what counts as expertise and to critically reflect on the circumstances under which scientists or expert bodies may become actors of ocean governance. This duty will become more pressing as global environmental changes increasingly affect the earth's oceans and seas.

Acknowledgements

This work was supported by the Research Council (ERC) under the European Union's Horizon 2020 research and innovation programme (grant agreement No 804599 – MARIPOLDATA – ERC-2018-STG).

References

Baker, M.C. et al. (2001) *The Status of Natural Resources on the High-Seas. Part I: An Environmental Perspective; Part 2: Legal and Political Considerations Gland.* CH. WWF-Worldwide Fund for Nature, Gland, Switzerland.

Barbier, M. et al. (2018) 'Ethical Recommendations for Ocean Observation', *Advances in Geosciences*, vol 45, pp 343–361.

Birnie, P. (1995) 'Law of the Sea and Ocean Resources: Implications for Marine Scientific Research', *The International Journal of Marine and Coastal Law*, vol 10, no 2, pp 229–251.

Campbell, L.M. Hagerman, S. and Gray, N.J. (2014) 'Producing Targets for Conservation: Science and Politics at the Tenth Conference of the Parties to the Convention on Biological Diversity', *Global Environmental Politics*, vol 14, no 3, pp 41–63.

Cash, D. et al. (2003) 'Knowledge Systems for Sustainable Development', *Proceedings of the National Academy of Sciences*, vol 100, no 14, pp 8086–8091.

Chasek, P.S. (2019) 'Linking Scientific Knowledge and Multilateral Environmental Governance', in: Peterson, M.J. (ed) *Contesting Global Environmental Knowledge, Norms and Governance*. Routledge, New York, pp 15–53.

Coleman, D. (2010) *Big Ecology: The Emergence of Ecosystem Science*. University of California Press Books, Berkeley.

Collins, H. and Evans, R. (2006) 'The Third Wave of Science Studies: Studies of Expertise and Experience', in: Selinger, E. and Crease, R. (eds) *The Philosophy of Expertise*. Columbia University Press, New York, pp 39–110.

Gieryn, T.F. (1995) 'Boundaries of Science', in A.I. Tauber (eds) *Science and the Quest for Reality: Main Trends of the Modern World*. Palgrave Macmillan, London.

Gonçalves, L.R. (2021) *Regional Fisheries Management Organizations: Are They Effective?* Springer, Cham.

Gray, N.J., Gruby, R.L. and L.M. Campbell. (2014) 'Boundary Objects and Global Consensus: Scalar Narratives of Marine Conservation in the Convention on Biological Diversity', *Global Environmental Politics*, vol 14, no 3, pp 64–83.

Haas, P. and Stevens, C. (2011) 'Organized Science, Usable Knowledge, and Multilateral Environmental Governance', in R. Lidskog and G. Sundqvist (eds) *Governing the Air: The Dynamics of Science, Policy, and Citizen Interaction*. MIT Press, Cambridge, MA.

Haas, P.M. (1992) 'Introduction. Epistemic Communities and International Policy Coordination', *International Organization*, vol 46, no 1, pp 1–35.

Haas, P.M. (2005) 'Science and International Environmental Governance', in P. Dauvergne (ed) *Handbook of Global Environmental Politics*. Edward Elgar, Cheltenham.

Hamblin, J.D. (2005) *Oceanographers and the Cold War: Disciples of Marine Science*. University of Washington Press, Seattle and London.

Hamblin, J.D. (2014) 'Seeing the Oceans in the Shadow of Bergen Values', *Isis*, vol 105, no 2, pp 352–363. https://doi.org/10.1086/676573.

Hannigan, J. (2016) *The Geopolitics of Deep Oceans*. John Wiley and Sons, Cambridge.

Hughes, H. and Vadrot, A.B.M. (2019) 'Weighting the World: IPBES and the Struggle over Biocultural Diversity', *Global Environmental Politics*, vol 19, pp 14–37.

IOC-UNESCO. (2017) *Global Ocean Science Report: The Current Status of Ocean Science Around the World*, L. Valdés (ed), United Nations Educational, Scientific and Cultural Organization, Paris.

IOC-UNESCO. (2020) *Global Ocean Science Report 2020-Charting Capacity for Ocean Sustainability* in K. Isensee (ed.), UNESCO Publishing, Paris.

IPSO. (2013) *Implementing the Global State of the Oceans Report*. Available at: http://www.stateoftheocean.org/

Jasanoff, S. (1990) *The Fifth Branch. Science Advisers as Policymakers*. Harvard University Press, Cambridge.

Jasanoff, S. (2004) *States of Knowledge: The Co-Production of Science and the Social Order*. Routledge, Abingdon.

Laloë, A.-F. (2016) *The Geography of the Ocean: Knowing the Ocean as a Space*. Routledge, Abingdon.

Latour, B. (1987) *Science in Action: How to Follow Scientists and Engineers through Society*. Harvard University Press, Cambridge.

Lidskog, R. and Sundqvist, G. (2002) 'The Role of Science in Environmental Regimes: The Case of LRTAP', *European Journal of International Relations*, vol 8, no 1, pp 77–101.

Litfin, K.T. (1984) *Ozone Discourses: Science and Politics in Global Environmental Cooperation*, Columbia University Press, New York City.

Lodge, M.W., Anderson, D., Lóbach, T., Munro, G., Sainsbury, K. and Willock, A. (2007) *Recommended Best Practices for Regional Fisheries Management Organizations: Report of an Independent Panel to Develop a Model for Improved Governance by Regional Fisheries Management Organizations*. Chatham House, London.

Lubchenco, J. (1995) 'The Role of Science in Formulating a Biodiversity Strategy', Science and Biodiversity Policy, a supplement to *BioScience*, vol 45, pp 7–9.

Lubchenco, J. et al. (2019) 'Connecting Science to Policymakers, Managers, and Citizens', *Oceanography*, vol 32, no 3, pp 106–115.

Mills, E.L.(1993) 'The Historian of Science and Oceanography After Twenty Years', *Earth Sciences History*, vol 12, no 1, pp 5–18.

Mossop, J. (2018) 'The Relationship between the Continental Shelf Regime and A New International Instrument for Protecting Marine Biodiversity in Areas beyond National Jurisdiction', *ICES Journal of Marine Science*, vol 75, no 1, pp 444–450.

Oreskes, N. (2003) 'A Context of Motivation: US Navy Oceanographic Research and the Discovery of Sea-Floor Hydrothermal Vents', *Social Studies of Science*, vol 33, no 5, pp 697–742.

Palmer, M.A. (2012) 'Socioenvironmental Sustainability and Actionable Science', *BioScience*, vol 62, no 1, pp 5–6.

Partelow, S. et al. (2020) 'Tropical Marine Sciences: Knowledge Production in a Web of Pat Dependencies', *PLos One*, vol 15, no 2, p e0228613.

Peterson, M.J. (2019) *Contesting Global Environmental Knowledge, Norms and Governance*. Routledge, New York.

Reidy, M.S. (2008) *Tides of History: Ocean Science and Her Majesty's Navy*. University of Chicago Press, Malden.

Rozwadowski, H.M. (2014) 'Focus: Knowing the Ocean: A Role for the History of Science', *Isis*, vol 105, no 2, pp 335–337.

Star, S. and Griesemer, J. (1989) 'Institutional Ecology, 'Translations' and Boundary Objects: Amateurs and Professionals in Berkeley's Museum of Vertebrate Zoology, 1907–39', *Social Studies of Science*, vol 19, no 3, pp 387–420.

Steinberg, P.E. (2001) *The Social Construction of the Ocean*. Cambridge University Press, Cambridge, New York.

Stephens, T. and Rothwell, D.R. (2015) 'Marine Scientific Research', in: D.R Rothwell, A.O. Elferink, K. Scott and T. Stephens (eds) *Oxford Handbook on the Law of the Sea*. Oxford University Press, Oxford.

Sullivan, J.M., Cerny-Chipman, E.B., Rosenberg, A.A. and Lubchenco, J. (2017) 'Bridging the Science-Policy Interface: Adaptive Solutions in the Anthropocene', in P.S. Levin and M.R. Poe (eds) *Conservation for the Anthropocene Ocean: Interdisciplinary Science in Support of Nature and People*. Academic Press, London.

Tessnow-von Wysocki, I. and Vadrot, A.B.M. (2020) 'The Voice of Science on Marine Biodiversity Negotiations: A Systematic Literature Review', *Frontiers in Marine Science,* vol 7, p 614282.

Till, M. Hillebrand, et al. (2018). 'Disciplinary Diversity in Marine Sciences: The Urgent Case for an Integration of Research', *ICES Journal of Marine Science*, vol 75, no 2, pp 502–509.

Tolochko, P. and Vadrot, A.B.M. (2021a) 'The Usual Suspects? Distribution of Collaboration Capital in Marine Biodiversity Research', *Marine Policy*, vol 124, no 2, p 104318.

Tolochko, P. and Vadrot, A.B.M. (2021b) 'Selective world-building: Collaboration and regional specificities in the marine biodiversity field', *Environmental Science & Policy*, vol 126, pp 79–89.

United Nations. (ed) (2021) *The Second World Ocean Assessment: World Ocean Assessment II*. United Nations, New York.

Vadrot, A.B.M. (2014) *The Politics of Knowledge and Global Biodiversity*. Routledge, London.

Vadrot, A.B.M. (2020) 'Multilateralism as a 'Site' of Struggle Over Environmental Knowledge: The North-South Divide', *Critical Policy Studies*, vol 14, no 2, pp 233–245.

Vanderpool, C. (1983) 'Marine Science and the Law of the Sea', *Social Studies of Science*, vol 13, no 1, pp 107–129.

Vermeulen, N. (2013) 'From Darwin to the Census of Marine Life: Marine Biology as Big Science', PLOS ONE 8(1): e54284, https://doi.org/10.1371/journal.pone.0054284.

Visbeck, M. (2018) 'Ocean Science Research Is Key for a Sustainable Future', *Nature Communications*, vol 9, no 1, p 690.
Wildavsky, A. (1979) *Speaking Truth to Power: The Art and Craft of Policy Analysis*. Little, Brown and Company, Boston.
Wright, G., Rochette, J., Gjerde, K.M. and Levin, L.A. (2018) 'Protect the Neglected Half of Our Blue Planet', *Nature*, vol 554, pp 163–165.

9

CIVIL SOCIETY

Nongovernmental organisations, public opinion, and individuals

Andrea Quirino Steiner, Elia Elisa Cia Alves, and Luís Paulo Santana

The spread of democracy and the increase in international flows of people, goods, services, and ideas in the 1990s boosted civil society's participation in policy formulation and implementation, and it transformed governance in terms of scope, function, and impact (Armstrong et al., 2010; Katz-Rosene, 2019). In the realm of marine governance, civil society action has evolved in extent, means, and content. According to van Tatenhove (2011: 87), marine governance can be defined as

> the sharing of policy-making competencies in a system of negotiation between nested governmental institutions at several levels (international, (supra)national, regional, and local) on the one hand and governmental actors, market parties and civil society organizations on the other to govern activities at sea and their consequences.

Civil society is key to specific aspects of marine governance because framing the political agenda to guarantee marine protected areas (MPAs) or ecosystem-based management (EBM), for instance, may require that groups such as families, social networks, and private businesses (see Chapter 5) play significant roles (McCay and Jones, 2011). Zondervan et al. (2013) argue there are three main challenges to marine governance in the Anthropocene: the rising pressures related to overexploitation of the oceans, the need for enhanced global coordination of government responses to such pressures, and the interconnectedness of marine systems. The authors also list five analytical problems that must be addressed regarding governance: (1) structures, (2) agency in its different levels and facets (from artisanal fisherfolk to private businesses and intergovernmental organisations, for example), (3) social adaptiveness to marine environmental change, (4) political and geographical accountability and legitimacy, and (5) allocation of and access to marine resources.

Within the context of global environmental change (GEC), especially involving climate change and the oceans (Harris, 2019), the literature reveals great advances and contributions from civil society towards the implementation and effectiveness of environmental public policies embedded to deal with this ongoing process (Katz-Rosene, 2019). Agenda 2030, adopted in 2015, is one of several international programmes that established 17 Sustainable Development Goals (SDGs) with civil society participation. The SDG dedicated to oceans (SDG 14) demands cross-sectoral cooperation to support the coordination of efforts

 DOI: 10.4324/9781315149745-12

to address ocean-related targets across shared sea basins and highlights the importance of civil society participation in ocean governance (Spalding and Ycaza, 2020; see Chapter 25). Some unsuccessful experiences of civil society participation in marine governance reveal persistent gaps. These gaps point to challenges to be tackled in the next decades, such as the eroding roles of traditional communities in coastal management (Reedy, 2020), difficulty in engaging with other types of stakeholders (Berkowitz et al., 2020), and the need for greater transparency, accountability (Cadman et al., 2020), and representativeness (Katikiro, 2020).

This chapter aims to better understand the actors representing civil society in marine governance amidst global environmental change by mapping their historical background as well as the roles, methods and topics related to their actions. Beyond theoretical insights, we illustrate civil society's role with specific cases in order to bring evidence of improvements in marine governance in the context of GEC challenges. The second section of the chapter focuses on core concepts necessary to understand the link between civil society and marine governance, such as nongovernmental organisations, public opinion, national and transnational networks, and civil society in its broader sense. The third section presents a brief account of the history of civil society actions regarding marine governance, and the fourth section reviews the topics that have been approached most frequently by civil society in this realm. The chapter then analyses the main roles civil society has been carrying out in marine governance worldwide, as well as the methods of action used to tackle issues related to marine conservation. In the sixth section we attempt to answer the following question: has civil society been successful in promoting marine governance? For this purpose, we briefly describe 13 cases from different regions of the world.

Core concepts related to civil society

Debates on civil society during democratisation have sought to define the space for human association without coercion: a set of relational networks, a place of social interaction and public activities, a place where citizens not only act individually but also collectively (Cohen and Arato, 1994). From this perspective, civil organisations fill a space that is not linked to the economy or to political power (Diamond, 1994). Putnam et al. (1993) argued that democratic governance and the performance of democratic institutions are conditioned by how much civic associations can socialise individuals to rules of reciprocity and trust. Putnam and his co-authors' key argument is that social contexts tend to generate cooperative attitudes. Literature on environmental politics has made significant contributions regarding the activity of several non-state actor groups in policy-related actions. This reflects the plurality of actors with a voice in democracies, with each actor pursuing its interests, whether that be better governance, more profit, defending a cause or guiding evidence-based policies. Additionally, globalisation has made environmental governance more difficult to achieve without the participation of non-state actors (Lorenzetti and Carrion, 2012). According to Jones and Long (2021: 4), 'coevolutionary governance in the shadow of hierarchy is subject to both the accountability of local initiatives to the state... and the accountability of the state to locals'. From this perspective, civil society can influence decisions, although decentralisation is conditional on specific conditions being fulfilled in order to provide for both effectiveness and equity.

When classifying actors that influence environmental issues and policies globally, O'Neill (2012) lists state and non-state actors. Whereas, in her typology, state actors include international intergovernmental organisations and states themselves, non-state actors encompass a much more diverse group: the global environmental movement, the corporate sector, expert groups, the population in general, and individual leaders. The global environmental movement, in turn, includes transnational networks of nongovernmental organisations (NGOs) and other groups. O'Neill (2012) also highlights that actors' roles in environmental issues – including civil society – are not static in any way. They can evolve in several aspects, including in what themes are dealt with and prioritised. Such change is not necessarily due to the actors' themselves, but (for instance) to structural modifications in governance, windows of opportunity and the actors' own evolving capacities and abilities as entrepreneurs of environmental change.

In this sense, there is no single definition of civil society. It usually refers to a sphere where non-profit organisations and social movements promote activities to meet social needs and common interests, in addition to bringing new themes and values to the public (Milani Filho and Viana, 2010). Despite the absence of consensus, there is convergence regarding the idea of a space where individuals and civil organisations articulate actions expressing their claims. At the empirical level, the debate can focus on which actors are inserted in this sphere, how they interact, and what is their power of influence on political institutions (Baccaro, 2001). We understand civil society as non-state actors that act at the local, regional and international levels through individual leadership (Miles et al., 2001); formally instituted groups, such as nongovernmental organisations (Parmentier, 2012); less institutionalised social movements (also called grassroots movements) that often have contributions from academics or other experts (Campbell et al., 2016; see also Chapter 8); public opinion (Hansen and Cox, 2015); and networks (such as the Climate Action Network) that encompass different types of groups sharing a common purpose (Parmentier, 2012; Gerhardinger et al., 2018).

Although literature usually presents these actors in different categories, the real picture is a much more complex net of interrelationships because they all play important roles in this polycentric arrangement towards marine governance (Rudolph et al., 2020). It is also possible to highlight the different roles played by NGOs and social movements. While NGOs offer services to the state by helping formulate, implement, and/or monitor public policies, social movements are usually more associated with bringing evidence on the existence or importance of public problems and demanding services and changes from the state. This difference also may appear between two types of NGOs: a first kind with a participatory, identity-based vision, and a more militant character; and a second that is more pragmatic, which usually acts strategically, without a defined ideological profile, using rational, goal-oriented, and market-oriented logic (Gohn, 2013). In the first case, an example would be an organisation based mainly on volunteer work and with strong (and specific) moral values. The second case includes NGOs that spend considerable amounts of resources to attract members and to develop lobbying strategies to put forth their policy goals, and which may ultimately be more successful in this sense (Rootes, 2013).

When identifying the determinants of international environmental regimes, Miles et al. (2001) highlight the importance of problem-solving capacity and the interplay between its different dimensions. Among several aspects of regime effectiveness, they focus on instrumental (goal-oriented) leadership. In such cases, noteworthy individuals (which may or may not be acting as part of civil society) can help (re)frame problems or solutions by overcoming eventual gridlocks around specific issues (Merrie and Olsson, 2014).

Since 1945, the role of NGOs has been established in the United Nations (UN) Charter as supporting governments in policy making (Article 71). In ocean governance, NGOs are recognised by the UN Convention on the Law of the Sea (UNCLOS, 1982) as entities that can pressure governments to take action such as the creation or the implementation of policies that promote marine conservation (Parmentier, 2012). Less institutionalised movements of civil society, often called grassroots organisations, also play an important role in marine governance, especially when considering small fishing communities that might be institutionalised in the form of associations (Brewer, 2012). This is the case of the associations that comprise the International Fisherfolk and Fisherworkers' Coalition (IWFC). Indigenous peoples have also been gaining strength in the twenty-first century, such as the recipients of the United Nations Development Program (UNDP) Equator Prize, which contributed to the conservation of biodiversity throughout the world using indigenous practices (von der Porten et al., 2019). Public opinion encompasses citizens that may try to influence decision-makers on specific issues in marine governance, and it considers the importance of mass-mediated forms of communication (and, more recently, social media). The Ocean Conservation Trust advocates are an example: according to their website, they 'use their online voices, through social media and beyond, to help people to Think Ocean, and to realise that making some small changes in their daily lives can have a huge positive impact'. Newer vehicles of digital communication may also bring environmental issues to the attention of the population and of policymakers through informational and educational campaigns, as those created by the Australian Marine Conservation Society and made available through their website.

Hansen and Cox's (2015) handbook on the role of communication in environment governance also identifies several initiatives linked to marine governance, such as citizen groups that collect data on cases of deep-water oil spills and nuclear accidents. Civil society can also act through networks, alliances, and coalitions, which include different types of groups that foster knowledge exchange among themselves. One example is Ang Pamalakaya, a Philippine federation of small fisherfolk organisations. Parmentier (2012) points to several factors linked to the formation of such networks: the need to build a strong critical mass that is difficult for policymakers to ignore; to deal with several issues simultaneously; to enhance expertise; to make sure the population remembers past issues; and to get financial support more easily. A challenge to marine governance, in this sense, is the competition among several networks and campaigns.

It is important to bear in mind that the role of civil society in marine governance results from a process that dates back centuries. The next section highlights some historical facts.

Historical background

Thurstan et al. (2020: 162) state that 'Without an appreciation of the past our perception of what is natural is wired to our early experiences of [a geographical] area'. Citizens tend to link the role of civil society in conservation to recent movements that are more evident in the media, such as the recent role of climate activist Greta Thunberg, or to large NGOs such as Greenpeace. Yet concern about environmental issues by civil society is not new and can be traced as far back as the ninth century in works by Arab philosophers on topics related to water pollution (Gari, 2002). Civil society's contribution to marine conservation and governance has been recorded by environmental historians. For example, the role of artisanal fisherfolk has been especially emphasised in the sense that they have always explored and been attached to marine landscapes (Payne, 2012). Johannes (1978), for instance, described a centuries-old system of fisheries conservation based on limited entry in Palau, Micronesia, a practice that was widespread in the Pacific Islands before Western influence. Paranteau (2004) highlighted the importance of fishing sportsmen and sporting organisations to the promotion of Canadian salmon regulations in the late nineteenth century. This was the first wildlife resource regulated in Canada (and maybe in all of North America). More recently, it is noteworthy that the first conference on marine protected areas was put forth by the International Union for the Conservation of Nature (IUCN) in 1975 (WCPA, 2010). Although this is an intergovernmental organisation with strong links to the UN, it is also composed of hundreds of nongovernmental member organisations, and an even greater number of conservation experts.

Similarly, following the overall growth in the modern environmentalist movement from the 1960s onward, several marine conservation NGOs have been created, such as the Sea Turtle Conservancy (1959), Save Our Surf (1964), the American Cetacean Society (1967), the Ocean Conservancy (1972), the Cousteau Society (1973), Sea Shepherd (1977), and the Antarctic and Southern Ocean Coalition (1978), among others. Early marine conservation movements were frequently linked to the protection of so-called charismatic or flagship species, such as whales and seals, which often diverted efforts from more comprehensive conservation actions (Radkau, 2012). Rieser (2012) has dedicated an entire volume to the history of the conservation of the green turtle, starting from the 1940s, in which she highlights the role of scientists and conservationists. The twentieth century also revealed several noteworthy individuals who, although sometimes connected to organisations, had a body of work related to marine conservation linked to their names. People like Rachel Carson (1907–1964), Jacques Cousteau (1910–1997), Sylvia Earle (1935-) and John Veron (1945-), all of them scientists as well as public persona, helped attract attention to urgent matters related to the oceans and ocean conservation, or to specific ecosystems like coral reefs (as in the case of John Veron), and generally inspired awe regarding marine environments.

During the UN Conference on Sustainable Development (Rio+20) in 2012, historic progress in public policies for the ocean was intensively evaluated. The international community recognised the need to improve governance, build capacities, and address new issues. A number of initiatives emerged to address the ocean in a more coordinated way, including the Global Ocean Partnership, the World Ocean Assessment, and the Global Ocean Commission. The need to coordinate efforts and produce action-driven knowledge has also been recognised by several other meetings, events and strategic documents produced by the UN after the Rio+20 Conference. Frameworks evolved to deal with new environmental problems that emerged in the context of global environmental change, as we will see in the next section.

Topics in marine conservation approached by civil society

According to Campbell et al. (2016), controversial topics related to international ocean governance, such as poverty, gender, human rights, and justice, are seldom discussed. When they are discussed, they represent countries' perspectives. Small-scale fisherfolk are an exception, as they are represented in international forums by organisations such as the International Collective in Support of Fishworkers or the World Fishermen's Forum. While ocean governance's social-related topics are often downplayed, Campbell et al. (2016) identified a growing recognition of the economic potential of oceans. This has followed, for instance, the discoveries of seabed minerals and offshore energy-generation opportunities, which the UN has labelled the Blue-Green Economy. Yet in the context of GEC, traditional categories of marine governance topics are being increasingly discussed transversally. The rising levels of greenhouse gas emissions and the consequent deoxygenation, warming and changes in ocean chemistry impact species and food webs throughout ecosystems. Meanwhile, industrialisation, energy generation, mining, tourism, and food production lead to resource depletion, overfishing, and pollution (Rudolph et al., 2020).

Campbell et al. (2016) listed five broad categories of activities that are commonly being tackled by marine governance: food production, biodiversity conservation, industrialisation, global environmental change, and pollution. Within these categories, the authors draw attention to small-scale fisheries, aquaculture, biodiversity conservation in the high seas, large marine protected areas, seabed mining, ocean acidification, blue carbon, and plastic pollution. Petersson (2022) showed the importance of civil society in calling for transparency in global fisheries governance. Other relevant topics that can be included here are toxic waste and oil pollution; whaling, finning, and the protection of flagship species; overfishing; advocacy for marine conservation policies, and general public awareness. Some of these issues are discussed below.

Civil society roles and methods of action

Although the formal politics of nation-states and multilateral institutions continue to play a central role in global ocean governance, the historical highlights presented previously suggest the role and importance of non-state actors. This includes multiple agents and actors involved in public agencies, universities, political parties, businesses, and NGOs, for instance (Alonso and Maciel, 2010). Despite this growing professionalisation in the way of dealing with environmental issues (Raustiala, 1997) and GEC, voluntary services predominate among most associations. Limited by scarce resources, information is these groups' most important tool. It is both produced for and disseminated among populations, and it provides voters with data on government actions around issues related to the environment. The idea is to constrain policymakers to be more responsive to the demands of civil society. Even with scarce financial resources, civil society groups may provide consulting services on funding criteria and achieve positions of influence over the strategic decisions of many governments. Bulkeley and Newell (2010) illustrate this with the Regional Greenhouse Gas Initiative (RGGI), which created a cap-and-trade program for CO2 emissions from power plants in order to guide states. Organisations also seek to stimulate awareness through environmental education and to help transform public demands into policy (Hansen and Cox, 2015). An example is the work of Reef Environmental Education Foundation (REEF) in the USA. Literature indicates different roles and means of action of civil society to deal with GEC. In the case of marine governance, Parmentier (2012) points to policy development and agenda setting, education and outreach, management, science, and watchdogging, usually used together. One example is that of Greenpeace and whale conservation. Images of whales being speared were systematically shown to the public in the 1970s and helped bring awareness to the problem to policy makers. The act of recording whale hunting was also a form of watchdogging. More recently, scientific information on climate change has been used to show that whales are prone to other threats in addition to hunting.

Corell and Betsill (2008) defined the influence of civil society and provided methodological approaches to measure it that consider the relationship between influence and power. According to them, actors may use their specialised knowledge to modify actions taken by decision-makers. One example is that of TRAFFIC, an NGO that gathers information and reports on the illegal trade of wildlife products such as tortoise shells. Persuasion is usually the most employed too and, more recently, transparency (as promoted by NGO Transparency International regarding challenges faced by the International Maritime Organization and SDG 14, for instance). Corell and Betsill (2008) also highlight that participation, whether related to activities, access to negotiations and/or resources, does not necessarily mean influence. One example is that of a multi stakeholder council in an MPA in southern

Brazil (Pretto and Marimon, 2017), which faces several challenges to influence the MPA's management. Thus, the effects of these organisations' actions are unpredictable in terms of the beneficial results they may bring about; much will depend on context and opportunities for their influence (de Vivero et al., 2008).

Has civil society been successful in promoting marine governance?

When discussing the role of non-state actors in international environmental politics, O'Neill (2012) proposed interesting questions. Has the increasing participation of such actors brought about greater influence in practice? Are certain kinds of groups more influential than others (and why)? Do certain types of knowledge privilege the capacity or action of specific groups? Are groups from the Global North more powerful and influential than those from the South? Success and effectiveness of civil society's actions to deal with GEC and marine conservation have varied, and those actions can be observed in terms of political and biological responses and data: from changes in procedures, political behaviour and issue framing to improvements in the environment itself (see Keohane et al. (1993) for the use of data in studying environmental policy). Here we briefly analyse successful and unsuccessful cases on different continents. The criteria used to measure success align with the indicators of NGO influence suggested by Correll and Betsill (2008): (1) observable change in procedural issues, (2) whether ideas put forth by NGOs are taken into consideration (as in the texts of international agreements), (3) the position of NGO actors in different phases of the negotiation of agreements, and (4) whether issue framing enables or constrain NGO activity. Our search for cases took three directions: (1) information on organisations found online, (2) cases found in the literature, and (3) informal conversations with activists for marine conservation.

As a case of simultaneous success and failure of a noteworthy individual, Boyan Slat built up a non-profit foundation to create a device that could clean the ocean. He used social media and digital communication platforms to advocate for his project. His first attempt to collect ocean garbage failed to hold onto the plastic debris it collected. Nevertheless, after adaptation of this prototype, in 2019 Ocean Cleanup announced that the system could successfully retain the debris captured. Whether this will have a decisive impact on the Great Pacific Garbage Patch remains an open question (O'Neill, 2020). This project indirectly addresses the major GEC issue of climate change, as plastic pollution has been linked to a decrease in the ocean's ability to capture CO_2 (Harris, 2019). However, Henderson and Green (2020) have shown how difficult it is to engage public opinion to comprehend connections between personal use of plastics and ocean pollution: practical initiatives to reduce single-use plastics reveal people's unawareness of the effects of microplastics on marine life and human health.

Climate change also makes the ocean more susceptible to radioactive contaminants (Karcher et al., 2010). Hansen and Cox (2015) have documented successful cases regarding public opinion in this arena. Ken Buesseler, when gathering data on radioactivity for the Woods Hole Oceanographic Institution, was able to convince citizens to collect samples of seawater from sites along the west coast of North America and send them to his research centre. The information was then made available online. In relation to grassroots organisations, successful and unsuccessful cases come from *Onda Limpa para Gerações Futuras* (Clean Wave for Future Generations) and its efforts to deal with another major GEC issue: biodiversity loss. This group acts on marine conservation issues along the southern coast of the state of Pernambuco, Brazil. Although it is now an NGO, for over ten years, it was an informal group made up of volunteers. In 2018, the group observed a reduction in the population of a type of ghost crab which inhabits the sand. A volunteer biologist suspected this was linked to the recent use of a tractor with a plough to clean the beach. The group presented the state prosecution office with a petition signed by the public and the equipment stopped being used. On the other hand, the group's attempt to investigate high rates of sea turtle deaths the following year did not succeed due to insufficient measures by public authorities.

Still regarding biodiversity loss, other cases include conservation efforts by indigenous groups. Von der Porten et al. (2019: 1) stated that 'the reinvigoration and reestablishment of indigenous ways of living [are linked to] the management, restoration, and conservation of marine and coastal ecosystems around the world'. Among the cases studied by these authors is that of the Andavadoaka, from Madagascar, which promote restrictions around species such as octopuses. Vierros et al. (2020) discussed the potential of indigenous communities towards the conservation of four groups of marine migratory species. In the case of leatherback turtles (and sea turtles in general) in the Pacific, whose populations have been continuously declining due to multiple human-caused impacts, the authors argue that involving indigenous communities in their international governance is crucial.

Berdej and Armitage (2016) showed the importance of NGOs in 'bridging' in Indonesia, that is, strategically connecting diverse actors and groups. Reef Check Indonesia, an NGO that seeks to conserve the biodiversity of reefs, was able to promote an increase in capacity and knowledge in community-based governance. Conversely, Balboa (2014) documented several failures from transnational NGOs. She presented the case of the Coral Milne Bay Project in Papua New Guinea, a success in terms of people involved, achievement of some goals and international recognition. The head organisation, Conservation International (CI), was able to promote important concepts through a high-impact *Nature* publication resulting from the initiative; this changed discourses on conservation and influenced institutions that create conservation rules and regulations across the globe. However, the project was brought to a halt two years before the end of phase one due to lack of funding. Balboa (2014) documented problems in the long negotiating process to obtain funding for the project (three years until the first payment) and the amount of bureaucracy imposed by the UN's Global Environment Facility.

The literature is packed with successful cases of networks and coalitions acting through advocacy and diplomacy by using technical information and education tactics and articulating several actors from society to address GEC. Examples include the Deep Sea Conservation Coalition (DSCC), which has raised awareness around the importance of the deep sea and monitored the implementation of United Nations General Assembly resolutions around the topic (Fuller et al., 2020). There are also several examples coming from countries of the Global South, such as the Brazilian Future Ocean Panel (Painel Mar, 2015), which deals with GEC and marine governance through a collaborative platform that gathers representatives from academia, civil society and governmental entities to articulate coastal-marine knowledge networks. Collaborators of *Painel Mar* have been the first to map these networks of interaction within Brazil, and they have identified 21 national nongovernmental, 15 governmental and 10 international networks up to 2017 (Gerhardinger et al., 2018). Nevertheless, although there have been advances, Gerhardinger et al. (2020) have discussed the failures of Brazilian marine spatial planning (MSP), a crucial tool to address the implications of GEC for marine conservation. In 2013, a special working group was officially designated within the Commission of Sea Resources to become the governmental think-tank for the country's MSP process. Since its creation in 2003, the commission has remained dormant despite the increase in the number and diversity of non-state agents in the MSP policy arena. Table 9.1 provides an overview of the cases.

Table 9.1 Successful and unsuccessful cases of civil society participation in marine governance

Type of civil society group or individual	*Outcome*	*Case/location*	*Source*
Noteworthy individual	Unsuccessful	Boyan Slat (new effort to clean up the Great Pacific Garbage Patch), Pacific Ocean	O'Neill (2020)
	Successful	Boyan Slat (first effort to clean up the Great Pacific Garbage Patch), Pacific Ocean	O'Neill (2020)
Public opinion	Unsuccessful	Knowledge gaps on the impact of personal plastic consumption and microplastics in the ocean, worldwide	Henderson and Green (2020)
	Successful	Public awareness of ocean radioactivity, North America	Hansen and Cox (2015)
Grassroots/informal organisation	Unsuccessful	*Onda Limpa para Gerações Futuras* (case of sea turtle deaths), Itapuama Beach, Pernambuco, Brazil	Data from the organisation itself
	Successful	*Onda Limpa para Gerações Futuras* (case of ghost crab deaths), Itapuama Beach, Pernambuco, Brazil	Data from the organisation itself
Indigenous group	Unsuccessful	Indigenous peoples and the conservation of migratory species in the northern Pacific (leatherback turtle)	Vierros et al. (2020)
	Successful	The Andavadoaka (restrictions on certain species), Madagascar	von der Porten et al. (2019)
NGO	Unsuccessful	Conservation International (Coral Milne Bay Project), Papua New Guinea	Balboa (2014)
	Successful	Reef Check Indonesia (bridging), Indonesia	Berdej and Armitage (2016)
Network/coalition	Unsuccessful	Special working group was officially designated within the Commission of Sea Resources (never fully activated), Brazil	Gerhardinger et al. (2020)
	Successful	Deep Sea Conservation Coalition followed the implementation of UN resolutions towards the protection of the deep sea, worldwide	Fuller et al. (2020)
	Successful	*Painel Mar* (mapped networks of interaction), Brazil	Gerhardinger et al. (2018)

Conclusion

Global environmental change has intensified challenges related to ocean governance in many ways. Pre-existent challenges such as pollution, navigation, and fishing overlap with both biophysical processes, such as acidification, deoxygenation, and warming, and socioeconomic-related pressures like degradation, inequality, mining, overfishing, energy production, and poverty. As the literature suggests a need to reframe ocean governance towards a global commons approach (instead of a state-centred one), the role of civil society, whether it be through individuals, public opinion, formal and informal organisations or networks, becomes even more relevant than it has been historically. Participation of civil society in ocean governance is centuries old, although it has only been formally recognised internationally in the last seven decades. Actions by civil society actors have proliferated in scope and number from the 1990s, yet there are several challenges to their potential influence, among them a shortage of financing, regulatory constraints, and the lack of technical capacity. On the other hand, positive examples proliferate across different continents, illustrating how civil society actors' goals have sometimes been attained. Greater success in the future will be important if marine governance is to successfully cope with the challenges of global environmental change.

Acknowledgements

We thank Estevão Paixão for providing information on the *Onda Limpa* cases.

References

Alonso, A. and Maciel, D. (2010) 'From protest to professionalization: Brazilian environmental activism after Rio-92'. *The Journal of Environment & Development*, vol. 19, no. 3, pp. 300–317.

Armstrong, D., Bello, V., Gilson, J. and Spini, D. (eds.) (2010) *Civil society and international governance: The role of non-state actors in global and regional regulatory frameworks*. London: Routledge.

Baccaro, L. (2001) *Civil society, NGOs, and decent work policies: Sorting out the issues*. ILO/International Institute for Labour Studies.

Balboa, C. M. (2014) 'How successful transnational non-governmental organizations set themselves up for failure on the ground'. *World Development*, vol. 54, pp. 273–287.

Berdej, S. M. and Armitage, D. R. (2016) 'Bridging organizations drive effective governance outcomes for conservation of Indonesia's marine systems'. *PLos One*, https://doi.org/10.1371/journal.pone.0147142.

Berkowitz, H., Crowder, L. B. and Brooks, C. M. (2020) 'Organizational perspectives on sustainable ocean governance: A multi-stakeholder, meta-organizational model of collective action'. *Marine Policy*, vol. 118, p. 104026.

Brewer, J. F. (2012) 'Revisiting Maine's lobster commons: rescaling political subjects'. *International Journal of the Commons*, vol. 6, no. 2, pp. 319–343.

Bulkeley, H. and Newell, P. (2010) *Governing climate change*. New York: Routledge.

Cadman, R., MacDonald, B. H. and Soomai, S. S. (2020) 'Sharing victories: Characteristics of collaborative strategies of environmental non-governmental organizations in Canadian marine conservation'. *Marine Policy*, vol. 115, no. 103862.

Campbell, L. M., Gray, N. J., Fairbanks, L., Silver, J. J., Gruby, R. L., Dubik, B. A., and Basurto, X. (2016) 'Global oceans governance: new and emerging issues'. *Annual Review of Environment and Resources*, vol. 41, pp. 517–543.

Cohen, J. L. and Arato, A. (1994) *Civil society and political theory*. Cambridge, MA: MIT Press.

Corell, E. and Betsill, M. M. (2008) 'Analytical framework: Assessing the Influence of NGO diplomats'. In: *NGO diplomacy: The influence of nongovernmental organizations in international environmental negotiations*. Cambridge, MA: MIT Press.

de Vivero, J. L. S., Mateos, J. C. R. and del Corral, D. F. (2008) 'The paradox of public participation in fisheries governance. The rising number of actors and the devolution process'. *Marine Policy*, vol. 32, no. 3, pp. 319–325.

Diamond, L. (1994) 'Rethinking civil society: toward democratic consolidation'. *Journal of Democracy*, vol. 5, pp. 4–18.

Fuller, S., Currie, D., Gianni, M., Goldsworthy, L., Rigby, C., Scheit, K., SImpfedorfer, C., Watling, L. and Weeber, B (2020) *Preventing Biodiversity Loss in the Deep Sea*. Amsterdam: DSCC.

Gari, L. (2002) 'Arabic treatises on environmental pollution up to the end of the thirteenth century'. *Environment and History*, vol. 8, no. 4, pp. 475–488.

Gerhardinger, L. C., de Andrade, M. M., Corrêa, M. R. and Turra, A. (2020) 'Crafting a sustainability transition experiment for the Brazilian blue economy'. *Marine Policy*, vol. 120, 104157. https://doi.org/10.1016/j.marpol.2020.104157

Gerhardinger, L. C., Gorris, P., Gonçalves, L. R., Herbst, D. F., Vila-Nova, D. A., De Carvalho, F. G.,... and Glavovic, B. C. (2018) 'Healing Brazil's Blue Amazon: The Role of Knowledge Networks in Nurturing Cross-Scale Transformations at the Frontlines of Ocean Sustainability'. *Frontiers in Marine Science*, vol. 4, no. 395, pp. 1–17.

Gohn, M. G. M (2013) 'Desafios dos movimentos sociais hoje no Brasil'. *Ser Social*, vol. 15, pp. 301–311.

Hansen, A. and Cox, R. (eds) (2015) *The Routledge handbook of environment and communication*. London: Routledge.

Harris, P. G. (2019) 'Climate change at sea: Interactions, impacts, and governance'. In: Harris, P. G. (ed.) *Climate change and ocean governance: Politics and policy for threatened seas*. Cambridge: Cambridge University Press, pp. 3–26.

Henderson, L. and Green, C. (2020) 'Making sense of microplastics? Public understandings of plastic pollution'. *Marine Pollution Bulletin*, vol. 152, 110908. https://doi.org/10.1016/j.marpolbul.2020.110908

Johannes, R. E. (1978) 'Traditional marine conservation methods in Oceania and their demise'. *Annual Review of Ecology and Systematics*, vol. 9, no. 1, pp. 349–364.

Jones, P. J. and Long, S. D. (2021) 'Analysis and discussion of 28 recent marine protected area governance (MPAG) case studies: Challenges of decentralisation in the shadow of hierarchy'. *Marine Policy*, vol. 127, 104362. https://doi.org/10.1016/j.marpol.2020.104362

Karcher, M., Harms, I., Standring, W. J. F., Dowdall, M. and Strand, P. (2010) 'On the potential for climate change impacts on marine anthropogenic radioactivity in the Arctic regions'. *Marine Pollution Bulletin*, vol. 60, no. 8, pp. 1151–1159.

Katikiro, R. E. (2020) 'The role of local institutions in governance of MPAs'. *Marine Policy*, vol. 117, no. 103906.

Katz-Rosene, R. M. (2019) 'The treatment of global environmental change in the study of international political economy: An analysis of the field's most influential survey texts'. *International Studies Review*, vol. 21, no. 3, pp. 477–496.

Keohane, R. O., Haas, P. M. and Levy, M. A. (1993) 'The effectiveness of international environmental institutions'. In: Haas, P., Keohane, R. O. and Levy, M. A. (eds) *Institutions for the earth: Sources of effective international environmental protection*. Cambridge: The MIT Press.

Kingdon, J. W. and Stano, E. (1984) *Agendas, alternatives, and public policies*. Boston: Little, Brown & Co.

Lorenzetti, J. V. and Carrion, R. M. (2012) 'Governança ambiental global: atores e cenários'. *Cadernos EBAPE.BR*, vol. 10, no. 3, pp. 721–735.

McCay, B. J. and Jones, P. J. S. (2011) 'Marine protected areas and the governance of marine ecosystems and fisheries'. *Conservation Biology*, vol. 25, no. 6, pp. 1130–1133.

Merrie, A. and Olsson, P. (2014) 'An innovation and agency perspective on the emergence and spread of Marine Spatial Planning'. *Marine Policy*, vol. 44, pp. 366–374.

Milani Filho, M. A. F. and Viana, M. S. R. (2010) 'Civil society and social capital in Brazil'. In: Anheier H. K. and Toepler S. (eds) *International encyclopedia of civil society*. New York, NY: Springer, pp. 218–223.

Miles, E. L., Andresen, S., Carlin, E. M., Skjærseth, J. B. and Underdal, A. (2001) *Environmental regime effectiveness: confronting theory with evidence*. Cambridge, MA: MIT Press.

O'Neill, K. (2012) 'Actors in international environmental politics'. In: O'Neill, K. (ed) *The Environment and International Relations*. Cambridge: Cambridge University Press.

O'Neill, K. (2020). 'Can the world win the war on plastic?' World Politics Review. Available at: https://www.worldpoliticsreview.com/articles/28590/having-polluted-the-oceans-plastics-are-now-facing-a-popular-backlash

Paranteau, B. (2004) 'A "very determined opposition to the law": Conservation, angling leases, and social conflict in the Canadian Atlantic Salmon Fishery, 1867–1914'. *Environmental History*, vol. 9, no. 3, pp. 436–463.

Parmentier, R. (2012) 'Role and impact of international NGOs in global ocean governance'. In: Chircop et al. (eds.), *Ocean Yearbook Online*, vol. 26, Leiden: Martinus Nijhoff Publishers, pp. 209–230.

Payne, B. (2012) 'Local economic stewards: The historiography of the fishermen's role in resource conservation'. *Environmental History*, vol. 18, no. 1, pp. 29–43.

Petersson, M. T. (2022) 'Transparency in global fisheries governance: The role of non-governmental organizations'. *Marine Policy*, vol. 136, 104128. https://doi.org/10.1016/j.marpol.2020.104128.

Pretto, D. J. and Marimon, M. P. C. (2017) 'Desafios à gestão participativa na perspectiva dos gestores e conselheiros da Reserva Biológica Marinha de Arvoredo, Santa Catarina'. *Desenvolvimento e Meio Ambiente*, vol. 42, pp. 328–344.

Putnam, R. D., Leonardi, R. and Nanetti, R. Y. (1993) *Making democracy work: Civic traditions in modern Italy*. Princeton, NJ: Princeton University Press.

Radkau, J. (2012) 'Religion and environmentalism'. In: McNeill, J. R. and Mauldin, E. S. (eds) *A companion to global environmental history*. West Sussex/UK: Blackwell Publishing Ltd.

Raustiala, K. (1997) 'States, NGOs, and international environmental institutions'. *International Studies Quarterly*, vol. 41, no. 4, pp. 719–40.

Reedy, K. (2020) 'Neoliberal Aleutians: Seeing like a fishing company, seeing like a coastal community'. *Marine Policy*, vol. 118, no. 103981.

Rieser, A. (2012) *The case of the green turtle: An uncensored history of a conservation icon*. Baltimore, MD: The Johns Hopkins University Press.

Rootes, C. (2013) 'Mobilising for the environment: Parties, NGOs, and movements', *Environmental Politics*, vol. 22, no. 5, pp. 701–705.

Rudolph, T. B., Ruckelshaus, M., Swilling, M., Allison, E. H., Österblom, H., Gelcich, S. and Mbatha, P. (2020) 'A transition to sustainable ocean governance'. *Nature Communications*, vol. 11, no. 1, pp. 1–14.

Spalding, A. K. and de Ycaza, R. (2020). 'Navigating shifting regimes of ocean governance: From UNCLOS to sustainable development goal 14'. *Environment and Society*, vol. 11, no. 1, pp. 5–26.

Thurstan, R. H., Buckley, S. M. and Pandolfi, J. M. (2020) 'Oral histories: Informing natural resource management using perceptions of the past'. In: Mañez, K. S. and Poulsen, B. (eds) *Perspectives on oceans past - A handbook of marine environmental history*. Dordrecht: Springer, pp. 155–174.

van Tatenhove, J. (2011) 'Integrated marine governance: Questions of Legitimacy'. *Maritime Studies*, vol. 10, no. 1, pp. 87–113.

Vierros, M. K., Harrison, A. L. Sloat, M. R., Crespo, G. O., Moore, J. W., Dunn, D. C., ... and Govan, H. (2020) 'Considering indigenous peoples and local communities in governance of the global ocean commons'. *Marine Policy*, vol. 119, 104039. https://doi.org/10.1016/j.marpol.2020.104039

von der Porten, S., Ota, Y., Cisneros-Montemayor, A. and Pictou, S. (2019) 'The role of indigenous resurgence in marine conservation'. *Coastal Management*, vol. 47, no. 6, pp. 527–547.

WCPA, I. (2010) *50 years of working for protected areas-A brief history of IUCN*. World Commission on Protected Areas. IUCN, Gland, Switzerland.

Zondervan, R., Gerhardinger, L. C., Noronha, I. T., Spalding, M. J. and Young, O. R. (2013) 'Ocean governance in the anthropocene'. *IGBP Global Change Magazine*, vol. 81, pp. 24–27.

10
PRIVATE GOVERNANCE
The case of Marine Stewardship Council certification in Russia

E. Carina H. Keskitalo, Dmitry L. Lajus and Lars H. Gulbrandsen

In the fields of global climate change and, more broadly, environmental sustainability, there has always been a strong emphasis on the notion that research should support change in governance systems (cf. Beck 2011, Reed et al. 2016). However, the way this governance change should be developed has often been given less attention (Beck 2011). In climate change literature, the focus has often been on 'mainstreaming': the idea that climate change would need to become integrated as a consideration in numerous other fields (Runhaar et al. 2018). This mainstreaming has often been assumed to progress relatively automatically, based on underlying assumptions that knowledge of the need for change should generate governance change (Beck 2011). In relation to such an assumption, climate change research has largely expressed a focus on the future (for instance, through scenario research) and on defining (for instance through modelling) the extent of natural changes that governance should need to relate to (Beck 2011). Contrary to this, social science studies, which have consistently been less emphasised in the climate change field, have stressed that knowledge cannot be assumed to result in changes in governance, and that implementation or mainstreaming cannot be taken as given (Beck 2011, cf. Runhaar et al. 2018). Instead, responding to global environmental change must be regarded as a broad institutional matter: different institutional configurations, comprising for instance different actors or organisations within a governance context, will respond differently to requirements for governance change, in relation to their institutionally embedded motivations and rationale (Young 1999). This type of wider institutional perspective might imply that not all organisations may respond directly to a need to adapt in relation to climate change, and among those who may recognise a need, the ways in which they do so may relate to existing motivational structure, organisational logics, or the means or instruments available for acting on this (cf. Naess et al. 2005, Thornton et al. 2012). Not all governance structures are easily revised – rather the opposite – and embedding environmental and longer-term aims into what many actors often consider primary economic 'bottom line' aims is itself a challenge, as these may contradict existing assumptions and organisational purposes. In addition, if new instruments or approaches are introduced in an attempt to change an existing structure, these may not always fit with the organisational logics of the case (Chuenpagdee and Song 2012, cf. Thornton et al. 2012).

This chapter discusses how change in a governance system can be understood and analysed, using illustrations from private governance, particularly the Marine Stewardship

 DOI: 10.4324/9781315149745-13

Council (MSC) certification system introduced in Russia (cf. Lajus et al. 2018). One of the most prominent private governance cases has been the voluntary market-based case of forest certification, particularly certification by the Forest Stewardship Council, followed by MSC fisheries certification. The implementation of these instruments illustrates the difficulties involved with re-orienting a fundamentally economically focused governance towards traceable environmentally oriented management (Auld et al. 2008, Gulbrandsen 2010).

In the MSC system, climate change considerations are taken into account while assessing the stock status of target and by-caught species, habitats, and ecosystems. The MSC standard requires an analysis of the effect of fisheries on all these components. All large-scale MSC-certified Russian fisheries are situated in high-latitude seas – Barents, Bering, and Sea of Okhotsk – and are therefore especially sensitive to global climate change. Currently, all of these fisheries are near historical maximums. This means that a decline in stocks due to further climate change is more likely than an increase. This decline may be caused either by direct effects of climate change on commercial fish species, or indirectly, by changes in the affected ecosystem mediated by complex biotic interactions with eventual effects on commercial species (Cheung et al. 2009). However, climate effects on ecosystems are very difficult to predict and incorporate into population models relevant to fishery management (King et al. 2015).

Participation in the MSC programme requires an understanding of its standards, which are based (explicitly or implicitly) on ecosystem-based management. However, implementing MSC also requires an understanding of standards and differences in institutional prerequisites for implementation in various areas. It is possible that the MSC process could offer a means to include the effects of climate change on fisheries management (McLeod and Leslie 2009, Lajus et al. 2017). However, climate change, rather than mainly providing a knowledge input into systems, may also result in multiple influences and increase conflict between actors over resources. As climate change is taking place in the present, with impacts filtered by existing structures, it must also be seen as an influence on governance regimes, with impacts on and through them.

Understanding private governance

Private governance is often seen as the quintessential new environmental governance instrument (Jordan et al. 2013). For instance, as they emerged, market-based voluntary initiatives were considered a response to the growing environmental crisis, based on the assumption that companies themselves, voluntarily and over time, would shift towards more sustainable means of undertaking business (Jordan et al. 2013). Initiatives in private governance include standard-setting initiatives intended to provide for 'constant improvement' and certification systems whereby a third party evaluates compliance with specific environmental and social requirements, often through the review of organisational guidance documents, routines, and case review (Keskitalo and Liljenfeldt 2012). The most prominent certification system has been the Forest Stewardship Council (FSC) system (Cashore et al. 2004), which also inspired the MSC system (Gulbrandsen 2010). The FSC system has often been regarded as a success, and in some markets it has almost become an export requirement, meaning that it may signify higher requirements for compliance than might otherwise be expected of a voluntary system (Keskitalo and Liljenfeldt 2012). However, FSC forest certification has also been criticised by environmental NGOs for whittling down environmental requirements through its council system, in which both environmental and business actors must agree on how international standards are modified for national cases (Counsell and Loraas 2002, Greenpeace

2021). For this reason, some environmental NGOs have left FSC national systems, whereas others have stayed on with the intention of trying to influence the systems from the inside. On the business side, by contrast, forest companies and forest owners have conceived of the changes brought on by certification as extensive, as certification schemes have required the inclusion of specific environmental aims, and mainstreaming of at least some environmental requirements as part of market requirements, which may not otherwise have taken place, absent a political situation in which legislative or economic instruments were enacted (Auld et al. 2008, Johansson and Keskitalo 2014).

Certification can be considered more or less successful depending on where the bar for the inclusion of environmental aims is placed: it is difficult to comprehensively assess the full environmental impacts of certification (Auld et al. 2008). Criticism has also been directed at certification systems with stringent standards that in practice require well-developed environmental management systems, including a high level of documentation of organisational routines, traceability through systems, and formalisation, meaning that it is difficult for smaller actors without this administrative capacity and established routines to manage. When it comes to marine systems, these are also inherently more difficult to manage than are forest systems: many fish stocks migrate quickly, and often unpredictably, and they are altered by climate change, whereas trees, and thereby also compliance with forest harvesting systems, are localised and thus more easily controlled. For forests, local communities and thus actors relevant for a social role in relation to forest are also more easily identified, while a social role for local communities (as actors who are to be consulted, for instance) has so far largely been absent from marine certification systems (Gulbrandsen 2010).

Private governance instruments like MSC are thus of a specific character and, similar to other voluntary instruments, they will only be successful if there are features in place (such as, for FSC, market requirements in certain areas) that support compliance. Voluntary instruments should also not be assumed to be effective on their own but may often be seen as complementary to, rather than replacing, more demanding instruments (such as legal or economic instruments) (Alberini and Segerson 2002). Hence, claims that private authority could serve as a substitute for public authority in areas of 'limited statehood' – areas where national governments have limited practical authority – particularly in developing countries (Börzel and Risse 2010), appear to overestimate the potential of certification as a governance tool. Indeed, as Gulbrandsen (2010: 182) concludes, 'there is a great deal of evidence to suggest that effective implementation of certification programmes requires well-functioning legal systems, property rights and national and local administrations'. It has also been noted that instruments cannot be considered neutral or possible to implement in similar ways in different cultural or other contexts. Instead, they should be regarded as being derived from and applicable to specific cultural and organisational contexts, and as carrying specific presumptions in relation to the structure or organisational style that is relevant to their usability, acceptability, and implementation (Lubatkin et al. 2005).

This is in line with a more institutionally focused understanding that implementation cannot be taken as given. Instead, how a specific challenge (or instrument) can be implemented requires a well-developed understanding or institutional analysis to be able to tailor instruments or communications to a specific context (Jordan et al. 2013). Research on certification also demonstrates that global standards are layered onto an existing matrix of institutions that can reconfigure the standards in myriad ways (Bartley 2011). In addition, it has been recognised that often assumedly neutral instruments such as certification or standard setting are in fact developed largely in relation to broadly Western or even North American contexts, and the specific interpretation of private governance has been given in this context

(Lubatkin et al. 2005). As a result, the implementation of certification or standard-setting systems is not necessarily possible to transfer to more locally based, less export-oriented, smaller-scale or more informal, or public rather than private, systems. As a result, in many cases it has been observed that the expansion of certification systems may result in small-scale and local producers be outcompeted by larger ones (Jacquet and Pauly 2008).

Implementing MSC certification in Russia

The complexities of implementing MSC certification can be illustrated by the Russian case. Here we draw upon a range of literature, including what was, to our knowledge, the first comprehensive study of all MSC-certified fisheries in Russia (Lajus et al. 2018). That study includes all three types of MSC certifications that exist in Russia: (1) large-scale (some hundreds of thousands of metric tonnes), high-seas industrial fisheries using mid-water or bottom trawls for walleye pollock in the Far East, Atlantic cod and codfish in the Barents Sea; (2) mid-scale (up to dozens of thousands of metric tonnes) Pacific salmon fisheries in the Far East that are in coastal areas or in rivers; and (3) small-scale (hundreds of metric tonnes) fisheries for freshwater perch and pikeperch using mostly gillnets and traps in various lakes and reservoirs. There are also high sea fisheries for Red King Crab, halibut, and Pacific cod in the Far East and the Barents Sea. The study was based on interviews and public certification reports (Lajus et al. 2018).

The Russian case is especially relevant due to the types of governance practised. Fisheries in Russia vary greatly in size and organisation, and include a number of small-scale fisheries, many of which are not very export-oriented. Russian fisheries also typically, both historically and presently, have quite a significant informal component in their management, even though the entire fisheries management system across the country is quite uniform. This holds true particularly for small-scale fisheries, which means that the inclusion of formal environmental management systems and standards in these cases has often been limited. Russian business contexts can also not be seen to have internalised the typically more North American or Western style of standard-setting and environmental management system implementation of predominantly private systems that constitutes the foundational logic of certification systems. The literature has noted that similar circumstances have arisen regarding the implementation of certification systems in, for instance, South American or broader developing country contexts (Gulbrandsen 2009, Jacquet et al. 2010).

The concerns regarding implementation in this case do not necessarily focus on limitations to steering per se, but rather on the types of steering and logics that are assumed to exist. While certification can be seen to be reliant on assumptions related to broader environmental management and private governance at large, with a notable North American basis, Russian systems may steer as strongly and clearly, but built on logics that are less directly compatible with a private governance system. For instance, a global survey of fisheries management systems that focused on the management of target species ranked Russia's fishery management fourth out of 28 counties surveyed (Melnychuk et al. 2017). However, the management of target species traditionally receives more attention in Russia than the ecosystem effects of fishing (Lajus et al. 2018) and is not per se necessarily dependent on the existence of the types of formalised systems that certification processes require: management can also rely on strong, informally developed processes (practices that are not formalised in writing or process management) or on formal processes that are not necessarily those required in environmental management systems (cf. Keskitalo and Andersson 2017). In addition, some of the factors noted for a certification system to be effective, such as environmental NGO, media, or consumer market pressure (Gulbrandsen 2010), have not necessarily been present.

In an appraisal of the early MSC certifications in Russia, Gulbrandsen and Hønneland (2014) found that consensus was emerging that the Russian national system for fisheries management fulfilled the MSC requirements. However, at the beginning of the process some of the certifications took a very long time. For instance, the Sea of Okhotsk walleye pollock fishery assessment, begun in 2008, was not certified until 2013. Other fisheries proceeded through the process far more quickly, possibly learning from experience with MSC fisheries assessments in Russia (Gulbrandsen and Hønneland 2014). Particularly export-oriented companies generally became certified, but there are also examples of fisheries withdrawing from certification processes. In the study by Lajus et al. (2018), all certified companies were export-oriented, operating on markets where certification was considered a must. Most fisheries were pushed by buyers to enter certification, and in some cases the role of environmental NGOs in supporting the processes was significant. In cases of strong market pressure, Russian small-scale fisheries, such as European perch fisheries, have also become certified (Lajus et al. 2020).

The domestic Russian market, on the other hand, was not seen to reward certification in relation to, amongst other things, governmental, cultural, and organisational features that impacted understandings of certification processes, purposes, and data availability for implementation. As a result, '[s]everal fisheries that passed pre-assessment with the support of NGOs pulled out of the process before entering full assessment' (Lajus et al. 2018: 111). This may contribute to a 'growing divide between larger-scale, internationally oriented, distant-water fishing companies and the smaller-scale coastal fisheries' (Pristupa et al. 2016: 128, cf. Keskitalo and Liljenfeldt 2014, Le Manach et al. 2020 for a comparative discussion in forestry; see Chapter 13). In relation to this, Lajus et al. (2018) note that despite general compliance with the MSC standards, the considerations around implementing certification can be seen to largely centre on institutional logics and requirements related to them. For instance, as noted above,

> Russian fisheries management traditions … differ notably from those in the West, on which MSC standards are built. For instance, in Russia less attention has traditionally been paid to the ecosystem effects of fisheries and to public involvement in management.
> *(Lajus et al. 2018: 109)*

These factors can be seen to be related to the role of environmental standards and potentially also to private standard-setting in a Russian context: not required domestically, it has mainly been pressed by export requirements and, secondly, NGOs. This may partly mean that the consumer foundation for domestically certified fish, and thereby the imperatives for certification on this market, may be lacking. Thus, there is a lack of the pressures that are generally assumed to be the underlying drivers of market-based certification and eco-labelling schemes.

The focus in certification on environmental factors, compared to the structure of the Russian market, consumer, and buyer preferences, has been made difficult by the limited emphasis on data availability. Lajus et al. note 'problems related to language and fisheries culture in Russia, which differ from those in the West, and [related] difficulties accessing scientific and management information that limit certification' (Lajus et al. 2018: 106). This can be seen to be related to the fact that much fisheries information has been regarded as a part of fisheries industry sector data in Russia, whereas more of this sort of data is readily available in North American or Western systems, for instance. This is because in Russia fish are considered a strategic resource, on which information is therefore intentionally limited, in line with Soviet-era traditions (Lajus et al. 2018). As a result, '[d]ifficulty with access to scientific information, which is crucial to certification, was mentioned by most of the interviewees ... [who] more often explained this in terms of bureaucratic difficulties, rather than the actual absence of data' (Lajus et al. 2018: 108). Again, the underpinnings for the certification system – here, availability of data – were thus limited due to historical and strategic factors.

These considerations, as well as others related to the organisation of Russian fisheries, were compounded by the perceived limited knowledge of these factors as well as of established local practices amongst assessment teams. Certification, based on assumptions similar to those in formal environmental management, typically requires formalised processes and formally traceable information. This places extensive requirements on administration and reporting, and while this is standard in large international companies, it can be very demanding for smaller or more informal (practice-based) management. Assumptions on what data may exist, and in what form, may thus differ significantly among those schooled in informal or formal systems, as well as in formal environmental management compared with other governance forms. In relation to this, it has been noted that '[a]lmost half ... of respondents reported communication problems with non-Russians in assessment teams' (Lajus et al. 2018: 110). Respondents also noted that non-Russians in assessment teams 'may lack familiarity with local fisheries and practices' (Lajus et al. 2018: 110). Examples of this included statements that 'illustrate genuinely different perceptions and assumptions about the processes and gear, which confuse the certifiers and companies, as well as perceived concerns with communication and clarity' (Lajus et al. 2018: 110).

As a result of these factors, communication issues in particular were highlighted in this first study of all certified Russian fisheries, compared to previous work (Lajus et al. 2018). Communication issues were seen to have

> at least two interrelated aspects. The first was purely translation and interpretation problems. These resulted in increased costs and a longer certification process during both the face-to-face and remote phases of assessment. The second was cultural differences, and differences in the traditions of fishery management.
>
> *(Lajus et al. 2018: 112)*

Here, communication components were seen to play a large role as

> the MSC process creates a new type of interaction between key participants – fishery clients, research institutions, governmental agencies, NGOs, and the general public. As in any new type of interaction, tensions are unavoidable because of lack of trust between stakeholders.
>
> *(Lajus et al. 2018: 112)*

'All of these issues may impact attitudes toward the certification process' (Lajus et al. 2018: 110).

For these reasons, one interviewee considered that 'it might be more effective to start the certification process with a site visit' (Lajus et al. 2018: 110). Through a site visit, a number of local practices, processes, and gear used could be made more apparent to assessment teams who fundamentally held other assumptions. A site visit can be used to highlight the roles of both management, culture, and motivation. The authors of the study concluded:

> Based on this analysis, three factors were identified in this study that shape the MSC certification process in Russia, and differentiate it from the general process and assumptions related to certification in the West: (i) linguistic and cultural differences between Russia and Western countries, where methods of assessing the sustainability of fisheries were developed, (ii) differences between the traditions of Western fishery management and Russian fishery management, which are adapted to situations in Russia, and (iii) the absence of an internal market for sustainable seafood, so that the only motivation for fisheries to obtain certification is to export their products.
>
> *(Lajus et al. 2018: 113)*

As a result of these types of institutional impacts, there may also be difficulties with adapting the systems directly to climate change and with implementing the systems in the way assumed in the originating context. For instance, teams may find it difficult to assess the types of unknown factors that certification systems require taking into account and that could impact target stocks, and that include not only climate change but anthropogenic non-fishery related factors. They may also find it difficult to define adjustments in ways that work well within fisheries systems. At the same time, comparing the private governance system of MSC with a more conservative state governance system, private governance may include a possibility to increase flexibility and include a direct focus on issues such as global environmental change.

The relationship of MSC certification in Russia to other cases

The results in the Russian case highlight many of the issues around MSC certification that have been noted in literature, such as the importance of international commercial and reputational benefit for certification. These results also highlight some of the potential institutional assumptions in different national cases that may make systems based on specific institutional assumptions – regarding how resource governance is undertaken or what information is available – difficult to directly apply. Related results have been found elsewhere. Key barriers to the limited uptake of the MSC beyond Europe and North America are limited market demand, lack of knowledge about certification, shortcomings of scientific data about fish stocks and ecosystem impacts, and financial costs of fisheries assessments (Gulbrandsen 2009, Jacquet et al. 2010, Carlson and Palmer 2016). It has also been noted that information related to certification is sometimes 'difficult to obtain, hard to understand, or not available in an appropriate language or format' (Stewart et al. 2003: 4–5, quoted in Carlson and Palmer 2016: 130–131). Similarly, while it has been noted that the Japanese case can be compared with cases in Europe or North America (Gulbrandsen 2009, Jacquet et al. 2010), Swartz et al. found that seafood sustainability in Japan is viewed through a community lens: 'the unique manifestation of the linkage between community and sustainability drives fisheries and seafood consumption in Japan' (Swartz et al. 2017: 188). As a result, they note that '[i]n addition to market limitations, the success of marine eco-labels in Japan is further limited by … differing perceptions of what "sustainable" seafood means' (Swartz et al. 2017: 188).

To some degree, these results can be seen to echo the illustration in the Russian case of the need to understand how fisheries management is conceived, and the role of environmental considerations in this. In the Russian case, national or societal understandings of fisheries, and what demands are placed on them, are coupled with management systems that do not directly comply with the assumption in MSC systems of how marine governance is structured. Despite the idea that private governance should not require government participation, it has thus been found that cases in countries whose national governments support adjusting management to MSC requirements, or may even be willing to do so, may find it easier to reach MSC compliance because many regulative assumptions are part of larger governmental assumptions (Gulbrandsen 2010, Blandon and Ishihara 2021, cf. Pristupa et al. 2016). Similar to much of the literature, Blandon and Ishihara conclude that, while certification schemes were originally seen as an alternative to governmental management, 'the MSC certification scheme cannot act as an alternative to government management, as it often directly assesses the fisheries management implemented by such authorities' (Blandon and Ishihara 2021: 8). Studies also highlight the limited possibilities for actors in fisheries to change management systems per se (Kaiser and Edward-Jones 2006). For instance, while the Japanese decentralised co-management system of fisheries could be considered close to an 'ideal' multi-layered co-management system, its composition means that MSC certification 'backfires … because the applicant may not have the authority to change all the measures in the coastal fisheries management system' (Blandon and Ishihara 2021: 7).

Due to concerns like these, particularly the issues of alignment to national systems and assumptions and the demands of formal environmental management systems that mean that certification may be less likely to be undertaken without significant market pressure, the *way* environmental issues are integrated may be just as significant as – or even more significant than – the *fact* that environmental issues are integrated. While market-based or private governance systems have often been regarded as a generalist way to integrate environmental considerations, this may rest on a misassumption: that different natural resource governance systems and also the market systems in which they act are more similar than they in fact are. Examples from markets in both Russia and Japan illustrate that domestic conceptions of seafood sustainability may relate to factors other than purely environmental ones, with systems unable to be tailored to represent consumer demand. In addition, the sorts of governance systems, the data available, and the type of organisation are not one-size-fits-all but are rather marked by national variation and varying assumptions.

In general, this supports a broader institutionalist understanding of the systems that are required to conceive of how environmental considerations may come to play a role in MSC certification. Much of the institutional perspectives in fisheries have drawn on rational choice theory, with a focus on property rights (Chuenpagdee and Song 2012), in relation to 'generic governance systems, or advocating what "should be" the governance arrangements' (Miller et al. 2010: 342), for instance developing measures for engaging multiple stakeholders in management (Miller et al. 2010). However, a broad conceptualisation of institutions that encompasses fisheries culture, organisation, state linkage, and the numerous assumptions existing in these systems illustrates that more generic organisation cannot be assumed to be implemented separately from existing institutional logics.

Attempts to develop steering systems for fisheries that are exposed to climate change and other environmental changes must thus include understanding of the systems in which governance change is to be implemented, and the motivating factors in those systems. As Miller et al. note: '[t]he fishery governance challenge is not new – and exists independently of the climate change reality – but the latter adds new complexity and urgency to dealing with the former' (Miller et al. 2010: 342). In addition, 'in fact there are always impediments to any change – notably adjustment costs, distributional impacts creating winners and losers, etc.' (Miller et al. 2010: 342). Climate change could even exacerbate conflicts among fishers (Mendenhall et al. 2020). In this context, it may be necessary to situate considerations related to environmental change in relation to national systems or societal assumptions, rather than assuming that one approach applies equally to all countries and situations.

Conclusion

Global environmental change increases the need to understand different institutional settings and their possibilities as well as implicit assumptions in governance instruments. As we may attempt to shift governance systems using standard-setting or certification systems, or other instruments, we need to understand specific built-in features that drive or motivate those systems, and we need to understand how they may differ between regional or national contexts (cf. Lubatkin et al. 2005, Jordan et al. 2013). The role of private governance and voluntary market-based instruments can be considered as cases in point. Although having largely been regarded as more general means that should be possible to implement based on private governance (outside of governmental involvement) as well as in different contexts, this chapter illustrates the context dependence of such systems and how they may in fact rest on specific assumptions of how different systems work (Pristupa et al. 2016, Swartz et al. 2017, Blandon and Ishihara 2021).

In responding to climate change, which is likely to impact existing systems by shifting and potentially increasing costs (Miller et al. 2010), understanding of the underlying systems becomes more important than ever. Changing governance towards a goal – in a more general way, whereby the focus is on the goal rather than the systems to be changed – may differ relatively strongly from conceiving of a goal in light of different systems and their requirements, instead placing the focus on how those systems can be motivated towards a goal. In a broader institutionalist understanding (Chuenpagdee and Song 2012), this highlights the need to understand different organisational contexts as well as the difficulties involved in assuming instruments to be neutral or taking for granted the mainstreaming of environmental goals, whether or not integrating climate change.

References

Alberini, A., & Segerson, K. (2002). Assessing voluntary programs to improve environmental quality. *Environmental and Resource Economics* 22 (1–2): 157–184.

Auld, G., Gulbrandsen, L. H., & McDermott, C. L. (2008). Certification schemes and the impacts on forests and forestry. *Annual Review of Environment and Resources* 33 (1):187–211.

Bartley, T. (2011). Transnational governance as layering of rules: Intersections of public and private standards. *Theoretical Inquiries in Law* 12 (2): 25–51.

Beck, S. (2011). Moving beyond the linear model of expertise? IPCC and the test of adaptation. *Regional Environmental Change* 11 (2): 297–306.

Blandon, A., & Ishihara, H. (2021). Seafood certification schemes in Japan: Examples of challenges and opportunities from three Marine Stewardship Council (MSC) applicants. *Marine Policy* 123: 104279. https://doi.org/10.1016/j.marpol.2020.104279

Börzel, T. A., & Risse, T. (2010). Governance without a state: Can it work? *Regulation & Governance* 4 (2): 113–134.

Carlson, A., & Palmer, C. (2016). A qualitative meta-synthesis of the benefits of eco-labelling in developing countries. *Ecological Economics* 127: 129–145.

Cashore, B., Auld, G., & Newsom, D. (2004). *Governing through markets – forest certification and the emergence of nonstate authority.* New Haven: Yale University Press.

Cheung, W. W. L., Lam, V. W. Y., Sarmiento, J. L., Kearney, K., Watson, R., & Pauly, D. (2009). Projecting global marine biodiversity impacts under climate change scenarios. *Fish and Fisheries* 10 (3): 235–251

Chuenpagdee, R., & Song, A. M. (2012). Institutional thinking in fisheries governance: Broadening perspectives. *Current Opinion in Environmental Sustainability* 4 (3): 309–315.

Counsell, S., & Loraas, K. T. (2002). *Trading in credibility. The myth and reality of the forest stewardship council.* London: The Rainforest Foundation.

Greenpeace. (2021). *Destruction: Certified. certification; not a solution to deforestation, forest degradation and other ecosystem conversion*. Amsterdam: Greenpeace International.

Gulbrandsen, L. H. (2009). The emergence and effectiveness of the Marine Stewardship Council. *Marine Policy* 33 (4): 654–660.

Gulbrandsen, L. H. (2010). *Transnational environmental governance: The emergence and effects of the certification of forests and fisheries*. Cheltenham: Edward Elgar.

Gulbrandsen, L. H., & Hønneland, G. (2014). Fisheries certification in Russia: The emergence of nonstate authority in a postcommunist economy. *Ocean Development & International Law* 45 (4): 341–359.

Jacquet, J., & Pauly, D. (2008). Funding priorities: Big barriers to small-scale fisheries. *Conservation Biology* 22 (4): 832–835.

Jacquet, J., Pauly, D., Ainley, D., Holt, S., Dayton, P., & Jackson, J. (2010). Seafood stewardship in crisis. *Nature* 467 (2 September): 28–29.

Johansson, J., & Keskitalo, E. C. H. (2014). Coordinating and implementing multiple systems for forest management: Implications of the regulatory framework for sustainable forestry in Sweden. *Journal of Natural Resources Policy Research* 6 (2–3):117–133.

Jordan, A. R., Wurzel, K.W., & Zito, A. R. (2013). Still the century of 'new' environmental policy instruments? Exploring patterns of innovation and continuity, *Environmental Politics* 22 (1): 155–173.

Kaiser, M. J., & Edwards-Jones, G. (2006). The role of ecolabelling in fisheries management and conservation. *Conservation Biology* 20 (2): 392–398.

Keskitalo, E. C. H., & Andersson, E. (2017). Why organization may be the primary limitation to implementing sustainability at local level. Examples from Swedish case studies. *Resources* 6 (1): 13ff.

Keskitalo, E. C. H., & Liljenfeldt, J. (2014). Implementation of forest certification in Sweden: An issue of organisation and communication. *Scandinavian Journal of Forest Research* 29 (5): 473–484.

King, J. R., McFarlane, G. A., & Punt, A. E. (2015). Shifts in fisheries management: Adapting to regime shifts. *Philosophical Transactions of the Royal Society B* 370: 20130277.

Lajus, D. L., Belyanin I. A., Ermolova E. P., & Golovin P. V. (2020). Certification of European perch from the Irikla Reservoir (basin of the Ural River) according to the standards of the Marine Stewardship Council. *Trudy VNIRO* 179: 124–148 [In Russian].

Lajus, D., Stogova, D., & Keskitalo, E. C. H. (2018). The implementation of Marine Stewardship Council (MSC) certification in Russia: Achievements and considerations. *Marine Policy* 90: 105–114.

Lajus, D., Stogova, D., & Lajus, J. (2017). Importance of consideration of climate change at managing fish stocks: A case of Northern Russian fisheries, In K. Latola, & H. Savela (Eds.), *The Interconnected Arctic — UArctic Congress 2016*. Cham, Switzerland: Springer Polar Sciences, pp. 127–134.

Le Manach, F., Jacquet, J. L., Bailey, M., Jouanneau, C., & Nouvian, C. (2020). Small is beautiful, but large is certified: A comparison between fisheries the Marine Stewardship Council (MSC) features in its promotional materials and MSC-certified fisheries. *PLoS One* 15 (5): e0231073.

Lubatkin, M. H., Lane, P. J., Collin, S. O., & Very, P. (2005). Origins of corporate governance in the USA, Sweden and France. *Organization Studies* 26 (6): 867–888.

McLeod, K. L., & Leslie, H. M. (2009) (eds.) *Ecosystem-based management for the oceans*. Island Press, Washington, DC.

Melnychuk, M.C., Peterson, E., Elliott, M., & Hilborn, R. (2017). Fisheries management impacts on target species status. *PNAS* 114 (1): 178–183.

Mendenhall, E., Hendrix, C., Nyman, E., Roberts, P. M., Hoopes, J. R., Watson, J. R., ... & Sumaila, U. R. (2020). Climate change increases the risk of fisheries conflict. *Marine Policy* 117: 103954.

Miller, K., Charles, A., Barange, M., Brander, K., Gallucci, V. F., Gasalla, M. A., ... & Perry, R. I. (2010). Climate change, uncertainty, and resilient fisheries: Institutional responses through integrative science. *Progress in Oceanography* 87 (1–4): 338–346.

Næss, L.O., Bang, G., Eriksen, S., & Vevatne, J. (2005). Institutional adaptation to climate change: Flood responses at the municipal level in Norway. *Global Environmental Change* 15: 125–138.

Pristupa, A. O., Lamers, M., & Amelung, B. (2016). Private informational governance in Post-Soviet waters: Implications of the Marine Stewardship Council certification in the Russian Barents Sea region. *Fisheries Research* 182: 128–135.

Reed, J., Van Vianen, J., Deakin, E. L., Barlow, J., & Sunderland, T. (2016). Integrated landscape approaches to managing social and environmental issues in the tropics: Learning from the past to guide the future. *Global Change Biology* 22 (7): 2540–2554.

Runhaar, H., Wilk, B., Persson, Å., Uittenbroek, C., & Wamsler, C. (2018). Mainstreaming climate adaptation: Taking stock about 'what works' from empirical research worldwide. *Regional Environmental Change* 18 (4): 1201–1210.

Swartz, W., Schiller, L., Sumaila, U. R., & Ota, Y. (2017). Searching for market-based sustainability pathways: Challenges and opportunities for seafood certification programs in Japan. *Marine Policy* 76: 185–191.

Thornton, P. H., Ocasio, W., & Lounsbury, M. (2012). *The institutional logics perspective. A new approach to culture, structure and process.* Oxford: Oxford University Press.

Young, O. R. (with contributions from Agrawal, A., King, L. A., Sand, P. H., Underdal, A., & Wasson, M., revised version prepared by H. Schroeder). (1999). *Institutional dimensions of global environmental change.* Science Plan, Bonn: IHDP.

11
PRIVATE-SECTOR INVESTORS
Climate action and blue carbon financing

Nata Tavonvunchai

There is increasing recognition of the significance of oceans in the international climate change agenda (Stephens, 2015), particularly through the United Nations (UN) 2030 Agenda for Sustainable Development and the Sustainable Development Goals (SDGs), which addresses oceans in SDG 14: Life Below Water (Johansen and Vestvik, 2020), and in the Intergovernmental Panel on Climate Change's *Special Report on the Ocean and Cryosphere in a Changing Climate* (Pörtner et al., 2019). Carbon sequestration is when atmospheric carbon is absorbed for long-term storage, and occurs as a result of the ocean's carbon cycle and flux (Hattam et al., 2015). Carbon in relation to these oceanic processes is called 'blue carbon' (Nellemann et al., 2009). Coastal ecosystems in particular have a significant role in absorbing and storing atmospheric carbon through transfer to sediment systems or conversion to biomass, thus can be known as 'blue carbon' ecosystems (Macreadie et al., 2019; Martin et al., 2016). The established blue carbon ecosystems include mangrove, saltmarsh, and seagrass ecosystems (Duarte et al., 2005) due to their high capability of storing carbon in the long term (Herr and Landis, 2016; Lovelock and Duarte, 2019; Nellemann et al., 2009). At least one of these blue carbon ecosystems is present in 151 countries, with all three present in 71 countries (Herr and Landis, 2016). These ecosystems cover less than 0.5% of the seabed but are estimated to contain 50–71% of all carbon stored in ocean sediment (Nellemann et al., 2009).

There has been increasing attention to how targeted investment in the conservation and restoration of these blue carbon ecosystems can support progress towards global carbon reduction and climate resilience (Macreadie et al., 2019; Pendleton et al., 2012). The Paris Agreement urges nations to make commitments to reduce their GHG emissions through Nationally Determined Contributions (NDCs), which are the voluntary commitments made by countries to address climate change and reduce emissions (Herr and Landis, 2016). Seventy percent of NDCs include reference to marine issues, including 27 NDCs referencing blue carbon strategies, although this includes more broadly wetlands and 'ocean carbon', in addition to the aforementioned blue carbon ecosystems (i.e., mangroves, saltmarshes, and seagrass), and 45 NDCs specifically referencing mangroves (Gallo et al., 2017).

Climate finance, broadly defined as investment to reduce the impact of climate change through mitigation or adaptation (Stadelmann et al., 2013), has contributed to increased investment in conserving and restoring blue carbon ecosystems (UNEPFI, 2021a). However, there is likely a funding gap in financing the broad-scale ambition of marine conservation,

DOI: 10.4324/9781315149745-14

understood here as activities taking place in the marine environment with the purpose of conserving its integrity. Funding needed to meet SDG 14, which entails but is not limited to blue carbon ecosystems, is estimated to be USD 174.52 billion, with a USD 149.02 billion attainment gap (Johansen and Vestvik, 2020: 5). This chapter focuses on private-sector investors and the climate financing models they use in relation to blue carbon ecosystems, which contribute to the wider aim of ocean protection for addressing climate change. For the purpose of this paper, private-sector investors are broadly defined as privately owned entities, including companies (i.e., corporations), financial institutions, and philanthropies (Lienert, 2009).

Marine ecosystem services and pressures

Ecosystem services are defined as the direct and indirect contributions of ecosystems to human well-being (Hattam et al., 2015: 64). Beyond their carbon storage potential, blue carbon ecosystems can also protect coasts (i.e., from extreme weather, coastal erosion, and sea-level rise) and support biodiversity (e.g., fish spawning grounds, improved water quality), thus contributing to local livelihoods (Herr and Landis, 2016; Macreadie et al., 2019). The marine environment provides an ecosystem service by regulating the climate through the hydrological cycle, temperature, and climate-influencing substances (Hattam et al., 2015: 65). The degradation of blue carbon ecosystems results in a decrease in the ocean's carbon storage capacity and contributes to the release of CO_2 (UNESCO, 2020; Pendleton et al., 2012). Blue carbon ecosystems are in decline (Duarte et al., 2005) as a result of the effects of climate change (i.e., ocean warming, sea-level rise, and acidification) (Hoegh-Guldberg et al., 2019), direct pressure from coastal development (e.g., aquaculture and port development), and diffuse pressure from pollution and erosion from land-based activities (e.g., agriculture, tourism industry-related development) (Bijoor, 2020; Moraes, 2019; Pendleton et al., 2012) and from offshore activity (e.g., drilling) (Bijoor, 2020).

Private-sector investors

A 'complex constellation of financial actors' is involved in the investment process that makes decisions about what and where investments are made (Glemarec, 2011: 77). Private-sector investments in blue carbon are increasingly involving the piloting innovative financing models and collaboration of a variety of investor types. More broadly, there have been commitments from some private-sector companies to achieve the Paris Agreement goals ten years early by regularly reporting, eliminating, and mitigating carbon and GHG emissions (The Climate Pledge, n.d.). Several factors are driving this shift, one of which can be explained by an increased awareness of the importance of a social licence to operate (SLO) (Hurst et al., 2020; Moffat et al., 2016). A SLO is interpreted as the broad societal acceptance of an activity or company, which is found in trust-based relationships between actors (Moffat et al., 2016). Social acceptance can also affect financial performance (Moffat et al., 2016) through the influence of consumer relationships, which can especially be seen in the case of consumers' increasing concerns for the environment (Holthus, 1999). Additionally, those investors who address climate resilience in their investments can reduce future costs from climate change-related damage. Thus, action to protect coastal ecosystems can be a form of 'insurance' (Rees et al., 2018; Thiele et al., 2020). The key private-sector investors relevant to financing blue carbon ecosystems include the following:

Multi-national corporations (MNCs) are companies (e.g., retailers and consumer goods companies) that operate in various countries. MNCs have been noted by the World

Benchmarking Alliance as essential 'keystone companies' for achieving the SDGs due to their size, power and large-scale impact on generating positive or negative environmental and social externalities.

(Virdin et al., 2021)

Commercial banks are entities which manage wealth via transactions, accounts, and loans.

Institutional investors include insurers/reinsurers and pension funds, which typically provide large-scale, low-risk investment over an extended duration (Glemarec, 2011). These investors invest within operational and regulatory constraints, such as having to consider the length of their investment horizon, level of risk and the information requirements of other entities they are affiliated with (Glemarec, 2011). Investment/asset managers may manage the assets of these institutional investors through various financial models in capital markets (Glemarec, 2011). With most institutional investors seeking to maximise investment returns (e.g., pension funds) (Glemarec, 2011), and investment in marine ecosystems not projected to have high investment returns, some institutional investors are less suited to funding blue carbon projects (Friends of Ocean Action, 2020). Conversely, some institutional investors (i.e., insurers) are especially well placed for climate finance and blue carbon investment, having long recognised the threat and associated insurance opportunities of, and being technically well-equipped to address, climate risks.

(EY, 2008)

Private equity investors are typically investment funds comprised of multiple entities, but also include venture capital (VC) (Friends of Ocean Action, 2020). Private equity and debt investments can be very conservative (i.e., less willing to take risk), thus they invest in low-risk and subsequently low-return investments. Examples of such investments are those in later-stage mature projects with more certain outcomes (Glemarec, 2011). On the other end of the spectrum, VC invests in many projects and products, with the idea of high-risk high-return in mind – meaning that there is an expectation that many of the projects will not generate returns, but that losses would be made up by those investments that succeed and bring sizeable returns.

(Friends of Ocean Action, 2020; Glemarec, 2011)

Philanthropic donors are entities that tend to invest without expecting return and can range from individuals to large-scale foundations. Because of this, philanthropies are especially helpful in providing early-stage funding to emerging projects, investing in new interventions, supporting, piloting, and scaling up innovation (e.g., providing essential technical assistance through aiding project preparation and strategic advisory) and developing a pipeline of investments.

(Blended Finance Taskforce and SYSTEMIQ, 2020; Dente et al., 2020)

Table 11.1 provides further details on private-sector investors and their relationship to blue carbon ecosystems. As blue carbon investment is a nascent are of investment, the highest level of implementation progress is the pilot phase. The stages of involvement of investors are also summarised. Further description is provided in Table 11.1 for the stages of the project investment lifecycle that each investor is likely to participate in. As the business models of pension funds and VC investors do not typically support blue carbon investment, and because there have been no case studies identified including these investors, these two investor types are excluded from the rest of this chapter's analysis.

Table 11.1 Private-sector investors and blue carbon ecosystems

Private-sector investor	*Relevance to blue carbon*	*Progress*	*Typical stage of involvement*	*Additional information*
Multinational corporations	★★★	★★	Early to later-stage/ mature projects	Blue carbon ecosystem support is gaining traction among MNCs, notably through blue carbon credits.
Commercial banks	★★★	★	Early to later-stage/ mature projects	Commercial banks have potential to be involved in blue carbon investment, but no concrete examples found to date.
Insurers	★★★	★★	Early-stage	While the relevance of and the narrative of oceans as a solution to climate risks is aligned with insurers' mandates, this investor type does not yet seem to be actively involved with blue carbon.
Pension funds	★	n/a	n/a	The financial model of pension funds is unsuitable for blue carbon ecosystems.
Asset managers	★★	★★	Early-stage	This investor type is involved in blue carbon.
Venture capital	★	n/a	n/a	The financial model of venture capital and its and investors' inclination for technological approaches are unsuitable for funding blue carbon ecosystems (Glemarec, 2011).
Investment funds	★★★	★★	Early to later-stage/ mature projects	This investors type is involved in blue carbon.
Philanthropic donors	★★★	★	Early-stage	Philanthropies have a long track-record in funding marine conservation projects, but the extent of their involvement in blue carbon is unclear.

Note: The stages of the 'relevance to blue carbon' and 'progress' columns are indicated by the following phases: limited applicability (★), some pilot projects (★★) and operational/mainstream (★★★).

Financial models for blue carbon investment

Financial models for blue carbon range from 'pure philanthropy', where no return of any kind is expected, to profit-seeking investments made with the primary goal of maximising financial return (Friends of Ocean Action, 2020). This section briefly explores private-sector financing models used for, or with potential for, blue carbon ecosystem protection and restoration, ordered by currently common practices. It should be noted that these cases are by no means exhaustive and are limited to the entities described above and the available literature on the topic at the time of writing. Additionally, private capital flows especially tend not to be disclosed voluntarily, making it challenging to understand completely the current state of private finance related to blue carbon and where the gaps in that finance may be (Friends of Ocean Action, 2020).

Grants

Grants allocated by public or private entities are commonly framed as 'impact only', meaning that there is no expectation of financial returns. Currently, marine conservation is typically financed by philanthropies and public-sector grants, but finance may also include that provided by the private sector (Bos et al., 2015; see Chapter 10). However, there does not seem to be evidence of grant funding specifically for the purpose of blue carbon, though it is assumed, with the definition of marine conservation, that grant funding would have also been allocated to ecosystems that would be classed as blue carbon ecosystems.

Corporate social responsibility

Corporate social responsibility (CSR) strategies are a common way in which MNCs self-regulate (as private businesses) and contribute more broadly to societal goals. CSR strategies may be aligned, for example, with goals to enhance the reputation of a company, reduce business and legal risk by taking responsibility for corporate actions, and/or driven predominantly by the ethical stance of the directors. However, the current literature suggests that there is a shift from a transactional to a social benefit-centred relationship (Moffat et al., 2016). With the realisation that natural resource management and environmental practice affect social perception (Moffat et al., 2016), CSR strategies can also contribute to social-ecological aims, including the protection or enhancement of blue carbon ecosystems. For some industries, the demonstration of commitment to societal goals has included blue carbon-credit investment (see the next paragraph). While the difference between CSR and pure grants seems to be that CSR often has underlying aims of benefitting investors (e.g., protecting them from being linked in some way to investors' supply chain or reducing investors' future risk exposure; see Box 11.1) (Friends of Ocean Action, 2020), CSR and pure grants are often similar in practice, thus have been considered together in this chapter.

Payments for ecosystem services and blue carbon credits

Payments for ecosystem services (PES), that is investment in ecosystems' protection and management, are commonly made by those directly benefitting from the ecosystem services in question (Bos et al., 2015). PES has high potential for marine conservation because of marine conservation's high investment risk and often limited immediate financial returns, which tend to be long-term and financially intangible, though bring social, cultural, and environmental benefits (Visbeck et al., 2014). Some examples of PES in marine conservation are fees for marine protected areas (MPA), tourism taxes, and community conservation agreements where local communities (investees) are funded or receive other forms of economic incentive for conservation-related activities (Bos et al., 2015), all of which can contribute to supporting blue carbon ecosystems. More recently, this has also included blue carbon credits (Thomas, 2014). Carbon credits are a type of offset in which the credits signify carbon emissions reductions (Ullman et al., 2013). One carbon credit is equivalent to an offset of one metric tonne of CO_2 or CO_2 equivalent (tCO_2e). 'Blue' carbon credits (i.e., ocean-based reductions) to date have been used for the conservation and rehabilitation of marine coastal ecosystems, specifically mangroves, saltmarshes, and seagrass. Projects must be able to demonstrate additionality and permanence (i.e., a direct link to additional carbon benefits as a result of the project and that emissions will remain stored) (UNESCO, 2020; Chapman and Wilder, 2013). Demonstrating benefits beyond carbon storage (co-benefits), such as positive biodiversity and community impact, is especially important for blue carbon-credit projects to differentiate themselves from other offset products (Chapman and Wilder, 2013) and appeal to corporate-offset credit buyers who are increasingly interested in projects that also generate additional, non-carbon-related benefits (Bijoor, 2020; Vanderklift, 2019).

Currently, carbon-credit trading mostly takes place in the voluntary carbon market and among corporate investors (Bonn et al., 2014). PES investors interested in investing in blue carbon are typically those that have high emissions levels (i.e., challenging-to-decarbonise MNCs) and are looking to 'offset' their emissions, while investees are typically communities or non-profit organisations that would steward blue carbon ecosystems through community conservation agreements (Bos et al., 2015). As most corporate buyers of carbon credits are looking to invest in credits that align with their company's mission (Bijoor, 2020), entities that are most likely to be investing in blue carbon credits related to ecosystems are those operating in tourism/recreation, energy (e.g., oil and gas), shipping (see Chapters 20 and 23), fishing and aquaculture, bioprospecting, and coastal development (Bijoor, 2020; Vanderklift, 2019; Thomas, 2014). Typically, investors 'offset' their emissions by purchasing a number of credits equivalent to emissions from a certain activity (see Box 11.1 for an example). While the blue carbon market is nascent, with few registered credit-issuing projects (Thomas, 2014), interest in blue carbon credits is growing as a result of the prominent climate change rhetoric and development, and subsequently increasing regulation, e.g., the Paris Agreement (Vanderklift, 2019; Thomas, 2014).

Box 11.1 Apple and mangrove restoration in Cispatá Bay, Córdoba, Gulf of Morrosquillo (Colombia)

Actors: MNC (Apple); NGO (Conservation International, Verra, Fundación Omacha); research institution (INVEMAR Research Institute); carbon project developer/consultancy (South Pole); government (Regional Autonomous Corporation of Sinu and San Jorge Valleys, Regional Autonomous Corporation of Sucre, Colombian General Directorate of Maritime Affairs); project validator (AENOR)

Aim: As part of its 2018 Earth Day campaign, Apple announced a partnership with Conservation International to protect mangroves, support communities, and fund blue carbon research. The initiative is expected to reduce at least 17,000 tCO_2e over its first two years, which is equivalent to the emissions of the fleet of vehicles needed to update Apple Maps over ten years. Apple's initiative is situated within the Conservation International project, 'Blue Carbon Project Gulf of Morrosquillo', developed with technical support from South Pole. The project is expected to store 1,221,717 tCO_2e over the project lifetime of 30 years in 7,561 hectares of mangroves, marshes, and streams. The Cispatá Bay project is the first blue carbon conservation project registered by Verra and issues its carbon credits through the Verified Carbon Standard, as well as first blue carbon project to quantify both the carbon stored in tree biomass and soil. In addition to ecosystem services, protection and carbon reduction, the project also engages local governments and communities in the project's management, protects endangered species, and encourages sustainable livelihoods and agriculture through bee-keeping, eco-tourism and community gardens. The sale of carbon credits is intended to provide the local community with financial security and livelihoods in place of mangrove deforestation activities, and to provide initial funding to develop an eco-tourism programme and build local capacity for sustainable fishing.

Project start date: 2015 (2018 announcement by Apple)

Sources: Verra, 2021; Friends of Ocean Action, 2020; Apple Inc., 2019; Conservation International, n.d.; Verra, n.d.

Insurance

The main objective of insurance is to reduce the risk of potential losses through an arrangement by which an entity (e.g., a company) undertakes to provide a guarantee of compensation for specified loss, damage, illness, or death in return for the payment of a specified premium. In relation to the environment, this can be done by placing the responsibility of mitigating environmental harms (as a loss) on entities negatively affecting the environment (Bos et al., 2015). For example, parametric insurance, a type of insurance whereby payout is determined by a 'triggering event', as defined by a certain numerical index, can be targeted towards improving climate change adaptation and resilience or de-risking climate investment (Friends of Ocean Action, 2020: 107) (see Box 11.2). Parametric insurance is particularly relevant for communities and locations affected by extreme weather because it transfers the climate-related financial losses to financially stable entities, in place of local communities that often have limited financial resources (Friends of Ocean Action, 2020). While parametric insurance has not yet been used for blue carbon ecosystems, it has been identified as highly suitable for them (Friends of Ocean Action, 2020). The following case study on coral reefs (Box 11.2), an ecosystem not widely recognised as a blue carbon ecosystem (Lovelock and Duarte, 2019), aims to shed light on the potential application of insurance to blue carbon ecosystems.

Impact investing

The Global Impact Investing Network (GIIN) defines impact investing as 'investments made into companies, organisations, and funds to generate [positive] social and environmental impact alongside financial return' (Bos et al., 2015: 122). Impact investing is still a relatively

Box 11.2 Quitana Roo Conservation Trust Fund, Mexico

Actors: private insurer (Swiss Re); corporations (local hotels); NGO (the Nature Conservancy); government (state government, National Parks Commission)

Aim: The Quitana Roo Conservation Trust Fund aims to manage, maintain, and restore the Quintana Roo state coastline and coral reefs through disbursed grants, which will go towards managing the Mesoamerican reef via parametric insurance. Grant money will come from collecting funding from a variety of sources, including private-sector user fees, paid by beachfront hotels using the beach, and philanthropic funding. The initiative, led by the Nature Conservancy with support from Swiss Re, is set up to restore reefs within municipalities covered by the Trust. Management of a payout for this purpose is determined by the Trust and a technical committee. Payouts are automatically triggered by high wind speeds (>100 knots) within the specified area to allow for rapid response (within the first 60 days in the event of extreme weather) and higher recovery outcomes for reefs. The maximum payout over a 12-month period is USD 3.8 million, with a typical investment of USD 100,000 to 150,000 each time. The Trust's activities also include coastal zone management and social development and security.

Start date: 2018 (legally established; not yet fully operational)

Source: Friends of Ocean Action (2020)

new field; only a small proportion of impact investments are environmentally focused, and even less so focused on the marine environment (Bos et al., 2015). Impact investing (i.e., impact funds) allows funds to be pooled for certain objectives from a variety of investor types (Friends of Ocean Action, 2020). Funds can also include carbon credits in its structure, and thus be linked to the carbon market, where carbon credits are traded (Vivid Economics and Environmental Finance, 2018). A global ocean fund that covers the costs of maintaining and protecting blue carbon ecosystems has been touted as an initiative with high potential (Virdin et al., 2021), with one involving blue carbon ecosystems recently established (see Box 11.3).

Box 11.3 Mirova Blue Carbon Facility/L'Oréal Fund for Nature Regeneration global funding facility

Actors: MNC (L'Oréal); asset manager (Mirova Natural Capital); carbon project developer/consultancy (EP Carbon)

Aims: Managed by Mirova, the over-arching objective of this impact fund facility is to reduce the impact of L'Oréal's supply chain on biodiversity in coastal, marine and forest ecosystems. The investment portfolio of this fund is envisioned to finance mangrove conservation and restoration carbon-credit projects, in addition to other objectives, including terrestrial forest and land restoration, sustainable agriculture and fisheries, and supporting rural livelihoods (via eco-tourism) and gender equality. For mangrove projects, the L'Oréal Fund targets early-stage innovative projects with local presence that preserve at least 5,000 hectares of mangroves are projected to generate a minimum of 100,000 carbon credits, and involve local communities and benefits. By 2030, the fund aims to have invested EUR 50 million (USD 59 million) into restoring 1 million hectares of degraded ecosystems and capturing 15–20 million tonnes of CO_2 emissions. For L'Oréal, this is part of a larger programme, 'L'Oréal for the Future', which is comprised of other impact funds. For Mirova, the Facility falls within its acquired Althelia Sustainable Ocean Fund, which covers other marine conservation topics. Partners of the fund include, but are not limited to, NGOs (Conservation International, Environmental Defence Fund, USAID); consultancies (Eunomia); asset managers (AXA Investment Managers, Caprock Group) and MDBs (European Investment Bank, Entrepreneurial Development Bank [FMO], the Inter-American Development Bank, *Bank aus Verantwortung* [KfW]).

Start date: 2021 (operational)

Source: Mirova, 2021; Blended Finance Taskforce and SYSTEMIQ, 2020; Mirova, 2020; L'Oréal Group, n.d.

Debt

Debt, typically provided by banks, is a financial loan which must be repaid, with its provision linked to investments' projected risk and return (Thiele et al., 2020). Key elements of debt instruments include a time-bound interest rate to the debtor, repayment agreements and often collateral (i.e., assets in which lenders can have claims over if debt is not repaid) (Friends of Ocean Action, 2020; Thiele et al., 2020). This means that debt has a low-risk profile and requires a positive financial return, which may explain private-sector investors' limited use of it for blue carbon: the challenge conservation and development financing pose for debt instruments is that its returns are often less clear and benefits less monetisable (Blended Finance Taskforce and SYSTEMIQ, 2020; Friends of Ocean Action, 2020). Despite this, there is an increasing market for sustainability-linked debt (Friends of Ocean Action, 2020), which is commonly linked to PES, in which lenders can finance sustainability initiatives by providing lower-priced loans (Blended Finance Taskforce and SYSTEMIQ, 2020; Friends of Ocean Action, 2020; Thiele et al., 2020). This chapter explores some potential applications of debt for private-sector investors, using examples from public-sector investment in blue carbon. Loans can be set up to increase access to capital for riskier investments as a revolving loan fund (RLF), which is often first funded by impact-only investment from philanthropies, the public-sector (e.g., governments; multilateral development banks [MDBs]) and private investors, and which are managed by local stakeholders (Friends of Ocean Action, 2020). RLF loans sit within a central fund; new loans are issued from the fund once loans are repaid. While not yet used for blue carbon ecosystems, RLFs may be suited for community-based efforts (e.g., MPA monitoring and mangrove restoration) and blue carbon conservation (Friends of Ocean Action, 2020).

Debt-for-nature swaps are types of debt instrument allowing indebted countries to 'swap' or redirect some or all of their debt into conservation initiatives via impact funds (Blended Finance Taskforce and SYSTEMIQ, 2020; Pascal et al., 2021). This mechanism 'restructures' debt, for example via reduced rates and more lenient repayment terms, in exchange for a binding commitment to the borrower to deliver the agreed socio-environmental outcomes rather than the original debt (McGowan et al., 2020). For example, in the case of a debt swap with Seychelles, this involved a philanthropy, Oceans 5 (The Nature Conservancy, n.d.). Typically, an intermediary, often a conservation organisation, contributes to restructuring the debt and supporting the institutional setup to manage the initiative, often in the form of trust funds but can also be in the form of funds from commercial banks (Friends of Ocean Action, 2020).

Impact bonds are used for specific and measurable positive impacts. Impact bonds can involve a variety of stakeholders and allow for initiatives that have limited but potential revenues to be financed through the transfer of repayments to third-party entities (Pascal et al., 2021; Friends of Ocean Action, 2020). There is an emerging market for 'blue' bonds. While private-sector investors have been involved in blue bonds, they have played a more supportive role in this process. As bonds often sit in the domain of the public-sector entities (e.g., sovereign states, MDBs, global funds), they have currently only been issued by the public sector (Shiiba et al., 2021). In 2018, Seychelles sold the world's first sovereign blue bond for marine protection and sustainable fishing practices (Wilkins and Gillespie, 2021). The bond of USD 15 million was placed with private-sector asset managers (Nuveen, Prudential and Calvert Impact Capital), with payments processed by commercial banks (The Standard Chartered Bank and Bank of New York Mellon) (Wilkins and Gillespie, 2021; Blended Finance Taskforce and SYSTEMIQ, 2020).

Table 11.2 summarises the financial models discussed above as they relate to blue carbon ecosystem protection and rehabilitation.

Further considerations for blue carbon finance

With respect to the literature, as summarised in Tables 11.1 and 11.2, MNCs, insurers, asset managers, and investment funds have progressed on schemes related to blue carbon the most, although this remains at the pilot stage. Private-sector investors seem to use several financial models, with PES, specifically blue carbon credits, most commonly used. This is followed by

Table 11.2 Implementation of financial models by private-sector investors for blue carbon

	Financial models				
Private-sector investor	*Grants / CSR*	*PES / blue carbon credits*	*Insurance*	*Impact funds*	*Debt*
MNCs	★★★	★★★	–	★★	–
Commercial banks	–	★★	★★	–	★
Insurers	–	★★★	★★★	–	–
Asset managers	★★	★★	–	★	★
Investment funds	★	★★	–	★★	–
Philanthropic donors	★★★	★★	–	–	★

Note: The current use of financial models by investors are indicated by the following phases: limited applicability (★), some pilot projects (★★) and operational/mainstream (★★★).

Table 11.3 Blended finance for blue carbon ecosystems

Financial models	*Blue carbon credits/offsets*	*Impact funds*	*Grants/CSR investment*	*Insurance*	*Debt*
Blue carbon credits/ offsets		✓	✓		
Impact funds	✓		✓		
Grants/CSR investment	✓	✓		✓	✓
Insurance			✓		
Debt/loans			✓		

Note: The check-mark symbol (✓) indicates the financing models that have been combined to date, and to the author's best knowledge, for blue carbon.

grants, CSR investment, insurance, and impact funds. These financial models are those typically expecting moderate returns, or where limited to no returns are anticipated. While financial models can be used independently, they are often 'blended' to achieve a common objective (see Boxes 11.1, 11.2, and 11.3) and combined with various other models and investors to de-risk investment that would otherwise be too risky for a single investor or investment model (Thiele et al., 2020). While blending finance to reduce investment risk and/or uncertainty has been critiqued as needing further examination (Christiansen, 2021), the idea is that this allows for new and higher impact initiatives (e.g., sustainability-linked or marine ecosystem investments) (Pascal et al., 2021; Thiele et al., 2020). In conservation, blended finance has been used the most in forest restoration, including that of mangroves, and as early-stage investment to provide technical assistance for project development and to build investment capacity (Thiele et al., 2020). Table 11.3 summarises the key financing models that have been combined for blue carbon.

Grants and CSR investments have been identified as relevant to all the other financial models. While these financial models are useful for ecosystem conservation, especially at the early/pre-investment stage, they are typically fixed-term with no guarantee for renewal (Friends of Ocean Action, 2020). Such funding is unreliable and can result in projects being 'fragmented' – projects may enter into several start-stop cycles and be susceptible to being modified according to each new investors' requirements (Bos et al., 2015: 118). This is not only problematic for blue carbon projects seeking to deliver on clear objectives and positive benefits, which can require decades to realise, but project resources may also be spent disproportionally on securing funding instead of on programme development and implementation (Bos et al., 2015). As fragmentation is not exclusive to projects mostly funded by grants and/or CSR, projects should have diverse revenue streams (i.e., sources and financial models), such as blended finance, for increased financial stability and generation of benefits (Bos et al., 2015).

Investment in marine conservation, including blue carbon, can be made more commercially viable by increased provision of public or philanthropic support (Pascal et al., 2021; Virdin et al., 2021; Vanderklift et al., 2019). This seems to be especially the case in emerging economies (Christiansen, 2021). In particular, private-sector investment can be facilitated through reduced uncertainty (Thomas, 2014), such as through standardised PES/impact metrics and legal and regulatory frameworks (Shiiba et al., 2021; UNEPFI, 2021b; Virdin et al., 2021; see Chapter 5). Uncertainty can also be reduced through tailoring investment to meet the demands of projects' socio-political contexts to ensure that they meet the demands and needs of those directly affected (Bos et al., 2015; see Chapters 12, 14 and 26). While these perceptions of seemingly reducing uncertainty are drawn from its quantification as risk, as seen in the framing of additional funding as a solution, risk reduction can be more nuanced and raises debates of who and how these risks are quantified (Christiansen, 2021). Aside from financial considerations, combining blue carbon investment with other marine conservation-related projects, such as fisheries, eco-tourism, and natural coastal infrastructure projects, may help reduce financial risks for investors (Bijoor, 2020).

There is a clear emerging link between blue carbon projects and community empowerment (UNESCO, 2020). Equitable involvement of communities is also a point raised by the literature on SLO (Moffat et al., 2016). Moreover, as the narrative of blue carbon has gained traction as a 'solution' towards addressing climate change, a similar idea of attaching quantifiable 'value' to the intangible, natural assets could be a useful entry-point for directing finance to the marine environment and may help leverage funds from investors and financing models that seek high returns (e.g., VC and debt instruments). For ecosystems, this can be through quantifying their economic and non-financial (e.g., cultural) value (Bos et al., 2015). However, substantial literature arguing against quantifying value must also be taken into consideration (Christiansen, 2021; Sullivan, 2017).

Conclusion

Blue carbon ecosystems can contribute to climate change adaptation by reducing atmospheric carbon as well as through climate change mitigation, through providing protection from extreme weather events, coastal erosion, and sea-level rise (Macreadie et al., 2019; Duarte et al., 2013). To close the funding gap, private-sector action is needed (Shiiba et al., 2021; Johansen and Vestvik, 2020; Bos et al., 2015). While investments may benefit investors' bottom lines (Spergel and Moye, 2004) or reduce climate-related costs (Thiele et al., 2020), private-sector investors are currently making investments in blue carbon, although this has yet to become mainstream. Financial models with low to moderate returns have been used, with blue carbon credits, most commonly used by private-sector investors, followed by grants, CSR investment, insurance, and impact funds. Although private-sector financing can be further facilitated by collaboration with other entities, most notably the public sector, it is important to make sure these investments are going towards the right projects in order to address climate change impacts and benefit local communities that rely upon a healthy, functioning marine environment for their well-being. More broadly, when considering blue carbon financing, it becomes apparent that effective marine environmental governance amidst climate change will very likely require substantial involvement of the private sector.

Acknowledgements

I am especially grateful to Siân Rees, Holly Niner, and Aymeric Reymond for their invaluable feedback in shaping this chapter. I would also like to thank Eric Zusman for sharing with me this publication opportunity.

References

Apple Inc. (2019) *Conserving Mangroves, A Lifeline for the World*. Available at: https://www.apple.com/newsroom/2019/04/conserving-mangroves-a-lifeline-for-the-world/ (Accessed: 1 July 2021).

Bijoor, S. (2020) *Mainstreaming Blue Carbon to Finance Coastal Resilience, Conservation Finance Network*. Available at: https://www.conservationfinancenetwork.org/2020/01/27/mainstreaming-blue-carbon-to-finance-coastal-resilience (Accessed: 12 August 2021).

Blended Finance Taskforce and SYSTEMIQ. (2020) *Better Finance, Better Food; Case Study Catalogue*. Available at: https://www.systemiq.earth/wp-content/uploads/2020/11/Better-Finance-Better-Food-Case-study-catalogue-2.pdf.

Bonn, A. *et al*. (2014) 'Investing in nature: Developing ecosystem service markets for peatland restoration', *Ecosystem Services*. Elsevier, 9, pp. 54–65. https://doi.org/10.1016/j.ecoser.2014.06.011.

Bos, M. *et al*. (2015) 'Marine conservation finance: The need for and scope of an emerging field', *Ocean and Coastal Management*. Elsevier Ltd, 114, pp. 116–128. https://doi.org/10.1016/j.ocecoaman.2015.06.021.

Chapman, S. and Wilder, M. (2013) 'Attracting private investment into REDD + projects: an overview of regulatory challenges', *International Journal of Rural Law and Policy*, (2013 Special Edition), pp. 1–15. http://doi.org/10.5130/ijrlp.i1.2013.3361.

Christiansen, J. (2021) 'Fixing fictions through blended finance: The entrepreneurial ensemble and risk interpretation in the Blue Economy', *Geoforum*. Elsevier Ltd, 120 (February 2020), pp. 93–102. https://doi.org/10.1016/j.geoforum.2021.01.013.

Conservation International (no date) *A Critical Investment in 'Blue Carbon'*. Available at: https://www.conservation.org/stories/critical-investment-in-blue-carbon (Accessed: 1 July 2021).

Dente, L. *et al*. (2020) *Sea beyond the Blue. A Guide on the Oceans for Philanthropists and Changemakers*. Available at: https://www.ubs.com/global/en/ubs-society/philanthropy/experiences/beyond-blue.html.

Duarte, C. M. *et al*. (2005) 'Major role of marine vegetation on the oceanic carbon cycle', *Biogeosciences*, 2, pp. 1–8. Available at: www.biogeosciences.net/bg/2/1/.

Duarte, C. M. *et al*. (2013) 'The role of coastal plant communities for climate change mitigation and adaptation', *Nature Climate Change*, 3(11), pp. 961–968. https://doi.org/10.1038/nclimate1970.

Ernst & Young (EY). (2008) *Strategic Business Risk Insurance 2008*. Available at: http://aaiard.com/11_2008/2008_Strategic_Business_Risk_-_Insurance.2.pdf.

Friends of Ocean Action. (2020) *The Ocean Finance Handbook*. Available at: http://www3.weforum.org/docs/WEF_FOA_The_Ocean_Finance_Handbook_April_2020.pdf.

Gallo, N. D. *et al*. (2017) 'Ocean commitments under the Paris Agreement', *Nature Climate Change*, 7(11), pp. 833–838. https://doi.org/10.1038/nclimate3422.

Gattuso, J. P. *et al*. (2018) 'Ocean solutions to address climate change and its effects on marine ecosystems', *Frontiers in Marine Science*, 5(OCT). https://doi.org/10.3389/fmars.2018.00337.

Glemarec, Y. (2011) *Catalyzing Climate Finance. A Guidebook on Policy and Financing Options to Support Green, Low-Emission and Climate-Resilient Development —Version 1.0*. United Nations Development Programme (UNDP). New York. Available at: https://www.undp.org/publications/catalyzing-climate-finance.

Hattam, C. *et al*. (2015) 'Marine ecosystem services: Linking indicators to their classification', *Ecological Indicators*, 49, pp. 61–75. https://doi.org/10.1016/j.ecolind.2014.09.026.

Herr, D. and Landis, E. (2016) *Coastal Blue Carbon Ecosystems. Opportunities for Nationally Determined Contributions. Policy Brief*. Gland, Switzerland. https://doi.org/10.2305/IUCN.CH.2015.10.en.

Hoegh-Guldberg, O. *et al*. (2019) *The Ocean as a Solution to Climate Change: Five Opportunities for Action*. Available at: http://www.oceanpanel.org/climate.

Holthus, P. (1999) 'Sustainable development of oceans and coasts: The role of the private sector', *UN Natural Resources Forum Journal*, 23(2), pp. 169–176. Available at: http://www.leisa-al.org/web/images/stories/revistapdf/vol22n2.pdf#page=30.

Hurst, B. *et al.* (2020) 'Engaging for a social licence to operate (SLO)', *Public Relations Review*. Elsevier, 46(4). https://doi.org/10.1016/j.pubrev.2020.101931.

Johansen, D. F. and Vestvik, R. A. (2020) 'The cost of saving our ocean - estimating the funding gap of sustainable development goal 14', *Marine Policy*. Elsevier Ltd, 112(103783). https://doi.org/10.1016/j.marpol.2019.103783.

L'Oréal Group (no date) *Investing for nature*. Available at: https://www.loreal.com/en/commitments-and-responsibilities/for-the-planet/investing-for-nature/ (Accessed: 1 July 2021).

Lienert, I. (2009) *Where does the public sector end and the private sector begin?*, *IMF Working Papers*. https://doi.org/10.5089/9781451872699.001.

Lovelock, C. E. and Duarte, C. M. (2019) 'Dimensions of blue carbon and emerging perspectives', *Biology Letters*, 15(3), pp. 1–5. https://doi.org/10.1098/rsbl.2018.0781.

Macreadie, P. I. *et al.* (2019) 'The future of Blue Carbon science', *Nature Communications*, 10(1), pp. 1–13. https://doi.org/10.1038/s41467-019-11693-w.

Martin, A. *et al.* (2016) *Blue Carbon - Nationally Determined Contributions Inventory Appendix to: Coastal blue carbon ecosystems, Opportunities for Nationally Determined Contributions*. Available at: https://www.grida.no/publications/378.

McGowan *et al.* (2020) 'Prioritizing debt conversion opportunities for marine conservation', *Conservation Biology*, 34(5), pp. 1065–1075. https://doi.org/10.1111/cobi.13540.

Mirova. (2020) *Mirova's Sustainable Ocean Fund Reaches a Final Close at $132 m of Commitments Exceeding its Target*. Available at: https://www.mirova.com/en/news/mirova-sustainable-ocean-fund-reaches-final-close-132m-commitments (Accessed: 1 July 2021).

Mirova (2021) *Call for Early-stage Blue Carbon Project Proposals*. Available at: https://www.mirova.com/en/news/call-early-stage-blue-carbon-project-proposals (Accessed: 1 July 2021).

Moffat, K. *et al.* (2016) 'The social licence to operate: a critical review', *Forestry*, 89(5), pp. 477–488. https://doi.org/10.1093/forestry/cpv044.

Moraes, O. (2019) 'Blue carbon in area-based coastal and marine management schemes – a review', *Journal of the Indian Ocean Region*. Taylor & Francis, 15(2), pp. 193–212. https://doi.org/10.1080/19480881.2019.1608672.

Nellemann, C. *et al.* (2009) *Blue Carbon. A Rapid Response Assessment*. Available at: http://www.grida.no/files/publications/blue-carbon/BlueCarbon_screen.pdf.

Pascal, N. *et al.* (2021) 'Impact investment in marine conservation', *Ecosystem Services*. Elsevier B.V., 48(101248). https://doi.org/10.1016/j.ecoser.2021.101248.

Pendleton, L. *et al.* (2012) 'Estimating global "blue carbon" emissions from conversion and degradation of vegetated coastal ecosystems', *PLoS One*, 7(9). https://doi.org/10.1371/journal.pone.0043542.

Pörtner, H.-O. *et al.* (2019) *IPCC Special Report on the Ocean and Cryosphere in a Changing Climate*. Geneva: IPCC Intergovernmental Panel on Climate Change.

Rees, S. E. *et al.* (2018) 'Bridging the divide: Social–ecological coherence in Marine Protected Area network design', *Aquatic Conservation: Marine and Freshwater Ecosystems*, 28(3), pp. 754–763. https://doi.org/10.1002/aqc.2885.

Shiiba, N. *et al.* (2021) 'How blue financing can sustain ocean conservation and development: A proposed conceptual framework for blue financing mechanism', *Marine Policy*. Elsevier Ltd, (104575). https://doi.org/10.1016/j.marpol.2021.104575.

Stadelmann, M. *et al.* (2013) 'Difficulties in accounting for private finance in international climate policy', *Climate Policy*, 13(6), pp. 718–737. https://doi.org/10.1080/14693062.2013.791146.

Stephens, T. (2015) 'Ocean acidification', in Rayfuse, R. (ed.) *Research Handbook on International Marine Environmental Law*. Cheltenham: Edward Elgar Publishing Limited, pp. 431–450. https://doi.org/10.4337/9781781004777.

Sullivan, S. (2017) 'Noting some effects of fabricating "nature" as "natural capital"', *The Ecological Citizen*, 1(1), pp. 65–73.

The Climate Pledge. (no date) *The Pledge*. Available at: https://www.theclimatepledge.com/us/en/the-pledge (Accessed: 9 August 2021).

The Nature Conservancy. (no date) *Debt Conversions for Marine Conservation and Climate Adaptation*. Available at: https://www.nature.org/en-us/about-us/who-we-are/how-we-work/finance-investing/naturevest/ocean-protection/ (Accessed: 1 July 2020).

Thiele, T. *et al.* (2020) Infrastructure Finance: A new approach, integrating Nature-based Solutions for coastal resilience. Gland, Switzerland. Available at: https://bluenaturalcapital.org/wp2018/wp-content/uploads/2020/03/Blue-Infrastructure-Finance.pdf.

Thomas, S. (2014) 'Blue carbon: Knowledge gaps, critical issues, and novel approaches', *Ecological Economics*. Elsevier B.V., 107, pp. 22–38. https://doi.org/10.1016/j.ecolecon.2014.07.028.

Ullman, R. *et al.* (2013) 'Including Blue Carbon in climate market mechanisms', *Ocean and Coastal Management*. Elsevier Ltd, 83, pp. 15–18. https://doi.org/10.1016/j.ocecoaman.2012.02.009.

UNESCO, 2020 *Custodians of the Globe's Blue Carbon Assets*. France. Available at: https://whc.unesco.org/en/blue-carbon-report.

United Nations Environment Programme Finance Initiative UNEPFI. (2021a) *Rising Tide: Mapping Ocean Finance for a New Decade*. Geneva, Switzerland. Available at: https://www.unepfi.org/publications/rising-tide/.

UNEPFI. (2021b) *Turning the Tide: How to Finance A Sustainable Ocean Recovery – A Practical Guide for Financial Institutions*. Geneva, Switzerland. Available at: https://www.unepfi.org/publications/turning-the-tide/.

Vanderklift, M. A. *et al.* (2019) 'Constraints and opportunities for market-based finance for the restoration and protection of blue carbon ecosystems', *Marine Policy*. Elsevier Ltd, 107(103429). https://doi.org/10.1016/j.marpol.2019.02.001.

Verra (20210) *Press Release: Verra Has Registered Its First Blue Carbon Conservation Project*.

Verra (no date) *BLUE CARBON PROJECT GULF OF MORROSQUILLO "VIDA MANGLAR"*. Available at: https://registry.verra.org/app/projectDetail/VCS/2290 (Accessed: 1 July 2021).

Virdin, J. *et al.* (2021) 'The Ocean 100: Transnational corporations in the ocean economy', *Science Advances*, 7(3), pp. 1–11. https://doi.org/10.1126/sciadv.abc8041.

Visbeck, M. *et al.* (2014) 'Securing blue wealth: The need for a special sustainable development goal for the ocean and coasts', *Marine Policy*. Elsevier, 48, pp. 184–191. https://doi.org/10.1016/j.marpol.2014.03.005.

Vivid Economics and Environmental Finance. (2018) *Sustainable financing mechanisms for Marine Protected Areas in North Devon*. Available at: https://www.vivideconomics.com/casestudy/sustainable-financing-for-marine-protected-areas-in-north-devon/.

Wilkins, R. C. and Gillespie, T. (2021) *Asia Primed for Rare Blue Bond Sales to Fund Sea Protection*. Available at: https://www.Wilkins and Gillespie.com/news/articles/2021–03-03/asia-primed-for-rare-blue-bond-sales-to-fund-ocean-protection?srnd=green (Accessed: 1 July 2021).

PART 4

Governing marine environments and regions

12
VULNERABLE NATIONS AND COMMUNITIES
Accounting for those most dependent on the seas

Greg Johnson, Amanda Alva, and Kelly Dunning

Perhaps nowhere is human dependence on nature more evident than in coastal and marine environments, where ecosystems and biodiversity are being lost and degraded at an alarming rate (MEA, 2005). Environmental governance systems can help prevent this loss by setting rules on how the extraction of natural resources occurs and by balancing competing economic uses to deliver value to society (Burroughs, 2011). For example, environmental governance systems can set limits on competing economic sectors such as energy production, fisheries, tourism, aquaculture, and conservation, with interests mediated by laws and regulations (Bellanger et al., 2020). Environmental governance can be defined as the processes and institutions (including cultural norms and rules) that enable effective decision-making related to protecting the environment (Pittman and Armitage, 2016). Major actors (i.e., decision-makers, nongovernmental organisations (NGOs), businesses, and scientists), institutions, power centres, and bodies of knowledge are all important parts of the process (Van Assche et al., 2020).

This chapter will survey environmental governance of coastal and marine systems in vulnerable nations, as well as the implications for the governance of global environmental change. Coastal communities are our focus because of the strong relationship between social systems (e.g., economies, food systems, and cities) and ecological systems (e.g., beaches, coral reefs, and wetlands). Governance cognizant of the relationship between social and ecological systems can enable communities to make decisions that steer themselves away from danger (Van Assche et al., 2020). Take a hypothetical example of a coastal policy created to prevent developers from building homes in a flood zone. The homes are exposed because they are adjacent to ecosystems like beaches that experience storms, floods, and erosion. Coastal policymakers can use historical flood data, predictions about future climate change impacts, and public preferences to determine when that exposure is too severe to permit development in certain zones on a landscape. This policy is, therefore, cognizant of the relationship between social systems of spatial planning, housing, and ecological systems of beaches and natural hazards. Governance limits human uses of resources, in this case, space on coastal landscapes, to steer home buyers and wider communities from dangers, such as losing lives and property in a flood.

Global governance systems are being created to respond to climate change impacts. Some of these impacts include rising seas, increasing storm intensity, threats to fisheries

DOI: 10.4324/9781315149745-16

and other coastal livelihoods, and the loss of coastal resources like wetlands that protect coastal populations. New ways of integrating different governmental agencies, scientific data, and grassroots actors, all for the common purpose of enacting governance at local and national scales, are being used worldwide to respond to global environmental change. In this chapter, we aim to provide an overview of the merits and challenges of these governance responses.

Vulnerable geographies

Vulnerability is caused by two factors: *exposure* to a potential hazard, such as communities existing in an area with frequent tropical storms, and *sensitivity* to hazards, or the degree of damage caused by hazards (Adger et al., 2006; Cutter et al., 2008). An example of sensitivity is how severely tropical storm damage impacts a community. Damages can range from minor wind damage as a result of stronger building standards to catastrophic damage in which a town becomes unrecognisable and basic services are lost. Resilience is the capacity of a social or ecological system to absorb this disturbance and retain its essential features (Adger et al., 2005). To select the case studies of coastal governance reviewed in this chapter, we used the United Nations Development Policy and Analysis Division's *World Economic Situation and Prospects* classification system, which is based on income and growth (United Nations, 2020). We focused primarily on low-income, least developed countries (LDCs) because of their resource-dependent communities, with high levels of vulnerability stemming from direct reliance on ecosystems (Fischer, 2018). These cases demonstrate the strong link between social and ecological systems and how governance cognizant of those links can reduce harm to coastal societies.

Africa

Given Africa's high concentration of LDCs, vulnerability is high, as low household incomes tend to increase vulnerability to hazards for much of the population. For example, low incomes may force large portions of the population into flood-prone areas with informal housing or settlements not in compliance with planning and building regulations. Informal housing, when exposed to hazards like major storms and floods, may suffer catastrophic, community-wide damage. One important way that African decision-makers are addressing vulnerability in coastal communities is through governance systems that focus on spatial planning for coastal zones. Often, this planning process is called *integrated coastal zone management* because it integrates, or considers, trade-offs between the conservation of natural resources and resource extraction for economic growth. Integrated coastal zone management must bring together (or integrate) different government agencies working towards a shared goal, different scales of government (community/local, regional, and national), and different types of knowledge (scientific and traditional), all the while working to use limited coastal space for numerous social and economic activities (Hassanali, 2015).

The integrated coastal zone management process begins when actors (such as governments), universities, and NGOs, study, and map natural resources (such as fisheries and habitats) and human social systems (such as low elevation housing vulnerable to storms and floods). In Central Africa, decision-makers are implementing ecosystem mapping of important coastal biodiversity areas to inform a spatial planning process (Ngoran and Xue, 2017; Trew et al., 2019). Similarly, Mauritania is using data collection to industrialise its coastal economy, and it is deploying a novel system of science-based fisheries management to accomplish this goal (Trégarot et al., 2020). This means that decision-makers are assessing the conservation status of key habitats and collecting biological data on commercially important species of fish to inform policy. In Mauritania, scientists are also collecting data to inform decisions on where to place marine protected areas (MPAs) or conservation zones in coastal and other marine environments, which limit the extraction of resources. For example, they are using data on the location of [fish] nurseries, like coral reef habitat, that sustain commercial fisheries (Trégarot et al., 2020). Thus, consideration of fishing-focused economic growth and protected areas for conservation are integrated. In another example, in Gabon, biodiversity-rich estuarine ecosystems located near major cities such as Libreville and Port Gentil are being mapped for the first time. Findings showcase an urgent need for officials to plan and implement protected areas as fewer than 33% of wetlands are protected and remain at risk of being lost to development (Aldous et al., 2021, p. 64).

In addition to examining important biodiversity areas, officials and researchers are also mapping human systems to study social vulnerability. In West Africa, a particularly vulnerable region where most countries are classified as LDCs, global environmental change is causing major disturbances that governance systems must address. For example, coastal erosion, worsened by sea-level rise and increased storm intensity from climate change, is creating significant vulnerability to flooding in places like Benin, Senegal, and The Gambia (Gomez et al., 2020; Ndour et al., 2018). Officials are beginning to understand the urgent need for governance responses to lessen this vulnerability. In The Gambia, for example, as many as 90% of households are vulnerable to coastal erosion, which is increasing in intensity due to climate change (Gomez et al., 2020, p. 9). With 74% of households lacking any capacity to bear these impacts of environmental change, erosion is threatening housing security and livelihoods by forcing a process known as adaptation, wherein locals are forced to alter their way of life to overcome unforeseen threats (Gomez et al., 2020, p. 9). Adaptation options could include the ability to transition into a new job or relocate to a less vulnerable home. Possible governance responses in The Gambia could include physical interventions like the construction of breakwaters and spatial planning interventions like integrated coastal zone management, both accompanied by increased education. Education and outreach could increase popular support for integrated coastal zone management efforts. Such programmes could focus on so-called *win-win policies* that protect natural resources like mangroves, which can protect coastal communities from erosion while increasing biodiversity.

Decision-makers can use maps and scientific data on coastal resources and social vulnerability to assess the impacts of the pressures we place on resources (such as fishing, oil and gas extraction, and coastal development). Then, they can create integrated coastal zone management plans with rules for resource extraction grounded in law. In North Africa, governments are balancing the interests of major economic sectors such as oil and gas (Maitig et al., 2018), water scarcity to meet the needs of a growing population (Soula et al., 2021), and the need for the development of pristine habitat to support tourism in coastal cities (Mohamed and Rachid, 2019). Balancing so many interests poses significant challenges to planning efforts in coastal zones, namely, how these plans are implemented. For example, coastal communities impacted by major planning decisions on Algerian coasts are asking for a greater role in decision-making, claiming their voices are not heard and that enforcement of laws and policies does not occur in practice (Khelil et al., 2019). In Ghana, by contrast, vulnerability to global environmental change enables some communities and grassroots actors to innovate and adapt in the face of hazards. Climate change is significantly impacting small-scale fisheries in Ghana, which increases the vulnerability of human communities due to lost livelihoods and diminished nutrition from declining catches (Freduah et al., 2019). In response, these communities are lobbying Ghana's government to enforce fishing laws, build local leadership focused on climate change impacts, and use of natural resources like sand to shore up defences against erosion. Similar grassroots efforts to implement coastal governance are happening in Tanzania, where beach management plans, MPAs, and protected estuarine areas are being created collaboratively with coastal communities and governmental actors (Katikiro et al., 2017).

In East Africa, integrated coastal zone management is taking place with some contrasting results. For example, Kenya is implementing a similar integrated coastal zone management plan to Ghana and Tanzania and faces similar challenges with grassroots inclusion in decision-making, the enforcement of laws, and a lack of data on how climate change impacts will look on the ground (Khelil et al., 2019; Ojwang et al., 2017). Key differences are seen in the depth and strength of Kenyan laws aimed explicitly at managing global environmental change. Specifically, Kenya has a Climate Change Response Strategy (2010), a Climate Change Act (2016), and a National Adaptation Plan (2016) with Climate Change Policy in the process of being created. However, these do not necessarily guarantee resilience at the local scale, where there exists a lag between climate change adaptation programmes and implementation (Ojwang et al., 2017).

Asia

Like in Africa, the greatest challenge for decision-makers involved in Asian coastal governance is responding to global environmental change. Questions on vulnerability in this region are focused on how nations can withstand the impacts of climate change on their rapidly urbanising coasts, which are home to over 300 million people in large population centres, termed 'mega-cities', such as Hong Kong and Singapore (Chan et al., 2018, p. 576). Mega-cities, like Singapore, are mitigating flash flood impacts through drainage projects by supplementing or replacing traditional concrete infrastructure with experimental 'green infrastructure.' Green infrastructure includes projects like vegetated drainage channels, constructed in place of concrete channels in the hopes of increasing plant and animal habitat and biodiversity. For example, the Kallang River was restored from a straight, open, concrete channel to a vegetated floodplain, with decision-makers targeting 100 additional projects like this to complete before 2030 (Chan et al., 2018, p. 584). Similar to Africa, Asia faces challenges with limited community participation in coastal governance. For example, as part of its integrated coastal zone management, Malaysia began creating MPAs in the 1990s. Local communities were not included, and fishing villages were stripped of their livelihoods overnight, which led to a lack of support in many communities adjacent to MPAs and a general scepticism for conservation (Dunning, 2018). Governance that enhances opportunities for grassroots community participation in MPAs can help remedy this by helping decision-makers create and enforce fisheries regulations, which increase compliance with the rules governing MPAs (Islam et al., 2017). Communities are also significant sources of novel governance ideas, or innovations that respond to global environmental change. For example, in 2016, community groups in Indonesian MPAs noticed increased coral reef stress from rising sea surface temperatures, a biological process known as coral bleaching (see Chapter 15). In response, they worked with local officials in the dive and tourism sectors to build artificial reefs for divers to visit, which reduced pressure on bleached, stressed reefs while increasing tourism opportunities (Dunning, 2021).

Uncertainty over the future impacts of global environmental change in some Asian countries has been shown to limit integrated coastal zone management efforts. For example, in South Asia, the most densely urbanised coasts have limited social and ecological data to inform planning efforts and governance responses. India's Andhra Pradesh coastline has a population of 50 million people with a high density of over 800 people per square mile (Kantamaneni et al., 2019, p. 393). People there have vulnerabilities to hazards like storm surges from cyclones, exacerbated by sea-level rise. Decision-makers have not implemented coastal governance strategies to cope with these vulnerabilities. Yet, data gathering techniques such as unmanned aerial vehicles are being tapped to study natural and social systems as the possible foundation for efforts aimed at coastal planning (Kantamaneni et al., 2019). Similarly, Iranian officials face an information gap regarding how to enact governance responses to climate change, namely sea-level rise. Climate change and sea-level rise may, according to Iranian land cover change models, convert immense stretches of mangrove forests into open water (Etemadi et al., 2018). This is worrisome to decision-makers because mangroves shelter human communities from coastal storms and floods while providing carbon storage to lessen the greenhouse gas emissions causing climate change. The Indian and Iranian cases show how gaps in data can create significant uncertainty that delays efforts at integrated coastal zone management.

Latin America and the Caribbean

Latin American officials are also responding to global environmental change through integrated coastal zone management, similar to those in Asia and Africa. Rapidly urbanising coasts and increased tourism throughout Latin America and the Caribbean have caused trade-offs between development and protecting natural resources such as water quality, fisheries, beaches, and MPAs (Banerjee et al., 2018; Cortés-Useche et al., 2021; Goulart et al., 2018; Hassanali, 2015). The region is wrestling with global environmental change as climate change crosscuts and increases problems and makes coastal populations vulnerable to hazards (Cortés-Useche et al., 2021; Goulart et al., 2018; Hassanali, 2015). Integrated coastal zone management attempts have been studied in Trinidad and Tobago, Ecuador, the Galapagos Islands, and Brazil, with varying results in implementation (Hassanali, 2015; Pazmiño Manrique et al., 2018). Two challenges have emerged in the region: the integration of agencies tasked with governance and a weak legal framework for coastal governance. Integration, defined here as coordination and communication between decision-makers, remains weak between agencies responsible for environmental regulations. This is specifically true for tourism regulations in Ecuador and MPA regulations in Mexico. These regulations should protect coastal systems from over development and overfishing, but they suffer lagging implementation (Mancha-Cisneros et al., 2018; Mestanza-Ramón et al., 2020). There is also a lack of appropriate laws to address environmental issues that impact vulnerable communities. For example, in Chile, there are no laws regulating water quality or invasive species (Anbleyth-Evans et al., 2020).

Challenges to grassroots participation in decision-making are present in this region. For example, Trinidad and Tobago's potential for integrated coastal zone management is limited by the lack of pathways for communities to participate in decision-making.

Integrating the actions of local and national governmental agencies is also lagging, lessening their ability to collaborate to create and implement policy (Hassanali, 2015). The U.S.-Mexico Joint Gulf of Mexico Assessment and Management Project provides an effective example of governance that enhances grassroots participation. The project includes mangrove wetland restoration through community engagement, education, and participation (Zaldívar-Jiménez et al., 2017). Sometimes, grassroots participation in governance is costly for vulnerable people as it takes away from time that could be spent fishing and earning an income. For example, small-scale fishers in Uruguay emphasised that costs outweighed the potential benefits from participating in meetings and decision-making processes, calling for capacity-building programmes led by state agencies to assist in expanding participation (Gianelli et al., 2018). Another noteworthy example, Ecuador, lacks a legal framework that would enable integrated coastal zone management and has weak institutions to implement it compared to the Kenyan case outlined above. However, there is potential for community participation, specifically by competent local governmental agencies that have shown administrative competence to enact some forms of integrated coastal zone management at the community level in Ecuador (Pazmiño Manrique et al., 2018).

A governance response to global environmental change worth noting is that of the large marine ecosystem (LME) governance frameworks that have been implemented in Latin America and the Caribbean. LME governance frameworks situate unsustainable fishing, pollution, and habitat degradation as major problems requiring coordinated governance responses, with climate change as a cross-cutting problem that worsens others. Ecosystems within these zones include the Humboldt Current LME in Mexico, the Gulf of Mexico LME, and the Wider Caribbean Region. These initiatives aim to address the interconnectedness of marine ecosystems, termed 'transboundary ecosystems', in order to make policies to recover fisheries and marine habitats undergoing environmental change (Fanning et al., 2021; Muñoz Sevilla and Le Bail, 2017; Villamizar and Cervigón, 2017; Zaldívar-Jiménez et al., 2017). These governance frameworks are an innovative way to strengthen management institutions for marine ecosystems. However, they face significant challenges such as a lack of funding, regional coordination, community engagement at global and local scales, and national-level supporters (Fanning et al., 2021).

Similar to African cases, one of the main governance issues in the Latin American and Caribbean region is enforcement and implementation of integrated coastal zone management policies, seen in the example of Guyana's Sustainable Development Goals. Guyana lacks the capacity for strengthened enforcement required to meet its obligations under the Convention on Biological Diversity, a treaty that requires setting aside important biodiversity areas for conservation (Elias-Roberts, 2020). An example of difficulties in implementation includes enforcing no fishing rules within MPAs. The Dominican Republic and Jamaica face similar issues (Chan et al., 2019; Cortés-Useche et al., 2021). One proposed solution is through a 'seascape approach', which is a novel institution where governments at different scales (national and local), businesses, NGOs, and communities work together in a network to make decisions for integrated coastal zone management, as has been seen in the Atlántida seascape in Honduras (Steadman, 2021). Another strategy in Costa Rica is through the government granting grassroots actors like communities the legal authority to form and implement Marine Areas of Responsible Fishing. These areas follow the Small-Scale Fisheries Guidelines outlined by the United Nations, but because of their robust inclusion of grassroots actors who actually fish the reserves, support for implementation is higher (Chavez-Carrillo et al., 2019).

Vulnerable groups

Climate change, refugees, and climate justice

Climate refugees and stateless people show how coastal and marine governance, global environmental change, and vulnerability intersect. Global environmental change and impacts such as heatwaves, droughts, and sea-level rise, among others, all play a role in what could become a humanitarian crisis for vulnerable communities in Asia, Latin America and the Caribbean, and Africa (Ahmed, 2017; Biermann and Boas, 2010). For example, when agriculturalists lose their livelihoods from drought and their homes and farms from sea-level rise, they may be forced to leave their country and seek refugee status elsewhere. Current institutions, organisations, and funding mechanisms are not prepared to respond to this emerging crisis (Biermann and Boas, 2010). Bangladesh is an example of the Asian climate refugee experience, namely in its densely populated urban centres along the coast. Recent estimates suggest that one in every seven people in Bangladesh will be displaced and forced to migrate either internally or internationally due to climate change (Government of Bangladesh, 2014, p. xvii). The refugee crisis in the Mediterranean region also demonstrates how low-lying coastal regions of Northern Africa, due to experiencing climate change, and currently embroiled in domestic strife, may produce refugees seeking safety in Europe (Biermann and Boas, 2010).

Gender, vulnerability, and marine governance

The importance of inclusion of gender in marine governance has become conventional wisdom among scholars and managers, yet it is notably lacking in documentation and implementation (Bradford and Katikiro, 2019; Koralagama et al., 2017; Lawless et al., 2021). Nowhere is the importance of gender inclusion in governance of marine resources more important than in small-scale fisheries, a sector employing millions of women all over the world (Koralagama et al., 2017). Small-scale fisheries are experiencing unprecedented and irreversible changes brought on by global environmental change (Hanich et al., 2018). The International Union for Conservation of Nature (IUCN) promotes the benefits of a form of integrated coastal zone management that includes stated goals of gender equality and sustainable small-scale fisheries management (Siles et al., 2019). The IUCN argues that decision-makers must create a gender action plan and strategise interventions to address gender gaps in fisheries (Siles et al., 2019). In African cases, the exclusion of women in decision-making led to discriminatory barriers for women in the fishing sector (Baker-Médard, 2017; Bradford and Katikiro, 2019). In Indonesia, although donor-funded development projects reached out to women to increase their participation, 40% of projects did not include people of all genders in implementation (Stacey et al., 2019, p. 366). Inclusive management, in which people of all genders are included in coastal resource management, is one possible solution. This means the decision-makers must (1) explicitly invite people of all genders to participate in decision-making, (2) create roles and activities for people of all genders related to decision-making, (3) create opportunities for economic empowerment, and (4) create opportunities for people of all genders to be involved in environmental protection (De la Torre-Castro, 2019).

Indigenous people, vulnerability, and coastal governance

Historically, in most countries, indigenous peoples have not been afforded explicitly recognised rights over marine areas, a lack of self-determination which makes them vulnerable to social and natural problems (Shapovalova, 2020). Indigenous communities have also experienced extended periods of overexploitation and contamination of their coastal areas due to overdevelopment through aquaculture and fisheries (Araos et al., 2020). Global environmental change is emerging as a major threat to indigenous ways of life. For example, in Bangladesh, indigenous communities such as the Jummas, face climate change impacts of drought, water shortages, pests, diseases, and human-wildlife conflict. Indigenous Bangladeshis are responding to global change by enacting their own governance responses, ranging from land management (e.g., changing drainage facilities for crops), to rebuilding infrastructure (e.g., roads after landslides), to water supply management (e.g., installing diesel-powered water pumps), and others. Scientists and decision-makers are beginning to study and adopt Bangladeshi indigenous knowledge as trusted governance responses to climate change (Rahman and Alam, 2016). Novel governance responses by decision-makers in Latin America and the Caribbean have given indigenous communities stewardship over coastal and marine environments. In Chile, for example, the creation of Marine and Coastal Areas for Indigenous Peoples (or MCAIPs) has resulted in 91 areas (totalling 32,000 square kilometres) being recognised as MPAs with cultural conservation components, with management rights held by indigenous communities (Hiriart-Bertrand et al., 2020, p. 4).

Conclusion

This chapter highlights the vulnerability of global coasts to a range of human and natural stressors. We have shown a wide range of governance responses intended to diminish vulnerability. In the African and Latin American and Caribbean contexts, despite resilient governance responses such as emerging integrated coastal zone management and LME management zones, challenges with the inclusion of grassroots level actors in decision-making remain. In both regions, where implementation and enforcement of governance responses meet comparable challenges, community inclusion may increase support for coastal regulations. Asian, East African, and South American cases of MPAs, where communities decide where MPAs are situated and how they are implemented, may be cases to learn from on this issue. On a global scale, marine and coastal systems face immense threats under global environmental change, demonstrating the need for new governance systems and institutional frameworks to inventory resources, study social vulnerability, and balance competing economic interests in the face of natural hazards. Governance responses to global environmental change are becoming increasingly complex, with outcomes that are difficult to predict. Climate change makes coastal communities, already characterised by high exposure and sensitivity, even more vulnerable. Thus, flexible governance systems for managing coastal ecosystems and human societies, especially those that draw on both scientific and traditional knowledge, such as that of indigenous communities in Bangladesh, are required. Marginalised groups, such as stateless people, refugees, indigenous communities, people of all genders, and others, will need to be included in a meaningful way if the world's most vulnerable people are to be part of the solution to marine governance amidst global environmental change.

References

Adger, W. N. (2006). Vulnerability. *Global Environmental Change*, 16(3), 268–281. https://doi.org/10.1016/j.gloenvcha.2006.02.006

Ahmed, B. (2017). Who takes responsibility for the climate refugees? *International Journal of Climate Change Strategies and Management*, 10(1), 5–26. https://doi.org/10.1108/IJCCSM-10-2016-0149

Aldous, A., Schill, S., Raber, G., Paiz, M., Mambela, E., and Stévart, T. (2021). Mapping complex coastal wetland mosaics in Gabon for informed ecosystem management: Use of object-based classification. *Remote Sensing in Ecology and Conservation*, 7(1), 64–79. https://doi.org/10.1002/rse2.161

Anbleyth-Evans, J., Leiva, F. A., Rios, F. T., Cortés, R. S., Vreni Häussermann, and Aguirre-Munoz, C. (2020). Toward marine democracy in Chile: Examining aquaculture ecological impacts through common property local ecological knowledge. *Marine Policy*, 113. https://doi.org/10.1016/j.marpol.2019.103690

Araos, F., Anbleyth-Evans, J., Riquelme, W., Hidalgo, C., Brañas, F., Catalán, E., Núñez, D., and Diestre, F. (2020). Marine indigenous areas: Conservation assemblages for sustainability in southern Chile. *Coastal Management*, 48(4), 289–307. https://doi.org/10.1080/08920753.2020.1773212

Baker-Médard, M. (2017). Gendering marine conservation: The politics of marine protected areas and fisheries access. *Society and Natural Resources*, 30(6), 723–737. https://doi.org/10.1080/08941920.2016.1257078

Banerjee, O., Boyle, K., Rogers, C. T., Cumberbatch, J., Kanninen, B., Lemay, M., and Schling, M. (2018). Estimating benefits of investing in resilience of coastal infrastructure in small island developing states: An application to Barbados. *Marine Policy*, 90, 78–87. https://doi.org/10.1016/j.marpol.2018.01.004

Bellanger, M., Speir, C., Blanchard, F., Brooks, K., Butler, J. R. A., Crosson, S., Fonner, R., Gourguet, S., Holland, D. S., Kuikka, S., Le Gallic, B., Lent, R., Libecap, G. D., Lipton, D. W., Nayak, P. K., Reid, D., Scemama, P., Stephenson, R., Thébaud, O., & Young, J. C. (2020). Addressing Marine and Coastal Governance Conflicts at the Interface of Multiple Sectors and Jurisdictions. *Frontiers in Marine Science*, 7. https://doi.org/10.3389/fmars.2020.544440

Biermann, F., and Boas, I. (2010). Preparing for a warmer world: Towards a global governance system to protect climate refugees. *Global Environmental Politics*, 10(1), 60–88. https://doi.org/10.1162/glep.2010.10.1.60

Bradford, K., and Katikiro, R. E. (2019). Fighting the tides: A review of gender and fisheries in Tanzania. *Fisheries Research*, 216, 79–88. https://doi.org/10.1016/j.fishres.2019.04.003

Burroughs, R. (2011). *Coastal Governance*. Island Press.

Chan, C., Armitage, D., Alexander, S. M., and Campbell, D. (2019). Examining linkages between ecosystem services and social wellbeing to improve governance for coastal conservation in Jamaica. *Ecosystem Services*, 39. https://doi.org/10.1016/j.ecoser.2019.100997

Chan, F. K. S., Chuah, C. J., Ziegler, A. D., Dąbrowski, M., and Varis, O. (2018). Towards resilient flood risk management for Asian coastal cities: Lessons learned from Hong Kong and Singapore. *Journal of Cleaner Production*, 187, 576–589. https://doi.org/10.1016/j.jclepro.2018.03.217

Chavez-Carrillo, I. I., Partelow, S., Madrigal-Ballestero, R., Schlüter, A., and Gutierrez-Montes, I. (2019). Do responsible fishing areas work? Comparing collective action challenges in three small-scale fisheries in Costa Rica. *International Journal of the Commons*, 13(1), 705–746. https://doi.org/10.18352/ijc.923

Cortés-Useche, C., Hernández-Delgado, E. A., Calle-Triviño, J., Blasco, R. S., Galván, V., and Arias-González, J. E. (2021). Conservation actions and ecological context: Optimizing coral reef local management in the Dominican Republic. *PeerJ*, 9. https://doi.org/10.7717/peerj.10925

Cutter, S. L., Barnes, L., Berry, M., Burton, C., Evans, E., Tate, E., & Webb, J. (2008). A place-based model for understanding community resilience to natural disasters. *Global Environmental Change*, *18*(4), 598–606. https://doi.org/10.1016/j.gloenvcha.2008.07.013

De la Torre-Castro, M. (2019). Inclusive management through gender consideration in small-scale fisheries: The why and the how. *Frontiers in Marine Science*, 6(156), 1–11.

Dunning, K. H. (2018). Managing coral reefs: An ecological and institutional analysis of ecosystem services in Southeast Asia. https://www.cambridge.org/core/product/identifier/9781783087976/type/BOOK

Dunning, K. H. (2021). How are managers responding to local and global ecological stressors? The case of Indonesian co-managed coral reefs in the Anthropocene. *Marine Policy*, 131, 104560.

Elias-Roberts, A. (2020). Balancing environmental protection and offshore petroleum developments in Guyana. *Global Energy Law and Sustainability*, 1(1), 1–27. https://doi.org/10.3366/gels.2020.0004

Etemadi, H., Smoak, J. M., and Karami, J. (2018). Land use change assessment in coastal mangrove forests of Iran utilizing satellite imagery and CA–Markov algorithms to monitor and predict future change. *Environmental Earth Sciences*, 77(5), 208. https://doi.org/10.1007/s12665-018-7392-8

Fanning, L., Mahon, R., Compton, S., Corbin, C., Debels, P., Haughton, M., Heileman, S., Leotaud, N., McConney, P., Moreno, M. P., Phillips, T., and Toro, C. (2021). Challenges to implementing regional ocean governance in the wider Caribbean region. *Frontiers in Marine Science*, 8, 667273. https://doi.org/10.3389/fmars.2021.667273

Fischer, A. P. (2018). Pathways of adaptation to external stressors in coastal natural-resource-dependent communities: Implications for climate change. *World Development*, 108, 235–248. https://doi.org/10.1016/j.worlddev.2017.12.007

Freduah, G., Fidelman, P., and Smith, T. F. (2019). Adaptive capacity of small-scale coastal fishers to climate and non-climate stressors in the Western region of Ghana. *The Geographical Journal*, 185(1), 96–110. https://doi.org/10.1111/geoj.12282

Gianelli, I., Horta, S., Martínez, G., de la Rosa, A., and Defeo, O. (2018). Operationalizing an ecosystem approach to small-scale fisheries in developing countries: The case of Uruguay. *Marine Policy*, 95, 180–188. https://doi.org/10.1016/j.marpol.2018.03.020

Gomez, M. L. A., Adelegan, O. J., Ntajal, J., and Trawally, D. (2020). Vulnerability to coastal erosion in The Gambia: Empirical experience from Gunjur. *International Journal of Disaster Risk Reduction*, 45, 101439. https://doi.org/10.1016/j.ijdrr.2019.101439

Goulart, F., Galán, Á. L., Nelson, E., and Soares-Filho, B. (2018). Conservation lessons from Cuba: Connecting science and policy. *Biological Conservation*, 217, 280–288. https://doi.org/10.1016/j.biocon.2017.10.033

Government of Bangladesh. (2014). *Trend and Impact Analysis of Internal Displacement due to the Impacts of Disaster and Climate Change*. ReliefWeb. Retrieved May 12, 2021, from https://reliefweb.int/report/bangladesh/trend-and-impact-analysis-internal-displacement-due-impacts-disaster-and-climate

Hanich, Q., Wabnitz, C. C., Ota, Y., Amos, M., Donato-Hunt, C., & Hunt, A. (2018). Small-scale fisheries under climate change in the Pacific Islands region. *Marine Policy*, 88, 279–284.

Hassanali, K. (2015). Improving ocean and coastal governance in Trinidad and Tobago - Moving towards ICZM. *Ocean and Coastal Management*, 106, 1–9. https://doi.org/10.1016/j.ocecoaman.2015.01.002

Hiriart-Bertrand, L., Silva, J. A., and Gelcich, S. (2020). Challenges and opportunities of implementing the marine and coastal areas for indigenous peoples policy in Chile. *Ocean and Coastal Management*, 193, 105233. https://doi.org/10.1016/j.ocecoaman.2020.105233

Islam, G. M. N., Tai, S. Y., Kusairi, M. N., Ahmad, S., Aswani, F. M. N., Muhamad Senan, M. K. A., and Ahmad, A. (2017). Community perspectives of governance for effective management of marine protected areas in Malaysia. *Ocean and Coastal Management*, 135, 34–42. https://doi.org/10.1016/j.ocecoaman.2016.11.001

Kantamaneni, K., Sudha Rani, N. N. V., Rice, L., Sur, K., Thayaparan, M., Kulatunga, U., Rege, R., Yenneti, K., and Campos, L. C. (2019). A systematic review of coastal vulnerability assessment studies along Andhra Pradesh, India: A critical evaluation of data gathering, risk levels and mitigation strategies. *Water*, 11(2), 393. https://doi.org/10.3390/w11020393

Katikiro, R., Namkesa, F., Ponte, S., and Minja, R. (2017). Sustainability partnerships for the governance of coastal resources in Tanzania. NEPSUS Working Paper 2017/5, 2017.

Khelil, N., Larid, M., Grimes, S., Le Berre, I., and Peuziat, I. (2019). Challenges and opportunities in promoting integrated coastal zone management in Algeria: Demonstration from the Algiers coast. *Ocean and Coastal Management*, 168, 185–196. https://doi.org/10.1016/j.ocecoaman.2018.11.001

Koralagama, D., Gupta, J., and Pouw, N. (2017). Inclusive development from a gender perspective in small scale fisheries. *Current Opinion in Environmental Sustainability*, 24, 1–6. https://doi.org/10.1016/j.cosust.2016.09.002

Lawless, S., Cohen, P. J., Mangubhai, S., Kleiber, D., and Morrison, T. H. (2021). Gender equality is diluted in commitments made to small-scale fisheries. *World Development*, 140, 105348. https://doi.org/10.1016/j.worlddev.2020.105348

Maitig, A., Lynch, K., and Johnson, M. (2018). Coastal Resources Spatial Planning and Potential Oil Risk Analysis: Case Study of Misratah's Coastal Resources, Libya. *International Journal of Earth, Energy and Environmental Sciences*. https://doi.org/10.5281/zenodo.1315968

Mancha-Cisneros, M. del M., Suárez-Castillo, A. N., Torre, J., Anderies, J. M., and Gerber, L. R. (2018). The role of stakeholder perceptions and institutions for marine reserve efficacy in the Midriff Islands Region, Gulf of California, Mexico. *Ocean and Coastal Management*, 162, 181–192. https://doi.org/10.1016/j.ocecoaman.2018.01.024

MEA. (2005). *Millenium Ecosystem Assessment: Ecosystems and human well-being*. Island Press. http://agris.fao.org/agris-search/search.do?recordID=XF2006408644

Mestanza-Ramón, C., Chica-Ruiz, J. A., Anfuso, G., Mooser, A., Botero, C. M., and Pranzini, E. (2020). Tourism in continental Ecuador and the Galapagos Islands: An integrated coastal zone management (ICZM) perspective. *Water*, 12, 1647. https://doi.org/10.3390/w12061647

Mohamed, H., and Rachid, E.-D. (2019). Assessing sustainable tourism: Trends and efforts in Essaouira in Morocco as a Coastal City. *International Journal of Sustainability Management and Information Technologies*, 5(1), 23. https://doi.org/10.11648/j.ijsmit.20190501.14

Muñoz Sevilla, N. P., and Le Bail, M. (2017). Latin American and Caribbean regional perspective on ecosystem based management (EBM) of large marine ecosystems goods and services. *Environmental Development*, 22, 9–17. https://doi.org/10.1016/j.envdev.2017.01.006

Ndour, A., Laïbi, R. A., Sadio, M., Degbe, C. G. E., Diaw, A. T., Oyédé, L. M., Anthony, E. J., Dussouillez, P., Sambou, H., and Dièye, E. hadji B. (2018). Management strategies for coastal erosion problems in West Africa: Analysis, issues, and constraints drawn from the examples of Senegal and Benin. *Ocean and Coastal Management*, 156, 92–106. https://doi.org/10.1016/j.ocecoaman.2017.09.001

Ngoran, S. D., and Xue, X. (2017). Public sector governance in Cameroon: A valuable opportunity or fatal aberration from the Kribi Campo integrated coastal management? *Ocean and Coastal Management*, 138, 83–92. https://doi.org/10.1016/j.ocecoaman.2017.01.006

Ojwang, L., Rosendo, S., Celliers, L., Obura, D., Muiti, A., Kamula, J., and Mwangi, M. (2017). Assessment of coastal governance for climate change adaptation in Kenya. *Earth's Future*, 5(11), 1119–1132. https://doi.org/10.1002/2017EF000595

Pazmiño Manrique, P., Barragán, J. M., and García Sanabria, J. (2018). Progress on coastal management in Ecuador (2007–2017). *Environmental Science and Policy*, 90, 135–147. https://doi.org/10.1016/j.envsci.2018.09.016

Pittman, J., and Armitage, D. (2016). Governance across the land-sea interface: A systematic review. *Environmental Science and Policy*, 64, 9–17. https://doi.org/10.1016/j.envsci.2016.05.022

Rahman, M., & Alam, K. (2016). Forest dependent indigenous communities' perception and adaptation to climate change through local knowledge in the protected area—A Bangladesh case study. *Climate*, 4(1), 12.

Shapovalova, D. (2020). Indigenous rights and resource extraction in Guyana: A learning opportunity for the new offshore sector? (SSRN Scholarly Paper ID 3555121). Social Science Research Network. https://doi.org/10.2139/ssrn.3555121

Siles, J., Prebble, M., Wen, J., Hart, C., and Schuttenberg, H. (2019). *Advancing gender in the environment: Gender in fisheries — A sea of opportunities*. IUCN and USAID. Washington, DC: USAID. 68pp. https://portals.iucn.org/library/sites/library/files/documents/2019-040-En.pdf

Soula, R., Chebil, A., McCann, L., and Majdoub, R. (2021). Water scarcity in the Mahdia region of Tunisia: Are improved water policies needed? *Groundwater for Sustainable Development*, 12, 100510. https://doi.org/10.1016/j.gsd.2020.100510

Stacey, N., Gibson, E., Loneragan, N. R., Warren, C., Wiryawan, B., Adhuri, D., and Fitriana, R. (2019). Enhancing coastal livelihoods in Indonesia: an evaluation of recent initiatives on gender, women and sustainable livelihoods in small-scale fisheries. *Maritime Studies*, 18(3), 359–371. https://doi.org/10.1007/s40152-019-00142-5

Steadman, D. (2021). Towards ecological and social impact through collaborative governance of a seascape of marine protected areas in Honduras. *ORYX*, 1–12. https://doi.org/10.1017/S0030605320001155

Trégarot, E., Meissa, B., Gascuel, D., Sarr, O., El Valy, Y., Wagne, O. H., Kane, E. A., Bal, A. C., Haidallah, M. S., Fall, A. D., Dia, A. D., and Failler, P. (2020). The role of marine protected areas in sustaining fisheries: The case of the National Park of Banc d'Arguin, Mauritania. *Aquaculture and Fisheries*, 5(5), 253–264. https://doi.org/10.1016/j.aaf.2020.08.004

Trew, B. T., Grantham, H. S., Barrientos, C., Collins, T., Doherty, P. D., Formia, A., Godley, B. J., Maxwell, S. M., Parnell, R. J., Pikesley, S. K., Tilley, D., Witt, M. J., and Metcalfe, K. (2019). Using cumulative impact mapping to prioritize marine conservation efforts in Equatorial Guinea. *Frontiers in Marine Science*, 6. https://doi.org/10.3389/fmars.2019.00717

United Nations. (2020). World Economic Situation and Prospects 2020 [Country classification tables as defined by United Nations], 41, 166–169.

Van Assche, K., Hornidge, A. K., Schlüter, A., and Vaidianu, N. (2020). Governance and the coastal condition: Towards new modes of observation, adaptation and integration. *Marine Policy*, 112, 103413. https://doi.org/10.1016/j.marpol.2019.01.002

Villamizar, G. E. Y., and Cervigón, F. (2017). Variability and sustainability of the Southern Subarea of the Caribbean Sea large marine ecosystem. *Environmental Development*, 22, 30–41. https://doi.org/10.1016/j.envdev.2017.02.005

Zaldívar-Jiménez, A., Ladrón de Guevara-Porras, P., Pérez-Ceballos, R., Díaz-Mondragón, S., and Rosado-Solórzano, R. (2017). US-Mexico joint Gulf of Mexico large marine ecosystem based assessment and management: Experience in community involvement and mangrove wetland restoration in Términos lagoon, Mexico. *Environmental Development*, 22, 206–213. https://doi.org/10.1016/j.envdev.2017.02.007

13

COASTLINES AND NEARSHORE HABITATS

Interactive governance in an era of global environmental change

Tony George Puthucherril

Coasts comprise transitional areas where the terrestrial meets the marine. The influence exerted by the ocean determines the landward spread of the coastal zone, and the perimeter of the marine segment is ascertained likewise. Because of this land-sea interface and tidal influence, the coastal zone is unique and highly dynamic. It holds some of the most remarkable biodiversity and economic resources, including estuaries, wetlands, mangroves, salt marshes, lagoons, deltas, coral reefs, beaches, rocky cliffs, aquifers, and seagrass meadows. More importantly, these places have some of the highest densities of human populations and are a platform that supports a diverse spectrum of economic activity, including marine transportation of goods, offshore energy drilling, resource extraction, fish cultivation, recreation, and tourism. Most of these cannot be replicated in the hinterland, and therefore, coastal lands and resources are critical to the economies of practically every country that borders the ocean (World Bank n.d.).

During the past several decades, a noticeable ongoing phenomenon has been the migration of people in large numbers to coastal regions due to economic and aesthetic reasons (National Academies of Sciences, Engineering, and Medicine 2010: 27–82). In several countries, rural coastal areas are quickly turning urban (Wolff et al. 2020: 1–11). Coastal urbanization has become one of the foremost sustainable development-related challenges (International Human Dimensions Programme on Global Environmental Change 2015: 6). Because of this high human presence, there is a profusion of development and settled space on the coastal fringe. Squatter settlements, lack of drinking water, unsanitary sanitation, and improper and untreated waste disposal are some of the significant challenges that have risen due to the increasing urbanization of coastal areas and the resulting need for services (Celliers and Ntombela 2016: 386–404). The increasing number of dead zones along coastal waters and algal blooms are symptomatic of these currently plaguing ills (IPCC 2019: 62).

Even though coastal ecological systems provide various services essential to humans, including storm surge protection and effluent filtration, they are often destroyed to make way for coastal urbanization without appreciating their value. Mangrove felling, mining, filling, and draining coastal wetlands for aquaculture and agriculture, marine pollution and eutrophication, and upstream diversion of waters are significant threats in this regard (Shine and Klemm 1999). However, due to the indispensable nature of these ecosystem services,

DOI: 10.4324/9781315149745-17

attempts to rebuild them are often made through artificial and costly technological interventions. Rarely do these attempts equal or exceed the services offered by natural ecosystems.

In short, coastal and marine ecosystems are in a precarious situation, and these stresses and changes can cause significant harm to coastlines on their own. However, all these pale when compared to the effects of the rapid rise in sea levels, storms, warming oceans, intense wave action, and other climate change consequences. Depending on a host of anthropogenic and natural factors like the rate of its rise and the nature of the coastal zone, the sea will wear away and swallow large chunks of coastal land and many of these nearshore ecosystems (Hudson 2012: 31–68). Ironically, even though they are under grave threat, as bio-shields, these nearshore habitats are also at the forefront of evolving strategies to adapt to many of these challenges wrought upon the coastal zone by climate change (IPCC 2019). In addition, the sequestration capabilities of these ecosystems will play an essential role in mitigating the increasing presence of atmospheric carbon, the primary cause of climate change (National Academies of Sciences, Engineering, and Medicine 2019: 87–136).

This chapter explains some of the unique features of nearshore habitats. It analyses the challenges that they face due to global environmental change and its implications for marine and coastal governance. The chapter posits that a broader, holistic, and interactive governance that considers the ecosystem's uniqueness and transcends sectoral approaches is needed to sustain nearshore habitats and related ecosystems. Since good governance is equated with sustainability, there is a need for sustainable coastal development and Integrated Coastal Zone Management (ICZM) to secure interactive governance in the service of coastal and nearshore habitats and their associated ecosystems.

Nearshore habitats and their features

Among the various nearshore habitats and related ecosystems of critical importance are mangroves, seagrass beds, and coral reefs, which are the focus of this chapter. Mangroves are unique salt-tolerant plants located in the intertidal zones throughout much of the world's tropical and subtropical coastal environs. Their spread ranges from sparse patches of stunted shrubs, often the last vestiges of what would once have been a luscious outgrowth cleared for development, to certain blocs where mangrove growth is intense. Here also, development threatens to overrun them. These ecosystems are among the most productive on earth, and they are repositories of biodiversity and critical habitats for an incredibly diverse range of species. For instance, in the case of the Sundarbans on the Bay of Bengal, these include fish (from bhetki to the Ganges shark), invertebrates (tiger prawns and crab), reptiles (from king cobra to the terrapin and estuarine crocodiles), birds (from kingfishers to osprey), primates (such as rhesus macaques) and the iconic semi-aquatic Bengal tiger (Rahman and Asaduzzaman 2010: 35–47). Mangroves are vital to human well-being. They provide various ecological and socio-economic services to benefit millions of people living in coastal areas, worth close to USD 800 billion per year (Save Our Mangroves Now! Initiative n.d.). As important nursery and feeding habitats for many fish, crab, shrimp, and mollusc species, they are a significant source of food and income for coastal communities. They also supply fuelwood and construction materials; the wood of many species is prized because it is resistant to fungal and insect pathogens.

Some of these mangrove ecosystems also have immense cultural and spiritual significance (Mitra 2020: 337–351). They also stand at the forefront of climate change impacts and can act as speed-breaker, dissipating wave energy (Hashim, Catherine, and Takaijudin 2013: 4483–4488). Often, they provide the first defence against shoreline erosion, storm surges, and flooding, and they are five times more cost-effective than grey infrastructure (IUCN Mangroves and Coast. Ecosystem n.d.). From a climate change perspective, mangroves also play a critical mitigation role apart from the adaptation dimension. They have tremendous carbon-absorbing capabilities and are part of the panoply of coastal blue-carbon ecosystems (Fourqurean et al. 2014). Mangroves can store as much or even more carbon per unit area as terrestrial forests. However, if these ecosystems are degraded or destroyed, their carbon sink capacity is lost or adversely affected. This can lead to releasing the stored carbon back into the atmosphere, thereby aggravating climate change. Other environmental services of mangroves include their ability to filter impurities and pollutants, thereby naturally maintaining water quality.

Well-managed, healthy mangrove forests are thus critical to sustainable development. Despite their immense value, almost everywhere they exist, mangroves are under considerable threat from anthropogenic sources. This is primarily because, for most people, mangroves conjure images of a dark and squalid swamp, a dense tangle of roots that provide them with an odd appearance, as if they were standing on stilts above the water. Their sometimes-foul-smelling environment is often perceived to be a haven for mosquitoes, snakes, and unwelcome reptiles. Mangroves are therefore felled to give way for aquaculture, agriculture, and urban and infrastructure development. In addition, over-grazing, pollution, diversion of upstream water sources, offshore mining, and land reclamation are other causes for their destruction. Between the 1980s and 1990s, approximately 35% of global mangrove cover was lost (Friess 2019: 95). While the loss rate has slowed in the past two decades — to an estimated 4% between 1996 and 2016 — many regions continue to remain hotspots for mangrove loss.

Seagrasses are monocotyledonous plants, meaning that, even though they have long strap-like leaves, they are not true grasses (Hogarth 2007: 1). They thrive in shallow brackish and marine waters, which may be intertidal or subtidal, down to about 50 metres, in diverse coastal environments ranging from the tropics to the Arctic Circle. Their evolution occurred over 100 million years ago, and there are approximately 72 different species of seagrass classified into four major groups (Reynolds 2018). In several ways, seagrasses are like terrestrial plants. They photosynthesize to produce their food, release oxygen, and provide shelter and food to an incredibly diverse community of animals, from tiny invertebrates to large fish, crabs, turtles, marine mammals, and birds. Seagrasses also provide many important services to humankind. They can dissipate wave energy and help protect vulnerable shorelines. They improve water quality and can store carbon to limit climate change. Globally, even though these marine plants occupy only 0.2% of the seabed, as a carbon sink, their sequestration capabilities are tremendous (UNEP n.d.). An acre of seagrass can store about three times as much carbon as an acre of rainforest, translating to roughly eight metric tonnes of carbon per year (The Blue Carbon Initiative n.d.). In fact, seagrasses hold more than 10% of the carbon buried in the ocean (UNEP n.d.). Despite being one of the most productive ecosystems globally, seagrasses have often received little attention, and many seagrass meadows have been lost because of human activities. Pollution, destructive fishing practices, and direct physical disturbance like dredging and boat strikes threaten seagrass beds. Marine heatwaves can also negatively impact seagrass meadows (United Nations 2020: 495). The Second World Ocean Assessment reports that since 1980, seagrasses have been disappearing at a rate of 110 km^2 per year (United Nations 2020: 355).

Coral reefs are often referred to as 'rainforests of the sea' (see Chapter 15). Even though they occupy less than 1% of the ocean floor, they are home to 25% of all marine species, making them one of the most diverse environments on the planet. They are a vital support system that sustains several fish species, a source of protein and livelihood for millions. Globally, more than 500 million people depend on corals for food, storm protection, jobs, and recreation, and the resources and services that they provide are worth an estimated 375 billion dollars each year (IUCN Coral Reefs n.d.). Recreational activities, tourism, coastal protection, and the maintenance of marine ecosystems are some of the other economic benefits provided by reefs. Furthermore, complex chemicals found in certain coral reefs, particularly deep-sea coral habitats, possess remarkable pharma potential.

Corals are colonial, two-layered invertebrates made of polyps (Allen and Steene 1994). These polyps absorb calcium carbonate from the sea and secrete it as a calcium carbonate cup, which provides anchorage for the polyps, and when threatened, the polyp retreats into the cup's safety. It is these calcium carbonate cups of the many billions of polyps that fuse to form coral reefs (Miththapala 2008: 1). This process takes considerable time; some corals grow at only about 3–20 mm per year. Accordingly, it has taken millions of years for some of the outstanding coral reef ecosystems that we see today to form. As corals grow and die, they leave behind their calcium carbonate skeletons, which facilitate the growth of new corals. With time, coral walls are created, which act as a barrier against waves and currents. Another essential feature of corals is that unicellular algae called zooxanthellae live within their tissue. The zooxanthellae and corals share a symbiotic relationship. While the zooxanthellae obtain nutrients, carbon dioxide, and a safe haven from their coral hosts, in exchange, they provide the coral reef with the colour of the reef, food, and oxygen (IUCN What is a Coral Reef n.d.). Generally, coral reefs habit tropical shallow and warm waters (less than 100 metres in depth and usually between 25° and 29°C). Coral ecosystems can also be found in the cold, dark, and deep waters of the ocean. These cold corals have a wide distribution in many parts of the world's oceans, including off the Antarctica coast. They are also epicentres of marine biodiversity. For instance, about 1,300 species have been recorded as living in cold-water coral reefs in the north-eastern Atlantic (ICRI 2004). Unlike shallow-water corals, deep-sea corals do not require sunlight. They obtain the energy and nutrients for survival by trapping tiny organisms in passing currents (Smithsonian Ocean n.d.).

Globally, coral cover is also fast diminishing. Corals require particular water conditions for optimal growth and development. If the conditions are adverse, for instance if there is a rise in ocean temperature, the zooxanthellae will produce free radicals, which will trigger the coral polyp to release its symbiont. In this phenomenon, known as coral bleaching, the coral turns white and transparent. If this situation persists, the coral will eventually die (IUCN What is a Coral Reef n.d.). Extractive activities, pollution and sedimentation, diseases like crown-of-thorns starfish outbreaks, and physical coral reef degradation have also taken a severe toll on coral health (United Nations 2020: 307). Nevertheless, the gravest threat that this ecosystem faces is from rising temperatures due to the excessive presence of greenhouse gases in the atmosphere. If countries fail to limit the global average temperature to well below 2°C above pre-industrial levels in line with the Paris Agreement, future generations may never be able to experience the wonders that corals afford.

Conserving mangroves and nearshore habitats: governance and management dimensions

From an insipid and banal understanding that equated governance with command and control, a top-down hierarchical model where the State, to the exclusion of other stakeholders, established standards and adopted decisions often backed up by sanctions, the idea of governance has evolved into a more open-textured concept. Presently, governance is no longer the prerogative of the State. Ever since the idea of good governance permeated the international development discourse, it has come to imply a process of opportunity creation and avoidance of conflicts where a host of actors – the State, the market, and civil society, have significant stakes and roles to play. Governance implies organizational effectiveness and efficiency concerning policy formulation and implementation to secure overall development and growth. The idea of governance is now increasingly linked to sustainability. So critical is this inter-relationship that sustainable development is essentially all about good governance, which in its turn is essentially based on a principled approach (Meadowcroft, Langhelle and Ruud 2012: 1–13). The measure of success or the ability to secure good governance is now assessed against adherence to principles like accountability, transparency, precaution, participation, openness, adaptiveness, inclusiveness, collaboration, and the rule of law. This implies that governance must be 'interactive', holistic, and integrated. Creative thinking, transdisciplinarity and its application, a change from problem-solving to opportunity-creation, and successful dispute resolution are central to this idea of interactive governance.

Interactive governance involves resolving and addressing problems while creating opportunities through interactions between civil, public, and private actors (Kooiman et al. 2008: 1–11). Jan Kooiman (2003: 133–170) has classified interactive governance in terms of three 'orders'. First-order governing is all about the nitty-gritty of governance. Here, public and private entities collaborate to solve problems and create new opportunities. All actors — State, market, and civil society — are jointly and interactively responsible for problem-solving and opportunity creation (Kooiman 2003). Public responsibilities are discharged considering private needs and capabilities, and similarly, private tasks are accomplished to further and supplement public ends. The focus of second-order governing is on the institutional arrangements that enable the implementation of first-order governing. Institutions can be defined as the laws, agreements, norms, customs, procedures, and organizations used by 'first-order governors' to take and implement decisions. They are the framework within which first-order governance occurs and often serve as the meeting point for those who rule and those who are ruled. The third-order or meta-governance is the pivot of the governing process. It is the adhesive that binds the entire governing process. The meta-order of governance is based on principles, norms, and values. It provides methods and the avenue to think beyond institutional frameworks, enabling the governance process to be dynamic, inventive, and creative. In this context, it must also be noted that while governance generally considers the larger picture, the long-term objectives, and goals, in relation to natural resources management, it is through the medium of management that governance is put into practice. Management addresses day-to-day problems and is the lynchpin that links daily implementation to overall governance parameters, more so, to the three orders.

Given the diversity, complexity, and scale issues that emerge in the context of nearshore habitats and related ecosystems and the challenges posed by climate change and sea-level rise (SLR), this idea of interactive governance, its three orders and ICZM assume tremendous significance. ICZM is an iterative management process highly relevant to the coastal zones. It helps secure interactive coastal governance by feeding into and interlinking the three orders integral to interactive governance. These aspects are elaborated on below.

For a long time, natural resource governance systems, including those responsible for the coasts and the oceans, were based on the assumption that resources were plentiful and offered a bountiful supply that could support unfettered use and universal exploitation. This led to a single-resource or single-sector approach to management. As a result, distinct activities in the coastal landscape were assessed independently of one another, and sectoral regulatory regimes emerged along these lines. However, as the number of people who rely on coastal and marine resources grew, it became evident that the silo-based sectoral approach was impractical and was producing substantially less than optimal results. Single-sector management strategies ignore the unique peculiarities of the coastal environment and the reality that actions by one sector may have adverse effects on others. For example, suppose the government declares a particular stretch of the coastal zone as a marine reserve but fails to coordinate with other stakeholders. In that case, it is possible that, in due course, the marine reserve can be threatened. For instance, a private company may decide to set up a port in the vicinity of the marine reserve, or the government may decide to set up a coastal highway that may have to pass through a mangrove forest (The staff of the Coral Reef Alliance (CORAL) 2003: 58). Such developmental plans may impair the integrity of the marine reserve or the mangrove forest. They may even pit the developers and those who rely on these resources for their sustenance against each other. Therefore, a completely different planning and management method is required to avoid such workings at cross-purposes. This can lead to integration and ICZM to harmonize targets, objectives, and all management-related efforts to provide a holistic response to coastlines' ills (Cicin-Sain and Knecht 1998: 10).

ICZM has been in vogue since the 1970s. Almost every coastal country has some kind of ICZM programme in place. It is a continuous and dynamic management and planning process that seeks to prevent coastal degradation by holistically addressing threats to the coastal environment. ICZM secures this objective by overcoming fragmentation inherent in the sectoral management approach. It brings together the various regulatory agencies that regulate coastal development and other players, including the private who operate on various fronts, based on an agreed-upon set of governance parameters, thereby reducing conflicts and maximizing the advantages obtainable from this nuanced approach to coastal management (Cicin-Sain and Knecht 1998: 39). Without a well-thought-out ICZM plan, implementing coastal governance can be problematic. ICZM aims to thwart working at cross purposes while aiming to secure the functional integrity of the coastal resource systems and change coastal governance from a top-down hierarchical model to a bottom-up, participatory, fair, inclusive, transparent, and accountable one. In summary, it is a practical tool for achieving long-term sustainable coastal sustainability (SCD). Presently, ICZM aspires to go above and beyond. It now has wide recognition as a valuable adaptive methodology for addressing some of the problems raised by SLR and other climate change impacts on coastal zones. ICZM's objectives are increasingly geared to reduce coastal areas' vulnerability to natural disasters, climate change, and SLR while preserving vital ecological processes, life support systems, and biological diversity (Cicin-Sain et al. 2000: 291).

ICZM is closely linked to the idea of interactive governance because it furthers the three orders that are central to it (Kooiman and Bavinck 2005: 11–24). As noted above, the first order deals with the day-to-day dimensions of governance. An ICZM programme also deals with the day-to-day aspects of coastal governance, where decisions are taken and implemented for the sustainable use, development, and protection of coastal and marine areas and resources (Puthucherril et al. 2020: 4). Its operational remit can be extensive, encompassing upland watersheds, shorelines, nearshore coastal and estuarine habitats and their waters. It may even include the ocean space affected by the coastal landmass. ICZM can also operate on shorter or longer time scales, or it can begin on a shorter time scale and, in due course, metamorphose into a project with a longer time scale.

As far as the second order is concerned, the creation of appropriate institutions, the role of law and legal frameworks are critical. Even though ICZM's core purpose is to ensure holistic management through integration, it does not eliminate sectoral management. Instead, it attempts to strengthen sectoral management regimes and ensure that they operate within a broader framework that treats the coast as a system where discrete sectoral management regimes and decision-making machinery are inter-linked to produce more consistent and efficient decisions. Accordingly, in creating institutions that comply with the mandate of the second-order, integration emerges crucial and is a condition precedent to ICZM implementation. This can take various forms, including creating a coordinating mechanism, such as an inter-ministerial council or commission comprised of public and private sector representatives concentrating on policy formulation and implementation, strategy development, planning, architecture, research oversight, and collaboration. Similarly, due to their regulatory, coercive, and predictable nature, laws and legal framework also play crucial roles in promoting the second order. A coastal law helps achieve sustainable coastal governance and associated management by balancing the competing human demands for the use of the coast and its resources by providing a framework to resolve conflicts between coastal users and coastal management organizations through collaborative and coordinated efforts. The law establishes principles for controlling coastal resource management to ensure that all stakeholders act rationally, free riders are eliminated, and resources are used sustainably for the present and in the future.

The ICZM process also secures the meta-order dimension. The meta-order involves the application of principles that provide the governance process with consistency and cogency. Typically, an ICZM programme provides a blueprint to manage coastal areas and their resources. Since there is no fixed template to ICZM, these plans are highly flexible and can be developed at various scales and can take several forms. Nevertheless, since ICZM is a primary pathway to sustainable coastal development, ICZM plans are necessarily based on sustainability principles, both substantive and procedural. The main substantive principles relevant to an ICZM process include the polluter pays, precaution, inter and intra-generational equity, increasing human capabilities and adaptive capacities. The main procedural principles are environmental impact assessments, creation of marine protected areas, ecosystem-based management, adaptive management, transparency and accountability, subsidiarity, and decentralization (Puthucherril 2014: 191). Figure 13.1 below depicts how ICZM, with its substantive and procedural principles, feeds into the three orders central to interactive governance. By doing so, ICZM promotes good coastal governance leading to sustainable coastal development.

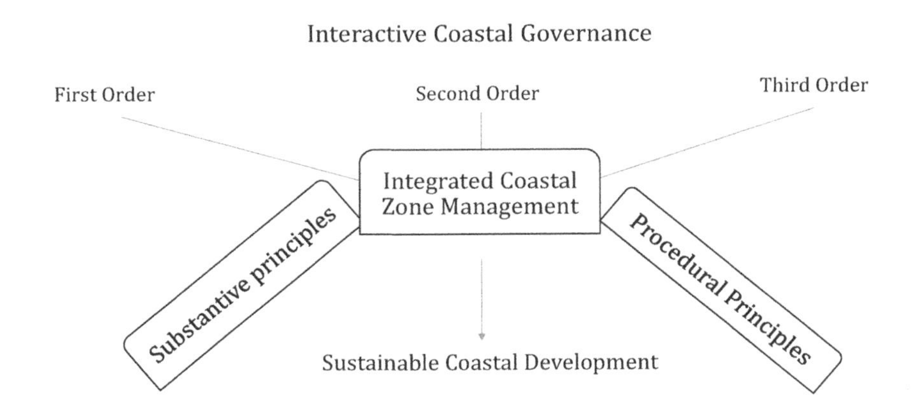

Figure 13.1 Interactive coastal governance and ICZM.

How has ICZM been employed to further interactive coastal governance in the backdrop of protecting nearshore habitats and related ecosystems? To answer this question, it is useful to highlight the experiences of certain coastal countries that have sought to protect nearshore habitats and related ecosystems in light of the challenges posed by climate change and SLR by effectuating interactive coastal governance through ICZM. For example, Belize possesses the second-longest uninterrupted reef system globally, namely the Barrier Reef Reserve, a world heritage site. Belize is also home to a broad range of habitats and marine attractions, including three main atolls, multiple coastal lagoons, mangrove forests, and more than 300 cays. The coastal zone is where much of the country's population (nearly 37%) lives and works (Azueta et al. 2020). Thirty percent of the country's gross domestic product is directly linked to the commercial activities in the coastal zone, which also supports more than 40% of its fisheries (Belize Coastal Zone Management Authority and Institute 2016). The potential of coastal-related tourism is also significant.

Given the possibility of conflicting interests between various sectors that depend on the coastal zone and its resources, Belize considers ICZM highly essential. To improve the governance of coastal and marine ecosystems and ensure their integrity and ability to provide services well into the future in light of the rising demand for coastal land, rapid development, overfishing, and population increase, Belize enacted its Coastal Zone Management Act (Government of Belize 1998). This statute reflects Belize's dedication to good coastal governance and management. It establishes the Belize Coastal Zone Management Authority (Belize CZMA), which is responsible for preparing and reviewing the coastal zone management plan. The law also establishes the Coastal Zone Management Institute, which assists the University of Belize in creating coastal zone management instructional programmes. The Act's core document is the 2016 Integrated Coastal Zone Management Plan, which was created in stages, starting with a strategy that outlined overarching national goals. The Plan envisions a 'sustainable future' in which healthy ecosystems support 'thriving local communities and a vibrant economy'. Adaptation to coastal climate change is an important part of the Plan. Climate change adaptation strategies are to be included in strategic management plans by all sectors, and inter-agency cooperation, public awareness campaigns, and the development of the Belize CZMA and the Institute's capabilities are also encouraged. The Plan also elaborates on mangrove protection and coastal habitat and species conservation.

The small island developing State of Barbados, with a coastline of 97 kilometres, is a leader in ICZM related best practices across the Caribbean. The coastline is this country's natural heritage and public asset. Barbados has more than 35 years of experience in implementing ICZM. This has enabled it to develop a sustainable coastal-based tourism industry, the primary lifeline of the economy (direct contribution of 13% to the total GDP in 2017 and indirect contribution of 40%). Nevertheless, the ever-expanding population (it is the 16th most densely populated country in the world, with a population density of 663 persons per square kilometre), economic needs, the impacts of climate change and SLR are placing more pressure on the coastal zones and nearshore habitats. There was a growing realization that Barbados needed a more modern coastal management delivery model to fully utilize the range of developmental opportunities that the coast provided. Accordingly, in 2020, Barbados revised the ICZM Plan. This revised plan integrates disaster management and climate change adaptation within the ICZM rubric to support effective coastal planning for better implementation of a climate and disaster risk resilient related approach. The emphasis here is on adaptation strategies to provide optimum response to factors that climate change exacerbates. This new delivery model seeks to introduce improved coastal risk understanding and procedures into infrastructure and non-structural adaptation measures designed to increase resilience to climate-induced coastal hazards.

ICZM implementation in Barbados furthers the three orders outlined above. Its ICZM Plan is a comprehensive document detailing all aspects relating to the practical operation of coastal zone management. The second key aspect of the ICZM process is the design of institutional processes that seek integration to overcome the fragmentation inherent in the sectoral management approach at the land-water interface. In this regard, the Coastal Zone Management Unit has the lead in developing and implementing ICZM. Public participation has also been identified as an essential element. Notably, the whole process has sufficient legal backing. The Plan also fulfils the meta-dimension and is based on nine principles: (1) sustainable development; (2) ecosystem-based management; (3) the use of sound science and best practice; (4) public and private participation; (5) Islands Systems management; (6) access and benefit-sharing; (7) good governance; (8) environmental Liability; and (9) gender equality and social inclusion. These nine principles reflective of international best practices in ICZM are supposed to support Barbados's Policy Framework delivery.

This idea of utilizing interactive governance and ICZM to afford protection to coastal resources and nearshore habitats, and to facilitate climate change adaptation, is fast gaining currency in several other coastal countries. Sri Lanka was one of the pioneers in implementing ICZM in Asia with its Coast Conservation Act of 1981. Based on this legislative mandate, the country also adopted its first Coastal Zone Management Plan (CZMP) in 1990 (first generation plan) (Sri Lanka Coastal Conservation Department 1990). With the conclusion of the civil war that ravaged the country for decades, Sri Lanka turned its attention to sustainably developing its coastal resources through ICZM. Accordingly, it amended the Coast Conservation Act in 2011. Alongside, it also revised and updated the CZMP in 1997 and 2004. In its more recent version of 2018, the Coastal Zone and Coastal Resource Management Plan seeks to ensure 'sustainable use of the coastal environment and its resources in the long term, consistent with the national development goals' (Ministry of Mahaweli Development and Environment (Sri Lanka) 2018: 2A). Given that Sri Lanka's coastline of 1,700 km includes diverse coastal habitats like mangroves, seagrass beds, coral reefs, estuaries and lagoons, the Plan focuses on the management and conservation of coastal habitats (Sri Lanka Coastal Conservation Department 1990: 33–50). Several strategies and actions have been identified to protect them, and institutionally, the implementation falls to the Coast Conservation and Coastal Resource Management Department.

As a country with a coastline of about 7000 km, India has also placed its faith in ICZM with a $ 400 million World Bank programme to strengthen ICZM implementation. Given the large population that lives in its coastal areas, the need to develop its blue economy, and the precarious nature of its coastal resources, India sought to regulate coastal development as early as 1991 through the Coastal Regulation Zone Notification. However, developmental considerations were often preferred over the integrity of the coastal environment, leading to rancorous legal battles that had to be settled by India's Supreme Court. This led the law to be amended several times, diluting its regulatory rigour. In its most recent iteration, the Coastal Regulation Zone Notification of 2019, despite its pro-development slant, has provisions to conserve ecologically sensitive areas such as mangroves, coral reefs, dunes, salt marshes, seagrass beds, and nesting grounds. The law classifies these ecosystems as coastal regulation zone (CRZ) CRZ-IA, and subjects them to a heightened degree of protection which requires developing a detailed environmental management plan. Subject to certain exceptions, there is a prohibition on the carrying out of a number of environmentally destructive activities in such areas (Ministry of Environment, Forest, and Climate Change (India) 2019: 31). In addition, certain critical coastal areas, such as the Sundarbans, the Gulf of Mannar, and the Bhitarkanika, have been classified as Critically Vulnerable Coastal Areas. Coastal communities, including fisherfolk whose livelihood depends upon the health of these sensitive ecosystems, have to be involved in their management (Ministry of Environment, Forest, and Climate Change (India) 2019: 32).

The current ICZM World Bank project in India essentially seeks to build upon a concluded bank-sponsored pilot project on ICZM implemented in Gujarat, Odisha, and West Bengal (Biliana et al. 2020). Based on the experience gained, the present project seeks to enhance coastal resources, protect coastal populations from pollution, erosion, and sea level rise, and improve livelihood opportunities (World Bank 2021). It will also assist states in developing state-level Integrated Coastal Zone Management Plans; provide long-term support to the Government of India in meeting national coastal and marine spatial planning needs; and develop and protect more intangible 'blue' resources, facilitate carbon sequestration, and develop coastal resilience.

Conclusion

Coastal habitats are among the world's most productive ecosystems, providing vital ecological services. As a source of food, supporting livelihoods, and contributing to development and poverty alleviation, the importance of nearshore habitats and coastal ecosystems can hardly be over-emphasized. However, the daunting challenge is that most of these ecosystems occur on or in parts of nearshore marine habitats most heavily influenced by human activities. These ecosystems are vanishing at a rapid rate. Marine pollution, urbanization, and the growing plurality of actors, competition with other sectors who wish to monopolize these habitats' resources, and the political, economic, and social marginalization of many communities and indigenous peoples who depend on these resources are significant issues. Also, many reports and studies categorically admit that several of the world's coastal environments are already and will continue to be detrimentally impacted by SLR and climate change. Since it is practically impossible to replace the ecosystem services that these coastal environments provide, we cannot afford to envisage a world without them. There is overwhelming consensus among academics and policymakers that with appropriate governance and management, much of this loss can be prevented and, in some cases, even reversed (Ricketts 2000: 291–359; Clark 1992: Ch. 2.2). Such a corrective course is extremely crucial given that one of the best antidotes to minimize or, in some cases, even to avert the impending climate change crises and many of the effects of SLR is by nurturing these nearshore habitats and related ecosystems.

Traditional ideas of governance regarding nearshore habitat resources, which prioritized coastal resources use through a sector-based approach, and enhanced user complementarities, have become intractable, complex, time- and cost-intensive. In this scenario, the idea of SCD, the interactive governance approach and ICZM emerge as highly relevant pathways to ensure that coastal development continues, resilience is strengthened, and nearshore habitats are protected. As noted, interactive governance provides a platform where government and nongovernmental actors can interact to resolve problems and create new opportunities more cooperatively. In its application to the coastal zones, interactive coastal governance establishes the framework within which management (i.e., ICZM) can take place. ICZM has emerged as the blueprint that lays out a framework for humans to interact with the coastal environment in environmentally sustainable ways. More importantly, there is a growing consensus that ICZM can support and include within its holistic and integrated framework climate change adaptation, which is highly critical for coastal countries (Cicin-Sain et al. 2020).

In sum, the idea of interactive governance and ICZM has finally come of age in this era of the Anthropocene, and it enjoys widespread popularity among experts and policy-makers. Coastal countries across the spectrum — economically developed, fast-emerging economies, least-developed, small-island developing states -- are all placing their faith in ideas of interactive coastal governance and ICZM. This is because they offer the best prospect to secure sustainable coastal development, reduce the vulnerability of coastal areas and their inhabitants to natural hazards and other climate change impacts, and help sustain essential ecological processes to make them available to future generations over longer time scales.

Acknowledgements

I dedicate this paper to the fond memory of the late Biliana Cicin-Sain, from whom I learned the fundamentals, and much more, of coastal and ocean management. Research assistance provided by Pushpit Singh, Symbiosis Law School, Hyderabad, is also acknowledged.

References

Azueta, James et al. (2020) 'State of the Belize Coastal Zone Report 2014–2018: Final Report', https://www.coastalzonebelize.org/wp-content/uploads/2021/02/The-State-of-the-Coast-Report-2014-2018-final-1-1.pdf

Belize Coastal Zone Management Authority and Institute. (2016) 'Belize Integrated Coastal Zone Management Plan', https://ambergriscaye.com/art4/BELIZE-Integrated-Coastal-Zone-Management-Plan.pdf

Biliana, Cicin-Sain et al. (2000) 'Education and Training in Integrated Coastal Management: Lessons from the International Arena', *Ocean and Coastal Management*, vol 43, p 291.

Celliers, Louis and Ntombela, Cebile. (2016) 'Urbanisation, Coastal Development and Vulnerability, and Catchments', in *Regional State of the Coast Report*. Nairobi, UNEP and WIOMSA.

Cicin-Sain, Biliana and W. Knecht, Robert. (1998) *Integrated Coastal and Ocean Management: Concepts and Practices*. Island Press, Washington, DC.

Cicin-Sain, Biliana et al. (2020) 'Volume 2 Review of International Experiences in Integrated Coastal Zone Management (ICZM) and Relevance to India: Emphasis on Federal/State/Local Interactions', https://www.ncscm.res.in/conference/images/ICZM%20REPORT%20Vol%202%20INTERNATIONAL%20EXPERIENCES.pdf

Fourqurean, James et al. (2014) 'Coastal Blue Carbon: Methods for Assessing Carbon Stocks and Emissions Factors in Mangroves, Tidal Salt Marshes, and Seagrass Meadows', https://www.iucn.org/sites/dev/files/english_blue_carbon_lr.pdf

Friess, Daniel A. et al. (2019) 'The State of the World's Mangrove Forests: Past, Present, and Future', *Annual Review of Environment and Resources*, vol 44, no 1, pp 89–115.

Allen, Gerald Robert, and Steene, Roger, Steens C. (1994) *Indo-Pacific Coral Reef Field Guide*. Tropical Reef Research, Singapore.

Government of Belize. (1998) 'Coastal Zone Management Act, 1998 (Act No. 5 of 1998)', http://extwprlegs1.fao.org/docs/pdf/blz13962.pdf

Hashim, Ahmad Mustafa, Catherine, Sim Mong Pheng and Takaijudin, Husna. (2013) 'Effectiveness of Mangrove Forests in Surface Wave Attenuation: A Review', *Research Journal of Applied Sciences, Engineering and Technology*, vol 5, no 18, pp 4483–4488.

Hogarth, Peter. (2007) *The Biology of Mangroves and Seagrasses*. Oxford University Press, New York.

Hudson, Blake. (2012) 'Coastal Land Loss and the Mitigation-Adaptation Dilemma: Between Scylla and Charybdis', *Louisiana Law Review*, vol 73, pp 31–68.

ICRI. (2004) 'Cold Water and Coral Reefs', https://www.icriforum.org/key-topics/cold-water-coral-reefs/

International Human Dimensions Programme on Global Environmental Change. (2015) *Coastal Zones and Urbanization*. IHDP, Bonn.

IPCC. (2019) 'Sea Level Rise and Implications for Low-Lying Islands, Coasts and Communities', in H.-O. Pörtner et al. (eds) *IPCC Special Report on the Ocean and Cryosphere in a Changing Climate*. The Press.

IUCN. (n.d.) 'Coral Reefs', https://www.iucn.org/theme/marine-and-polar/get-involved/coral-reefs

IUCN. (n.d.) 'Mangroves and Coastal Ecosystems', https://www.iucn.org/theme/marine-and-polar/our-work/climate-change-and-ocean/mangroves-and-coastal-ecosystems

IUCN. (n.d.) 'What is a Coral Reef', https://www.iucn.org/downloads/what_is_a_coral_reef__1.pdf

Kooiman, Jan. (2003) *Governing as Governance*. Sage, London.

Kooiman, Jan and Bavinck, Maarten. (2005) 'The Governance Perspective', in Jan Kooiman et al. (eds) *Fish for Life Interactive Governance for Fisheries*. Amsterdam University Press, Amsterdam.

Kooiman, J. et al. (2008) 'Interactive Governance and Governability: An Introduction', *Journal of Transdisciplinary Environmental Studies*, vol 7, no 1, pp 1–11.

Meadowcroft, James, Langhelle, Oluf, and Ruud, Audun. (2012) 'Governance, democracy and sustainable development: moving beyond the impasse', in Meadowcroft, James, Langhelle, Oluf, and Ruud, Audun (eds) *Governance, Democracy and Sustainable Development: Moving Beyond the Impasse*. Edward Elgar Publishing, Cheltenham.

Ministry of Environment, Forest, and Climate Change (India). (2019) 'Notification G.S.R. 37(E)', https://www.ncscm.res.in/pdf_docs/crz-2019.pdf

Ministry of Mahaweli Development and Environment (Sri Lanka). (2018) 'Sri Lanka Coastal Zone and Coastal Resource Management Plan – 2018', http://www.coastal.gov.lk/images/pdf/acts/czcrmp_2018_gazette_2072_58_e.pdf

Mitra, Abhijit. (2020) *Mangrove Forests in India: Exploring Ecosystems Services*. Springer Nature, Switzerland.

Puthucherril, Tony George. (2014) *Towards Sustainable Coastal Development: Institutionalizing Integrated Coastal Zone Management and Coastal Climate Change Adaptation in South Asia*. Martinus Nijhoff, Leiden.

Puthucherril, Tony George, et al. (2020) 'Volume 1 towards the Development of the Next Phase of Integrated Coastal Zone Management in India: Application of Lessons Learned from an International Review of Implementation of Integrated Coastal Zone Management (ICZM)', https://www.ncscm.res.in/conference/docs/VOL-1-INDIA-RELEVANCE%20OF%20INT%20PRACTICES%20AND%20POSSIBLE%20DIRECTIONS.pdf

R. Clark, John. (1992) 'Integrated management of coastal zones: FAO Fisheries Technical Paper 327', http://www.fao.org/3/t0708e/t0708e00.htm

Rahman, M. R. and Asaduzzaman, M. (2010) 'Ecology of Sunderban, Bangladesh', *Journal Science Foundation*, vol 8, no 1 and 2, pp 35–47.

Reynolds, Pamela L. (2018) 'Seagrass and Seagrass Beds', https://ocean.si.edu/ocean-life/plants-algae/seagrass-and-seagrass-beds

Ricketts, P. J. (2000) 'Integrated Coastal Zone Management in Atlantic Canada: Looking toward the Third Millennium', *Ocean Yearbook Online*, vol 14, no 1, pp 291–359.

Save Our Mangroves Now! Initiative. (n.d.) 'Why Is There a Need for Protection', http://www.mangrovealliance.org/wp-content/uploads/2019/12/SOMN_flyer_2019_ENG.pdf

Shine, Claire and De Klemm, Cyrille. (1999) 'Wetlands, Water and the Law. Using Law to Advance Wetland Conservation and Wise Use', https://portals.iucn.org/library/efiles/documents/eplp-038.pdf

Smithsonian Ocean. (n.d.) 'Deep-Sea Corals', https://ocean.si.edu/ecosystems/coral-reefs/deep-sea-corals

Sri Lanka Coastal Conservation Department. (1990) 'Coastal Zone Management Plan', https://www.crc.uri.edu/download/SLCZMPlan90.pdf

Sriyanie Miththapala. (2008) 'Coral Reefs: Coastal Ecosystems Series (Volume 1)', https://portals.iucn.org/library/sites/library/files/documents/CES-001.pdf

The Blue Carbon Initiative. (n.d.) 'About Blue Carbon', https://www.thebluecarboninitiative.org/about-blue-carbon

The National Academies of Sciences, Engineering, and Medicine. (2010) 'What We Know About Climate Change and Its Interactions with People and Ecosystems', in *Advancing the Science of Climate Change*. The National Academies Press, Washington DC.
The National Academies of Sciences, Engineering, and Medicine. (2019) 'Terrestrial Carbon Removal and Sequestration', in *Negative Emissions Technologies and Reliable Sequestration: A Research Agenda*. The National Academies Press, Washington, DC.
The staff of the Coral Reef Alliance (CORAL). (2003) 'Introduction to Coral Reef Ecosystems, Threats, and Solutions', https://www.sprep.org/att/IRC/eCOPIES/Global/264.pdf
UNEP. (n.d.) 'Seagrass—Secret Weapon in the Fight against Global Heating', https://www.unep.org/news-and-stories/story/seagrass-secret-weapon-fight-against-global-heating
United Nations. (2020) 'The Second World Ocean Assessment World Ocean Assessment II Volume I', https://mail.google.com/mail/u/1/?ogbl#inbox?projector=1
Wolff, Claudia et al. (2020) 'Future Urban Development Exacerbates Coastal Exposure in the Mediterranean', *Science Report*, vol 10, no 14420, pp 1–11.
World Bank. (2021) 'Integrated Coastal Zone Management', https://projects.worldbank.org/en/projects-operations/project-detail/P097985
World Bank. (n.d.) 'Oceans, Fisheries and Coastal Economies', https://www.worldbank.org/en/topic/oceans-fisheries-and-coastal-economies

14
ISLANDS
Rising seas, vulnerable shorelines, and territorial integrity

Anemoon Soete

Islands deal with a number of local, regional, and global environmental issues. Locally, for example, overpopulation can lead to waste disposal issues and a shortage of resources, in particular arable land and potable water, as has happened in Pacific Island Countries and Territories such as Kiribati and Tuvalu (Secretariat of the Pacific Regional Environment Programme, 2012). Another issue is the need to garner more income for the islands, which can lead to timber being harvested by offshore companies, as has happened in Papua New Guinea and the Solomon Islands (UNODC, 2016), resulting in a loss of habitat for fauna and flora. Runoff from deforested land makes for polluted rivers and muddied coastal waters, which can negatively impact the growth of coral, as has happened in the waters of Great Keppel Island, Australia (Fredston-Hermann et al., 2016). This adds to a number of historical issues with which many islands are already dealing with, such as barren land in Nauru due to phosphate mining by colonizing powers, and nuclear testing, which has left its radioactive footprint on Pacific islands such as those of French Polynesia and the Marshall Islands.

Another key issue plaguing islands is overfishing. In a 2020 report on the state of the world's fisheries and aquaculture, it was concluded that, worldwide, 'the percentage of stocks fished at biologically unsustainable levels increased [...] from 10% in 1974 to 34.2% in 2017' (Food and Agriculture Organization of the United Nations, 2020: 47). For the population of islands in particular, fish represents economic revenue as well as a source of food. The importance of fisheries for small island States was specifically recognized under the United Nations Sustainable Development Goals in Target 14.7, which is focused on increasing the economic benefits from the sustainable use of marine resources for small island developing States by 2030 (United Nations, 2015). Overfishing has many causes. One of the most important is the still rather unregulated fishing policy on the high seas (see Chapter 21). This remains the case despite efforts by regional fisheries management organisations, such as the Western and Central Pacific Fisheries Commission (WCPFC), to manage shared stocks of highly migratory species such as tuna. One of the greatest challenges lies in the increased demand for seafood worldwide and the difficulty of combatting illegal, unreported, and unregulated fishing (IUU) (High Seas Task Force, 2006).

The issue stretches beyond the high seas to the exclusive economic zones (EEZs) of island states, such as Nauru, which heavily rely on the revenues from selling fishing licences to foreign operators, something that puts further pressure on fish stocks (Fenner, 2018).

 DOI: 10.4324/9781315149745-18

One initiative which tries to tackle the overfishing of tuna is the 'Palau arrangement', a sub-regional agreement between Pacific small island States which provides the legal backdrop for parties to adopt management measures pertaining to the regulation of effort, capacity, and seasonal and area closures for their tuna fishery, among other things. As population growth continues to intensify on islands, so does the need for food. As fish is an important food source, increased local fishing puts further pressure on fish stocks which are already plagued by increased destruction of habitat (e.g., harvesting or clearing of mangrove forests, which act as nurseries for fish) and invasive species (e.g., invasion of the Indo-Pacific lionfishes in the Caribbean that is decimating small, reef-associated bony fishes) (Pippard et al., 2017). It is against this backdrop that global environmental changes are impacting islands, with the chief concern being climate change.

Climate change as a threat multiplier

Climate change causes a multitude of impacts on marine life and governance, and it exacerbates already existing issues for fish-dependent populations found in many island States. The Intergovernmental Panel on Climate Change (IPCC) has confirmed that the global ocean has warmed unabated since 1970, and marine heatwaves have doubled in frequency since 1982 and are still increasing in intensity (IPCC, 2019). In addition, the oceans have already taken up between 20–30% of total anthropogenic carbon dioxide emissions since the 1980s, resulting in ocean acidification. At the same time, the IPCC has demonstrated that there has been an oxygen loss of 0.5–3.2% over the oceans' upper 1,000 metres between 1970 and 2010, and that ocean currents are shifting (IPCC, 2019). The combination of these phenomena causes severe changes in marine and coastal ecosystems which lead to a decline of fish stocks and migratory shifts of fish away from warming waters. This in turn has a tremendous impact on fish populations and productivity of fisheries for those nations most dependent on fisheries (IPCC, 2019).

The effects on coastal ecosystems are furthermore intensified by an increase of extreme weather events that, for example, add to the harm caused by already bleached coral reefs and impact coastal mangroves and seagrass beds, which form vital habitats (Pippard et al., 2017; IPCC, 2019). On the island of Moorea of French Polynesia, seven coral bleaching events (1984, 1987, 1991, 1994, 2002, 2003, and 2007), two cyclones (1991 and 2010), and outbreaks of invasive species (notably the crown-of-thorns starfish) (1980–1982 and 2006–2010) took place during the last four decades. Given these continued hits suffered by the coral reefs of Moorea, it has been estimated that a decrease of coral cover on reef slopes, from 49% in 2005 to less than 1% in 2010, has taken place. This is one of the most extreme cases of reef destruction so far documented in the Indo-Pacific (Andréfouët and Adjeroud, 2018).

A final consequence of climate change impacting especially island nations is sea-level rise. The global mean sea level is rising, with acceleration in recent decades due to increasing rates of ice loss from the Greenland and Antarctic ice sheets, as well as continued glacier mass loss and ocean thermal expansion. Increases in tropical cyclone winds and rainfall, and increases in extreme waves, combined with relative sea-level rise, further exacerbate extreme sea-level events and coastal hazards. The impacts of sea-level rise on coastal ecosystems include habitat contraction, geographical shift of associated species, and loss of biodiversity and ecosystem functionality. Unfortunately, sea-level rise is projected to continue beyond 2100 in all scenarios of global temperature rise (Pippard et al., 2017). Therefore, absent protective measures, islands will be continuously and increasingly confronted with creeping oceans gnawing at their coasts, even leading to full submergence of low-lying islands.

Sea-level rise and the adage that 'the land dominates the sea'

Low-lying islands face an uncertain future as sea-level rise hangs as a sword of Damocles over their terrestrial territorial integrity. Under current international law, the loss of land will inevitably impact these States' sovereign grasp on their maritime zones and their ability to benefit from the economic resources that their waters and the seabed have to offer. This threat of partial or total loss of maritime 'territory' is essentially legal in its origin. It flows from the age-old, yet still dominant, adage that the land dominates the sea. In order to obtain and maintain sovereignty or sovereign powers over maritime zones, a State must command territorial sovereignty. Territorial sovereignty supersedes maritime sovereignty and both types of sovereignty are obtained – and lost – in a dissimilar fashion as is discussed hereafter.

The adage that the land dominates the sea reflects the idea that there exists a 'close dependence of the territorial sea upon the land domain' (*Fisheries Case*, 1951), dictating therefore that maritime territory has only a secondary reason to exist. The adage can be traced back to the need and desire of a territorial State to defend its land and coast. In order to fulfil this need for defence, States assumed control over the waters adjacent to their land territory. This led to defining the breadth of the territorial sea as being equal to a cannon-shot's range (i.e., the area of waters that could be controlled militarily) (*North Atlantic Coast Fisheries Case*, 1910). Though the military purpose of control of maritime zones remained intact, it was eventually joined by an economic objective of controlling maritime resources (see Chapter 2).

The land-first adage was already confirmed in case law in 1909 by the *Grisbådarna Award*. The case concerned the status of waters surrounding land territory ceded to Sweden by Norway in 1658. The question posed was whether the maritime waters surrounding the ceded land had been impacted by the shift in sovereignty over the land territory. The tribunal straightforwardly put it that such a decision needed to be based on 'fundamental principles of the law of nations, both ancient and modern, according to which a maritime territory is a dependency of a terrestrial territory, from which it is spawned' (*Grisbådarna case*, 1909). Given this finding, the tribunal concluded that 'the radius of the maritime territory forms an inseparable dependency of the terrestrial territory' (*Grisbådarna case*, 1909), and therefore that the status of the ceded land determined the status of the surrounding waters. The adage was later recognized and applied by the International Court of Justice (ICJ) in the *Fisheries Case* (1951), the *North Sea Continental Shelf Cases* (1969), the *Aegean Sea Continental Shelf Case* (1978), the *Case concerning the Continental Shelf (Tunisia/ Libyan Arab Jamahiriya)* (1982), the *Case concerning the Continental Shelf (Libyan Arab Jamahiriya/Malta)* (1985), the *Maritime Delimitation and Territorial Questions between Qatar and Bahrain* (2001; *Territorial and Maritime Dispute between Nicaragua and Honduras in the Caribbean Sea*, 2007) and the *Maritime Delimitation in the Black Sea* (2009; *Territorial and Maritime Dispute*, 2012; *Delimitation of the Maritime Boundary in the Bay of Bengal*, 2012).

It is not surprising that the land-first adage also underpins the 1982 Law of the Sea Convention (LOSC) in much the same way as it had already served as a building block for the 1958 Convention on the Territorial Sea and the Contiguous Zone (see Chapter 2). The adage in particular forms the cornerstone of the measurement of entitlements to maritime zones as these are measured from the point where sovereignty over land territory stops and the seas begin (i.e., at the coastline). It is the coastline or low-water line that represents the normal baseline from which all maritime zones are measured under the LOSC and thus where sovereign powers can be exercised (Article 3 LOSC). The land-first adage is therefore crucial to understanding the manner in which sovereignty over the seas is obtained. According to the LOSC, an entitlement to maritime zones relates to the status of the land territory; rocks (Article 121 LOSC) and low-tide elevations (Article 13 LOSC) cannot generate all types of maritime zones (1982).

The ways in which land territory can be acquired cannot be duplicated for maritime territory. Land can historically be obtained by a State in five different manners: accretion, cession, conquest, effective occupation, and prescription. Today only two real means remain to obtain land: accretion and cession, as conquest, usually through war, is no longer a legally accepted method of acquiring land. Though effective occupation is technically still possible, it requires *terra nullius* (i.e., unclaimed land), of which none exists. Prescription often cannot offer certainty of title over land. To obtain land through prescription, a State must govern the land over which the title is actually in the hands of another State for an uncertain yet significant amount of time. In order to have certainty over land acquired by prescription, one would *de facto* need a total lack of protest and full acquiescence by the titleholder or a formal renunciation of sovereignty over the land by the latter – a rather unlikely event. Acquiring land through accretion occurs through a natural, slow process of addition of soil to land on which title is already held. Just as accretion can increase the amount of land over which sovereignty is established, the process of erosion, which naturally and slowly washes away land, can take away this sovereignty. Finally, cession can procure land when one State transfers full sovereignty over its land to another State. Such a transfer can be made in the form of an agreement or treaty. Cession of land was in the past often a crucial part of peace treaties, yet it can also be found in treaties of cession or of delimitation. Likewise, a court or tribunal with the relevant capacity can authoritatively solidify a title to sovereignty in deciding a territorial dispute (Sharma, 1997).

It is important to realize the different mechanisms underpinning the acquisition of land territory or maritime territory so as not to confuse them. Firstly, when territorial disputes arise before a court or tribunal, the State with the better title, for example obtained through unchallenged effective occupation or proof of historical title, will obtain a favourable judgement. Such historic elements proving the better title of one party do not play a role for maritime delimitation (Weil, 1989). Maritime delimitation is based on contemporary facts and cuts into the entitlements of both parties, as both entitlements can be equal. This was captured by Judge Ndiaye in his separate opinion to the *Bangladesh/Myanmar* judgement where he quoted an excellent description of the difference between land and maritime delimitation: 'Whereas territorial delimitation has as its objective to determine and allocate each party's distinct and entire share according to its title, maritime delimitation is condemned to amputate parts of each party's legitimate title to maritime territory' (*Delimitation of the Maritime Boundary in the Bay of Bengal*, 2012).

Secondly, whereas title confirmed in treaty or judgement can set land boundaries in stone, these instruments cannot have the same effect on maritime boundaries. In order to come to this conclusion, one must consider the true meaning of Article 62 of the Vienna Convention on the Law of Treaties (VCLT) which deals with fundamental changes of circumstances. Paragraph 2 of this article reads that '[a] fundamental change of circumstances may not be invoked as a ground for terminating or withdrawing from a treaty [i]f the treaty establishes a boundary'. Case law appears to support the view that this exception of boundary treaties applies to land and maritime boundaries alike. The *obiter dicta* of the ICJ in the *Aegean Sea Continental Shelf Case* reads that '[w]hether it is a land frontier or a boundary line in the continental shelf that is in question, the process is essentially the same, and inevitably involves the same element of stability and permanence, and is subject to the rule excluding boundary agreements from fundamental change of circumstances' (*Aegean Sea Continental Shelf Case,* 1978). In the *Bay of Bengal Maritime Boundary Arbitration*, the Arbitral Tribunal noted that

> maritime delimitations, like land boundaries, must be stable and definitive to ensure a peaceful relationship between the States concerned in the long term. As the ICJ noted in its decision in the *Temple of Preah Vihear Case*, "[i]n general, when two countries establish a frontier between them, one of the primary objects is to achieve stability and finality". The same consideration applies to maritime boundaries.
>
> *(Bay of Bengal Maritime Boundary Arbitration, 2014)*

Importantly in light of sea-level rise, the Arbitral Tribunal added that 'neither the prospect of climate change nor its possible effects can jeopardize a large number of settled maritime boundaries throughout the world. This applies equally to maritime boundaries agreed between States and to those established through international adjudication' (*The Bay of Bengal Maritime Boundary Arbitration,* 2014).

The International Law Commission (ILC), which prepared the 1982 Draft Articles on the Law of Treaties between States and International Organizations or between International Organizations, discussed the potential impact of the VCLT on maritime boundaries. The ILC doubted the application of paragraph 2 of Article 62 VCLT by noting that '[l]ines of maritime delimitation (not to mention the delimitation of air space) may in fact have special features and it is possible that the stabilizing effect of Article 62 does not extend to certain lines of maritime delimitation, even if, to all intents and purposes, they constitute true boundaries.' (*The Bay of Bengal Maritime Boundary Arbitration,* 2014). This doubt is understandable when considering the *travaux préparatoires* to paragraph 2 of Article 62 VCLT. In the *travaux,* paragraph 2 was highlighted as 'an exception to the rule [of fundamental change of circumstances], because otherwise the rule, instead of being an instrument of peaceful change, might become a source of dangerous frictions.' (ILC, 1966). Paragraph 2 therefore aims to prevent conflict between States due to a contestation of borders originating from a change of circumstances. However, given the fact that the adage of the land dominating the sea and the LOSC determine that sovereignty over maritime zones originates from adjacent territory, the view that a lack of territory does not hinder the existence of sovereignty over maritime territory will be a source of dangerous friction between States rather than the application of the universally accepted principle of land dominating the sea.

The stability necessary for land boundaries would rather be counterproductive for maritime delimitation lines (Árnadóttir, 2016). As maritime boundaries depend on maritime entitlements, which are inherently unstable and moveable due to their dependency on the coastline of land territory, which has the potential to shift, it would appear absurd to freeze maritime delimitation lines decided upon in treaties or awards, which could be completely devoid of purpose in the event that the entitlements have completely retreated landward and away from the delimitation line. What is more, in the event one State's maritime zones remain capped by the delimitation line whereas the other State's entitlements having retreated from it entirely, the delimitation line would no longer fulfil its purpose of establishing an equitable result. Therefore, despite the presence of case law dictating stability of maritime boundaries, this stability cannot be maintained. Paragraph 2 of Article 62 of VCLT does not apply to maritime delimitation lines once this line no longer serves to delimit opposing or adjacent States' maritime entitlements as established under the LOSC. It appears therefore that when sea-level rise takes away land territory of a State by permanently flooding it and adding it to the seas, international law as it stands today will not prevent the relocation of the baseline from which maritime zones are measured. This in turn means that when the baseline of an island retreats inland, its maritime zones will shrink. For low-lying island States which are threatened to lose all land territory, the stakes are even higher as the loss of land might likewise procure a loss of statehood (for a detailed examination of this topic, see Soete, 2021).

Managing interconnected local, regional, and global issues of marine governance

Island States have turned to several measures to turn around environmental threats originating from outside their territory and which impact their maritime zones. It is important to understand which measures are taken to combat local, regional, as well as global threats to marine governance of islands, as these threats can influence one another.

For example, regionally, the Convention on the Conservation and Management of Highly Migratory Fish Stocks in 2000, which established the WCPFC, was a huge step towards improving conservation, management, and sustainable use of migratory tuna resources in the Western and Central Pacific Ocean. Yet issues remain. A persistent one lies in the difficulty of combatting IUU due to transfers of catches between vessels on the high seas. Such transfers enable fishing vessels to avoid having to visit ports to offload their catches. But when this happens, there is very little monitoring of the activity, allowing fish that have been caught in compliance with regulations to be mixed in with fish caught illegally. The WCPFC has prohibited transfers at sea by purse seine vessels. It also prohibits transfers for longline and non-purse seine vessels unless a WCPFC member determines that bringing catches to port is 'impracticable'. Definitive guidelines remain to be concluded on the meaning of what constitutes impracticable circumstances. Only interim guidelines are available at the moment, indicating that the circumstances are impracticable when they would cause 'significant economic hardship' or require a vessel to make 'significant and substantial changes to its historical mode of operation' (WCPFC, 2009: §§33–38). Unfortunately, several WCPFC members treat this exemption as the rule, thus invalidating this effort to combat IUU and overfishing in the region (Wold, 2019).

Bilaterally, States at times join forces in order to patrol their EEZs as well as the high seas just outside their EEZs. Australia and France have, for example, teamed up as a result of the Treaty on cooperation in the maritime areas adjacent to the French Southern and Antarctic Territories, Heard Island, and the McDonald Islands, and to regularly patrol the EEZs in the Indian Ocean in search of IUU fishing vessels. These joint patrols enable larger swathes of their EEZs to be patrolled all the while eliminating duplication of activities. Pooling their resources has increased the likelihood of apprehending IUU fishing vessels (Australian-French Treaty on cooperation in maritime areas, 2003). Monitoring vast EEZs remains a costly endeavour, however.

It is the inability to sufficiently monitor waters that remains a sticking point in the effective implementation of Marine Protected Areas (MPAs), a tool increasingly used by island nations to enable sustainable use of coastal and marine areas. An MPA does not have a universal definition, but the Convention for the protection of the marine environment of the Northeast Atlantic (OSPAR, 1992) states that an MPA is 'an area within the maritime area for which protective, conservation, restorative or precautionary measures, consistent with international law, have been instituted for the purpose of protecting and conserving species, habitats, ecosystems, or ecological processes of the marine environment' (OSPAR Commission, 2003: 1–2). The Maldives has 42 MPAs of which only one is monitored by rangers to protect the manta rays and whale sharks that visit during monsoon season. Pairing monitoring with a stringent approach to rulebreakers can lead to positive results to conserve coastal fauna and flora which in turn makes a coast more resilient to, for example, extreme weather events by conserving coral reefs, which act as a barrier between land and sea. An example of a stringent approach to combat overfishing is found in the 'burning boat policy' used in Sulawesi, Indonesia, which targets foreign flag fishing vessels engaged in IUU (Ambo-Rappe and Moore, 2018). – though the policy has led to clashes between nations, especially with China which claims part of the waters patrolled by Indonesia as Chinese fishing grounds (BBC, 2016).

To a certain extent, several island States have tackled the issue of lack of proper implementation of larger-scale MPAs by resorting to locally Managed Marine Areas (LMMAs), which have been set up to prevent overfishing and IUU. LMMAs are community-led approaches to conservation. LMMAs can be a valuable tool for islands with vast coastlines and geographic isolation of fishing communities paired with the lack of availability of sufficient funds and capacity to implement and follow up on large-scale conservation programmes (Harding, 2018). In Madagascar, for example, LMMAs have been successfully used to prevent illegal fishing and trading of shark, sea cucumber, and octopus during closed fishing seasons. However, this local approach highlights a flaw of MPAs and LLMAs. MPAs and especially LMMAs tend to cover smaller areas within the national jurisdiction of States, leaving two-thirds of the oceans unprotected as areas beyond national jurisdiction (see Chapter 21). Fortunately, there exist regional treaties which have installed MPAs outside the boundaries of States' EEZs and the ongoing negotiations within the United Nations on a novel convention to address a range of topics related to the conservation and sustainable use of marine biological diversity are underway.

MPAs and LMMAs can have significant value to prevent overfishing and IUU when implemented and monitored effectively. Yet even though these spatial measures can allow for the restoration of coastal ecosystems and habitats for example by restoring coral reefs, which make a coast more resilient to the worst effects of climate change, such as sea-level rise, they will most likely not provide sufficient protection against the effects of climate change. When it comes to dealing with the effects of climate change, island States, and the international community have tended to invest in monitoring and early warning systems, such as the programme for 'Enhancing Climate Information and Knowledge Services for resilience in five island countries of the Pacific Ocean' which has received funding through the Green Climate Fund and is a project of the United Nations Environmental Programme (UNEP) covering the Cook Islands, Niue, Palau, the Republic of the Marshall Islands and Tuvalu, or the programme 'Strengthening Hydro-Meteorological and Early Warning Services in the Pacific' of the World Meteorological Organization (WMO) covering 12 small island developing States in the Pacific (UNEP, 2020; WMO, 2021). Though extremely useful to assist the local populations in anticipating and preparing for the effects of climate change, such as extreme weather events, these measures cannot halt the primary effects of climate change. Sea-level rise will destroy coastal ecosystems and take away arable land and allow the maritime zones of islands to shrink by relocating the coastline inwards whether or not monitoring and early warning systems are in place.

To tackle the issue of sea-level rise, island States are charting the coordinates of their maritime zones in national legislation. Publishing the geographic coordinates of the baseline, as well as the outer limits of all maritime zones, has become a well-established practice of the threatened island States, as exemplified in legislation of the Marshall Islands,

Kiribati, and Tuvalu, all of which are fixing their maritime zones (United Nations Division for Ocean Affairs and the Law of the Sea; Pratt and Govan, 2010; International Law Association Committee on International Law and Sea Level Rise, 2018). However, this strategy is flawed in the same manner that judicial decisions and multilateral agreements cannot fix baselines or maritime boundaries to the extent that they no longer match the reality on land. Unilateral chartering or legislating of maritime boundaries cannot simply set aside the LOSC, which still respects the moveable nature of baselines in light of the adage that land dominates the sea (ILA Committee on Baselines, 2012). This is very clearly reflected in the LOSC which reads that 'the normal baseline for measuring the breadth of the territorial sea is the low-water line along the coast' (Article 5 LOSC). In the past for example, the USA has legitimately protested the continued use by Comoros of a submerged maritime feature as a basepoint from which it measured its maritime zones (USA Department of State, 2014).

Therefore, there is a great need for measures protecting the coastlines of islands as they exist today. Only the retention of baselines (or base points) will allow an island State to retain its maritime zones of old under the LOSC. Adaptation measures can be kept to the bare minimum, but they can also be quite comprehensive. Measures can range from planting sea grasses as beach nourishment to stabilize sediments or constructing sea walls several metres high to keep the rising tide at bay (Freestone, 1991). Hard measures, such as erecting sea walls, can however prove problematic. As opposed to soft measures, for example planting sea grasses, they can distress coral reefs and diminish calcification rates, leading to even greater erosion for an unprotected part of the shoreline due to less coral build-up by altering the natural sediment flow. This can jeopardize in particular coral islands' natural ability to adapt to sea-level rise (Schofield and Freestone, 2013). Furthermore, the costs of erecting sea walls can be too high for small island States alone. Even if costs are not too high and the erection of sea walls is possible, costly sea walls could shift funds to outer, uninhabited parts of the island simply in order to hold on to basepoints and baselines from which maritime zones can be claimed – when those funds may be put to better use to assist populated areas to adapt to climate change.

Given these downsides to erecting sea walls, it may be more constructive to reassess the reigning rule of moveable baselines and consider whether the LOSC should allow for the 'freezing' of current maritime entitlements of island States. After all, as is clear from cases of maritime delimitation, the LOSC does not shy away from equitable solutions which permit State and context-specific solutions. For example, when special circumstances, such as historic rights to fishing grounds, are present in a case of maritime delimitation, deviation is allowed from the normally automatic application of the rule of equidistance, which would simply divide the maritime zone between two parties in half starting from the respective baselines, because the LOSC dictates 'an equitable solution' must be achieved in cases of delimitation (Article 74 LOSC). This equitable solution was, for example, applied in a maritime delimitation case between Norway and Denmark (*Maritime Delimitation in the Area between Greenland and Jan Mayen*, 1993).

Given that many island States have contributed very little to climate change and sea-level rise, such a consideration may favour the retention of the *status quo* for these threatened islands. Maritime entitlements could be frozen in distinct ways by freezing either the baseline or rather the outer limits of the current maritime zones. Fixing the outer limits of maritime zones would entail a loss of internal waters and an expansion of the territorial sea as the baseline moves land inwards (Grote Stoutenburg, 2015). In choosing between both options, it is the latter that has more merit. It is true that a State has the highest degree of sovereignty in internal waters, as opposed to its powers in other maritime zones, yet this expanded area of internal waters would simply replace the equally far-reaching sovereignty the State enjoyed over its land territory. Such a change in legislation could come about in several ways. For starters, the LOSC itself could be amended. Unfortunately, amending the LOSC would mean opening up the entire agreement to debate. Given the nine years that it took to negotiate the initial LOSC, it is rather certain no State would be excited to open up this can of worms again (Barnett and Adger, 2003; Arsana and Schofield, 2012). Alternatively, a distinct implementing agreement could be added to the LOSC, as was done for the Agreement relating to the Implementation of Part XI of the LOSC, which reads that '[i]n the event of any inconsistency between this Agreement and Part XI [of the LOSC], the provisions of this Agreement shall prevail' (Agreement of Part XI of the LOSC, 1994).

Conclusion

Global environmental issues such as overfishing and the effects of climate change on fish and their habitats through ocean acidification, ocean warming, extreme weather events and sea-level rise, have and will continue to have a tremendous impact on island nations as these are often highly reliant on fish for sustenance and for economic reasons. Yet, especially daunting is the reality that current international law allows sea-level rise to gnaw at the maritime zones of island States due to the retreat of the baseline from which coastal entitlements of States are measured. These impacts are furthermore heaped on top of already existing local issues related to overpopulation, ineffective waste management, unfit crop management, pollution from the mining industry and the negative effects of having resources extracted during colonization. It is difficult to assign marine environmental decline to a single local, regional, or global cause. Different local and global issues, such as increased stress on resources due to overpopulation on the one hand and climate change on the other hand, can be viewed as interconnected, and in many cases, the different local, regional, and global types of drivers need to be addressed in order to achieve sustainable governance of marine areas surrounding islands, both near shore in coastal waters and offshore at the edges of EEZs and continental shelves.

That is not to say that valuable efforts to tackle environmental issues plaguing the islands' seas have not been made. Especially, the creation of MPAs or LMMAs has proven to be valuable to combat overfishing and IUU by foreign-flagged fishing vessels, but only when the planned measures in these areas are sufficiently monitored and enforced, which can be very difficult due to limited budgets and capacity. These spatial measures will unfortunately not do very much to alleviate the effects of climate change. Therefore, island States have turned to adaptive measures of precaution, such as monitoring and early warning systems to signal coastal populations of sudden-onset harmful events triggered by climate change. Despite these actions, it will remain crucial for islands to invest in hard protective measures to defend their coastlines from rising seas in order to retain their economically valuable maritime entitlements to the fullest. Given the high cost of such measures, the time has come to abandon the adage that land dominates the sea, and the assumption that baselines are inherently moveable under the LOSC. If the fixing of baselines would be made possible under the LOSC, at least international law would not be the origin of further aggravation of the effects of climate change. Such an outcome would be a reasonable request of those that are hardest hit by climate change, and which have contributed very little to causing it.

References

Aegean Sea Continental Shelf (Greece v Turkey), Merits, [1978] ICJ Rep 3.
Agreement relating to the Implementation of Part XI of the United Nations Convention on the Law of the Sea of 10 December 1982, 28 July 1994, 1836 UNTS 3.
Ambo-Rappe, R. and Moore, A.M. (2018) 'Sulawesi seas, Indonesia', in C. Sheppard (ed) *World seas and environmental evaluation: Volume II: The Indian Ocean to the Pacific.* Academic Press, Oxford.
Andréfouët, S. and Adjeroud, M. (2018) 'French Polynesia', in C. Sheppard (ed) *World seas and environmental evaluation: Volume II: The Indian Ocean to the Pacific*, Academic Press, Oxford.
Árnadóttir, S. (2016) 'Termination of maritime boundaries due to a fundamental change of circumstances', *Utrecht Journal of International and European Law*, vol 32, pp 94–111.
Arsana, A. and Schofield, C. (2012) 'Climate change and the limits of maritime jurisdiction', in R. Warner and C. Schofield (eds) *Climate Change and the Oceans – Gauging the Legal and Policy Currents in the Asia Pacific and Beyond*, Edward Elgar, Cheltenham.
Australian Customs Service. (2008) 'Illegal, unreported and unregulated fishing in the southern ocean', https://www.anao.gov.au/sites/default/files/ANAO_Report_2008-2009_06.pdf.
Barnett, J. and Adger, W. N. (2003) 'Climate dangers and atoll countries', *Climatic Change*, vol 61, pp 321–337.
BBC. (2016) 'Indonesian navy fires on Chinese fishing boat in disputed waters', https://www.bbc.com/news/world-asia-36573291.
Continental Shelf (Libyan Arab Jamahiriya/Malta), Merits, [1985] ICJ Rep 13.
Continental Shelf (Tunisia/Libyan Arab Jamahiriya), Merits, [1982] ICJ Rep 18.
Convention for the Protection of the Marine Environment of the Northeast Atlantic, 22 September 1992, 2354 UNTS 67 (entered into force 15 March 1998).
Convention on the Territorial Sea and the Contiguous Zone, 29 April 1958, 516 UNTS 205 (entered into force 10 September 1964).
Delimitation of the Maritime Boundary in the Bay of Bengal (Bangladesh/Myanmar), Merits, [2012] ITLOS Rep 4.
Fenner, D. (2018) 'Nauru', in C. Sheppard (ed) *World Seas and Environmental Evaluation: Volume II: The Indian Ocean to the Pacific.* Academic Press, Oxford.
Fisheries Case (United Kingdom v Norway), Merits, [1951] ICJ Rep 116.
Food and Agriculture Organization of the United Nations. (2020) 'The state of world fisheries and aquaculture: sustainability in action', www.fao.org/3/ca9229en/ca9229en.pdf.
Fredston-Hermann, A., Brown, C. J., Albert, S., Klein, C. J., Mangubhai, S., Nelson, J. L., Teneva, L., Wenger, A., Gaines, S. D. and Halpern, B. S. (2016) 'Where does river runoff matter for coastal marine conservation?', *Frontiers in Marine Science*, vol 3, pp 1–10.
Freestone, D. (1991) 'International law and sea level rise' in R. Churchill and D. Freestone (eds) *International Law and Global Climate Change*, Graham & Trotman, London.
Grisbådarna case (Norway v Sweden), PCA Award [1909], RIA XI 147.
Grote Stoutenburg, J. (2015) *Disappearing Island States in International Law*, Brill, Leiden.
Harding, S. (2018) 'Madagascar', in C. Sheppard (ed) *World seas and environmental evaluation: Volume II: The Indian Ocean to the Pacific,* Academic Press, Oxford.
High Seas Task Force. (2006) 'Closing the net: Stopping illegal fishing on the high seas', www.oecd.org/sd-roundtable/papersandpublications/39375276.pdf.
International Law Association Committee on International Law and Sea Level Rise. (2018) 'Sydney Conference Report', www.ila-hq.org.
International Law Association on Baselines under the International Law of the Sea. (2012) 'Sofia Conference Report', www.ila-hq.org.
International Law Commission. (1966) 'Draft Articles on the Law of Treaties with commentaries' in *Yearbook of the International Law Commission*, vol. II. United Nations, New York.
IPCC. (2019) 'Summary for Policymakers' in *IPCC Special Report on the Ocean and Cryosphere in a Changing Climate.* IPCC, Geneva.
Maritime Delimitation and Territorial Questions between Qatar and Bahrain, Merits, [2001] ICJ Rep 40.
Maritime Delimitation in the Area between Greenland and Jan Mayen (Denmark v. Norway), Merits, [1993] ICJ Rep 38.
Maritime Delimitation in the Black Sea (Romania v Ukraine), Merits, [2009] ICJ Rep 61.
North Atlantic Coast Fisheries Case (Great Britain v United States), Arbitral Award [1910] RIAA XI 167.

North Sea Continental Shelf (Federal Republic of Germany/Netherlands and *Federal Republic of Germany/Denmark),* Merits, [1969] ICJ Rep 3.

OSPAR Commission. (2003) 'OSPAR Recommendation 2003/3 on a Network of Marine Protected Areas', https://www.ospar.org/documents?d=32571.

Pippard, H., Ralph, G.M., Harvey, M.S., Carpenter, K.E., Buchanan, J.R., Greenfield, D.W., Harwell, H.D., Larson, H.K., Lawrence, A., Linardich, C., Matsuura, K., Motomura, H., Munroe, T.A., Myers, R.F., Russell, B.C., Smith-Vaniz, W.F., Vié, J.-C., Thaman, R.R., Williams, J.T. (2017) *The conservation status of marine biodiversity of the Pacific islands of Oceania,* IUCN, Switzerland.

Pratt, C. and Govan, H. (2010) 'Our Sea of Islands, Our Livelihoods, Our Oceania', www.forumsec.org/wp-content/uploads/2018/03/Framework-for-a-Pacific-Oceanscape-2010.pdf.

Schofield, C. and Freestone, D. (2013) 'Options to protect coastlines and secure maritime jurisdictional claims in the face of global sea level rise', in M. B. Gerrard and G. E. Wannier (eds) *Threatened island nations legal implications of rising seas and a changing climate,* Cambridge University Press, Cambridge.

Secretariat of the Pacific Regional Environment Programme. (2012) 'Pacific Environment and Climate Change Outlook', https://wedocs.unep.org/bitstream/handle/20.500.11822/9001/-Pacific%20Environment%20and%20Climate%20Change%20Outlook-2012PEECO.pdf?sequence=3&%3BisAllowed.

Sharma, S. P. (1997) *Territorial acquisition, disputes, and international law,* Martinus Nijhoff Publishers, The Hague.

Soete, A. (2021) *The international legal personality of island states permanently submerged due to climate change effects,* Maklu, Antwerp.

Stevens, G.M.W. and Froman, N. (2018) 'Maldives', in C. Sheppard (ed) *World seas and environmental evaluation: Volume II: The Indian Ocean to the Pacific,* Oxford, Academic Press.

Territorial and Maritime Dispute (Nicaragua v Colombia), Merits, [2012] ICJ Rep 624.

Territorial and Maritime Dispute between Nicaragua and Honduras in the Caribbean Sea, Merits, [2007] ICJ Rep 659.

The Bay of Bengal Maritime Boundary Arbitration (Bangladesh v India), Merits, [2014] PCA Award.

Treaty between the Government of Australia and the Government of the French Republic on cooperation in the maritime areas adjacent to the French Southern and Antarctic Territories (TAAF), Heard Island and the McDonald Islands, 24 November 2003, ATS 6 (entered into force 1 February 2005).

UNEP. (2020) 'New UNEP programme to support climate resilience in Pacific Islands through early warning systems', www.unep.org/news-and-stories/press-release/new-unep-programme-support-climate-resilience-pacific-islands.

United Nations. (2015) 'Sustainable development goals', https://sdgs.un.org/goals.

United Nations Convention on the Law of the Sea, 10 December 1982, 1833 UNTS 3 (entered into force 16 November 1994).

United Nations Division for Ocean Affairs and the Law of the Sea, 'Deposit of charts/lists of coordinates under the Convention', www.un.org/Depts/los/LEGISLATIONANDTREATIES/depositpublicity.htm.

United Nations Office on Drugs and Crime. (2016) 'Transnational organized crime in the Pacific: A threat assessment', https://www.unodc.org/documents/southeastasiaandpacific/Publications/2016/2016.09.16_TOCTA_Pacific_web.pdf.

United States of America Department of State. (2014) 'Comoros: Archipelagic and other maritime claims and boundaries', https://2009-2017.state.gov/documents/organization/224317.pdf.

Vienna Convention on the Law of Treaties, 23 May 1969, 1155 UNTS 331 (entered into force 27 January 1980).

Weil, P. (1989) 'Délimitation maritime et délimitation terrestre', in S. Rosenne, Y. Dinstein and M. Tabory (eds) *International Law at a Time of Perplexity: Essays in Honour of Shabtai Rosenne.* Martinus Nijhoff Publishers, Dordrecht.

Western and Central Pacific Fisheries Commission. (2009) 'Conservation and management measures on regulation of transhipment', www.wcpfc.int/doc/cmm-2009-06/conservation-and-management-measure-regulation-transhipment-0.

WMO. (2021) 'Strengthening hydro-meteorological and early warning services in the pacific', https://public.wmo.int/en/projects/strengthening-hydro-meteorological-and-early-warning-services-pacific-crews-pacific-sids-20.

Wold, C. (2019) 'The impracticability exemption to the WCPFC's prohibition on transhipment on the high seas', *Environmental Law,* vol 49, no 1, pp 131–186.

15
CORAL REEFS
The case for ecological reflexivity

Pedro Fidelman

In the context of global environmental change, the Anthropocene refers to the current geologic epoch that is characterised by unprecedented influence of human activities on the functioning of the Earth system (Steffen et al., 2011). Under it, coral reefs are rapidly shifting into degraded states driven by climate change and other anthropogenic drivers unlike anything seen in human history (Hughes et al., 2017a). On Australia's Great Barrier Reef (GBR), for instance, recurring bleaching events due to global warming have resulted in mass mortality of coral reefs (Hughes et al., 2017b) and, consequently, a staggering 1,000 km-scale transformation of coral assemblages (Hughes et al., 2018). In addition, sediments, nutrients, and pesticides from inadequate agricultural and land management practices are transported by land-based runoff to the GBR, negatively influencing its resilience. Nutrients derived from catchment runoff are also associated with outbreaks of crown-of-thorns starfish, which cause significant impacts on coral reefs (GBRMPA, 2019a). Under climate change, heavy rainfall events are predicted to increase, which, in turn, are likely to increase catchment runoff, compounding water quality issues in inshore areas (Wooldridge, 2009).

These complex drivers and impacts associated with climate change and local pressures pose profound challenges to coral reef governance, and environmental governance more broadly (Clement and Standish, 2018, Morrison et al., 2020). The GBR is exemplar of many coral reefs around the world, such as those listed under the UNESCO World Heritage Convention, where similar threats outpace conventional governance (Morrison et al., 2020). The following quote from the 2019 Great Barrier Reef Outlook Report is illustrative: 'threats to the Great Barrier Reef are multiple, cumulative and increasing... management agencies are constantly being challenged... to keep pace with a rapidly changing Reef' (GBRMPA, 2019a: iv). In fact, the existing arrangements for coral reef governance have been developed in the context of the relative stasis that characterised the Holocene (the geologic epoch that precedes the Anthropocene) (Dryzek, 2016). They have aimed to maintain historic ecological baselines in the face of conventional threats (e.g., pollution and overfishing). The Anthropocene, however, requires innovative forms of governance capable of responding and adapting to change, complexity, and uncertainty (Clement and Standish, 2018, Dryzek, 2016, Pickering, 2018, Morrison et al., 2020).

By drawing on the scholarship on governance for the Anthropocene, this chapter explores ecological reflexivity as an emerging concept with the potential to better equip coral reef

DOI: 10.4324/9781315149745-19

governance in the Anthropocene. It combines the concept of ecological reflexivity with elements of institutional analysis to provide insights into how ecological reflexivity may be operationalised in the context of coral reef governance.

Governance challenges on the Great Barrier Reef

This section draws on GBR to illustrate governance challenges to coral reefs in the Anthropocene. The governance of the GBR is complex. It involves intricate networks of arrangements (international treaties, domestic laws, policies, plans, and decision-making processes) and entities (e.g., government agencies, industry, environmental groups, and scientists) spanning multiple governance levels (from local to international) (Fidelman et al., 2019). However, the primary responsibility for the protection of the GBR lies with the Great Barrier Reef Marine Park Authority, an independent federal statutory authority. Under an intergovernmental agreement, the Authority shares responsibility for day-to-day planning and management of activities within the Marine Park with relevant State of Queensland government agencies. These arrangements are complemented by several state and federal agencies with mandates over resources and sectors both on the water and in the GBR catchment (GBRMPA, 2019a). The GBR is also listed as a World Heritage Area under the World Heritage Convention. Accordingly, the World Heritage Committee plays an international oversight and assistance role under this convention. Its decisions and recommendations also affect the governance of the GBR (Fidelman et al., 2019).

For over four decades, the Australian and Queensland governments, together with a range of stakeholders, have invested considerable resources in managing the GBR (Australia and Queensland, 2018). Since 2014, in response to a range of anthropogenic threats, the GBR and its catchment have been the target of significant management efforts. These include the implementation of the *Reef 2050 Long-term Sustainability Plan* (Reef 2050 Plan) and the *Reef 2050 Water Quality Improvement Plan*. The former aims to address cumulative impacts and increase the resilience of the reef to long-term threats, including climate change, by setting actions, targets, objectives, and outcomes for managing the GBR until 2050. The latter identifies how the water quality component of the Reef 2050 Plan is delivered with the aim of improving water quality from catchment runoff entering the GBR. These plans have, to some extent, helped enhance management arrangements in the GBR region. For example, the Reef 2050 Plan has improved jurisdictional consistency, coordination and resourcing across many management issues; encouraged and mandated positive engagement and collaboration between government agencies, partners and stakeholders; strengthened a number of policies (e.g., water quality and indigenous engagement in ecosystem management); and, encouraged increased consideration of cumulative impacts and community benefits (Leverington et al., 2019).

Management efforts have been accompanied by substantial investments. In 2016, the Reef 2050 Plan investment framework identified a five-year commitment worth $1.28 billion from both the Australian and Queensland governments. Since then, the GBR has received additional funding, which includes the Australian government's allocation of $443 million to the Great Barrier Reef Foundation to deliver a range of programmes aligned with the Reef 2050 Plan (Australia and Queensland, 2018). Nevertheless, management effectiveness for major threats to the GBR remains poor and, in the case of climate change, declining. As a result, the resilience of the region is degrading, with the outlook for the GBR recently downgraded to 'very poor' (GBRMPA, 2019a). Several compounded factors may help explain this situation. The resilience of ecological systems, such as the Great Barrier Reef, results from intricate interactions between internal and external structures and processes

(Ostrom, 2009). Accordingly, effectiveness in managing complex, large-scale issues in the GBR, such as climate change, results from the interplay between management arrangements and geographic, environmental, economic, and political drivers of change (Morrison, 2017). Such drivers involve several policy areas and geographical and jurisdictional scales (from local to global) (Fidelman et al., 2013). For example, on the one hand, considerable efforts have been made by the federal and state governments to protect the GBR; on the other hand, these same governments continue to provide support to the fossil fuel industry. Importantly, Australia is yet to develop comprehensive and coherent policies to reduce carbon emissions to meet its 2030 Paris emissions reduction target (Crowley, 2021).

Governance for the Anthropocene: ecological reflexivity

In the face of the Anthropocene, innovative approaches to science, management and policy are increasingly being called upon. Such approaches, the literature suggests, should be capable of embracing change, complexity, and uncertainty (Hughes et al., 2017a, Clement and Standish, 2018, Graham et al., 2014, Hobbs et al., 2011, Graham and Hicks, 2015, Hobbs et al., 2009, Seastedt et al., 2008). However, the literature offers limited substance on what these innovative approaches should specifically look like. By drawing on the environmental governance scholarship, this section seeks to contribute more specific guidance on this topic.

For Sørensen (2017), to qualify as innovative, governance approaches must be qualitatively new and produce some form of qualitative change. Rather than being 'never seen before', innovation may consist of adaptations developed elsewhere. Accordingly, invention and diffusion are regarded as key attributes of such innovative approaches. Together with performance (i.e., the impact of innovation), these attributes define governance innovation (Jordan and Huitema, 2014). In this context, a fundamental question is thus what kind of innovative approaches to coral reef governance are required in the Anthropocene. Pickering (2018) summarises desired qualities that could better equip environmental governance in the Anthropocene into four concepts. These can be conceptualised as follows:

1 *Social-ecological orientation* refers to the recognition of the interactions between humans and ecosystems. For example, anthropogenic drivers, such as fishing and pollution – and in the context of the Anthropocene, climate change – greatly influence the functioning of coral reef ecosystems; on the other hand, communities and industries directly or indirectly depends on goods and services provided by these ecosystems for livelihood, food security, recreation, tourism, and social and cultural services (Hughes et al., 2017a).
2 *Complexity orientation* refers to the ability to consider the intricate nature of human-ecosystem interactions. For instance, coral reefs comprise complex social-ecological systems involving multiple drivers, linkages, and feedbacks (Cinner et al., 2016). Further, there is considerable uncertainty in monitoring and predicting complex human-ecosystem interactions. An example is the significant knowledge gaps that exist in relation to how shifts in species composition due to climate change and other anthropogenic drivers will affect ecosystem functions (Hughes et al., 2017a).
3 *Dynamic orientation* refers to the capacity for identifying pathways for institutional change that can help steer society away from practices that undermine ecological conditions. For example, resilience-based management identifies management and policy interventions to address the cumulative impacts of local and regional anthropogenic activities on coral reefs (Anthony et al., 2015). Likewise, significant efforts to curb carbon emissions are imperative to slow down the effects of climate change on these ecosystems.

4 *Political orientation* refers to the ability to consider contested values associated with ecosystem conservation in the Anthropocene. The notion of managing altered ecosystem states (rather than returning these states to historic baselines) is illustrative as it involves highly contested ecosystem and social values (Clement and Standish, 2018, Hobbs et al., 2011). Similarly, addressing climate change involves multiple and, often, conflicting responses driven by divergent values, which reconciliation requires political contestation (Hulme, 2020).

Many approaches, such as resilience, adaptive, and transformative governance, social learning, experimental governance, and anticipatory governance, reflect, to some extent, the attributes outlined above. However, a comparative analysis reveals that governance underpinned by the notion of ecological reflexivity is one approach that holds all such qualities (for details of this analysis, see Dryzek, 2016, Pickering, 2018, Dryzek and Pickering, 2018). For Dryzek (2016:938), ecological 'reflexivity proves to be the primary requirement for [governance] institutions in the Anthropocene.'

Ecological reflexivity refers to the capacity to recognise social-ecological dynamics, rethink core values and practices and, in this light, respond accordingly by transforming such values and practices (Dryzek and Pickering, 2018, Pickering, 2018, Dryzek, 2016). It is a composite concept comprising three core elements: recognition, reflection, and response (Pickering, 2018). *Recognition* requires individuals and organisations involved in coral reef governance to be aware of changes in the social-ecological system, monitor risks and impacts of climate change and other anthropogenic drivers on this system, and anticipate future conditions and impacts. For example, the Great Barrier Reef Outlook report is prepared every five years to examine the Reef health, pressures, and likely future. It is one of the key mechanisms to track and predict changes in the GBR. *Reflection* requires individuals and organisations to combine learning from past experiences, to rethink core values and practices considering lessons learned and feedback from the social-ecological system, and to envision possible futures. Reflection on the deteriorating outlook for the reef has underpinned the Reef 2050 Long-term Sustainability Plan, which outlines measures for managing the GBR over a 35-year period. *Response* requires them to rearticulate core values, aims, and discourses and reconfigure existing functions and practices. The guidelines on permit applications for restoration and adaptation interventions to improve coral reef resilience prepared by the Great Barrier Reef Marine Park Authority, for example, represent a shift towards considering the adoption of novel and unconventional interventions in response to the impacts of climate change on the GBR (Dryzek and Pickering, 2018). Recognition, reflection, and response entail an iterative process whereby successive phases of recognition and reflection are informed by previous phases of response (Pickering, 2018).

Operationalising ecological reflexivity

Regarded as the primary requirement to address the governance challenges associated with the Anthropocene, ecological reflexivity is a powerful concept. Given the innovative nature of such a concept, a fundamental question is how ecological reflexivity may be translated into practice. This section combines the concept of ecological reflexivity (Dryzek and Pickering, 2018) with elements of institutional analysis (Ostrom, 2011) to provide insights into how ecological reflexivity may be operationalised in the context of coral reef governance.

Institutional analysis is an attempt to examine the rules – such as regulations, policies, norms and decision-making procedures (also known as institutions or institutional arrangements) – adopted by individuals and/or organisations to address their problems, leading to desired outcomes (Imperial, 1999). There are many ways of doing institutional analysis; several social science disciplines have their own approaches to investigate institutions (Hollingsworth, 2000). The Institutional Analysis and Development (IAD) framework (Ostrom, 2005) is one of the most well-established approaches used in institutional analysis (Blomquist and DeLeon, 2011, Heikkila and Andersson, 2018). Of particular interest to operationalising ecological reflexivity is the IAD's concept of action situation. It refers to the social space where a mix of individuals and organisations interact, make decisions, and realise outcomes from their interactions (Ostrom, 2011). In the context of this chapter, the action situation is the locus where government agencies and relevant stakeholders make decisions and implement actions in response to anthropogenic threats to coral reefs. Such an action situation is influenced by institutional arrangements, such as policies, regulations, and decision-making processes. In terms of the IAD framework, these institutional arrangements may be classified into seven types of rules:

- *Position rules* specify the participants and their roles in the coral reef governance
- *Boundary rules* specify how participants can take part in such governance
- *Choice rules* specify the actions participants can take as part of their roles
- *Aggregation rules* specify decision-making procedures, including arrangements to aggregate the preferences of participants
- *Information rules* specify the arrangements for information exchange between participants,
- *Payoff rules* specify the incentives and disincentives (e.g., resources available) to support governance
- *Scope rules* specify the functional scope (e.g., water quality, climate change, etc.) and geographic domain (catchment, marine park) that can be affected by coral reef governance (Ostrom, 2011).

In influencing the action situation, the seven types of rules also affect the conditions under which ecological reflexivity may (or may not) emerge as part of coral reef governance. Insights from the use of the IAD framework to examine collective problems (like those of governing coral reefs in the Anthropocene) emphasise the importance of certain attributes in creating enabling conditions for solving such problems. For example, in the context of co-management of marine resources, these attributes include involvement of various stakeholders, perspectives and solutions and learning capacity (Fidelman et al., 2017). Similarly, Dryzek and Pickering (2017) propose that ecological reflexivity involves the participation of various individuals and organisations, utilisation of expert and non-expert knowledge (see Chapter 8), and learning mechanisms within and across multiple levels of governance (Dryzek and Pickering, 2017). Accordingly, the operationalisation of ecological reflexivity in reef governance should be underpinned by institutional arrangements that enable these attributes. Such arrangements should, therefore, be capable of fostering participation, different types of knowledge, and linked networks. These are elaborated in terms of rule types and in the context of the GBR, as follows:

Participation of a range of individuals and organisations who affect and/or are affected by coral reef management. This includes partnership (position and boundary) arrangements involving the Great Barrier Reef Marine Park Authority, Traditional Owners, government agencies, reef-dependent industries, community groups, and the scientific community. Local Marine Advisory Committees, for example, are voluntary community-based groups that provide advice and a wide range of perspectives on local issues relating to the management of the Marine Park. Each of these committees comprise between 10 and 20 members, representatives of a diversity of interests, such as: agriculture, aquaculture, business, charter boat/

fishing guide, commercial fishing, conservation, diving and snorkelling, education, industry, natural and cultural resource management, ports and shipping, recreational fishing, research, sailing, boating, and kayaking, tourism, and other regionally specific reef interests. Representatives from state and federal government agencies and traditional owners may also be part of these committees (GBRMPA, 2021). The participation of such variety of interests entails (aggregation and scope) arrangements that support the consideration of multiple aspects of coral reef governance, e.g., protection, conservation, cultural values, as well as diverse perspectives on problem framing and solutions. It also entails (information) arrangements that facilitate the use of different types of knowledge.

The use of both technical information and local knowledge to inform decision-making. The participation of a range of interest groups in coral reef governance contributes different types of knowledge. In addition to the Local Marine Advisory Committees described above, other (information and aggregation) arrangements help contribute knowledge to the governance of the GBR. They include Traditional Use of Marine Resources Agreements and Reef Advisory Committees. The former consist of community-based plans for traditional resources management, which is based on both cultural knowledge and contemporary science. The latter consist of four committees (catchment and coastal, ecosystem, indigenous and tourism and recreation), which role is to provide advice to the Great Barrier Reef Marine Park Authority Board on the management of the Marine Park. Committees are competency based and their members are appointed on the basis of their experience, knowledge, skills, and networks with relevant groups and sectors (GBRMPA, 2019b).

Linked networks of institutional arrangements and entities. As mentioned previously, the GBR governance involves intricate networks of arrangements (international treaties, domestic laws, policies, plans, and decision-making processes) and entities (e.g., government agencies, industry, environmental groups and scientists) spanning multiple governance levels (from local to international) (Fidelman et al., 2019). Such networks give rise to a polycentric system featuring institutional gaps and overlaps (Fidelman et al., 2013). In this context, collaborative arrangements (e.g., information and aggregation rules) are imperative to minimise conflicting approaches and enable sharing of learnings between networks. Various such arrangements are identified in the GBR governance (Fidelman et al., 2013, Fidelman et al., 2019). For example, the Reef 2050 Plan is underpinned by Great Barrier Reef Intergovernmental Agreement between the Australian and Queensland governments. Such an agreement supports a unified and collaborative approach (associated with choice and scope rules) between the two governments to the management of the GBR and its catchment. Furthermore, while providing an overarching framework for the protection and management of the GBR, the Plan seeks to integrate and guide the actions of managing agencies, traditional owners, industry, resource users, researchers, and the community (Australia and Queensland, 2018).

Conclusion

Challenges to coral reef governance are unprecedented in the Anthropocene, requiring innovative approaches capable of embracing change, complexity, and uncertainty. This chapter identified the concept of ecological reflexivity as a promising such approach. It involves the capacity to recognise social-ecological dynamics, rethink core values and practices and, in this light, respond accordingly by transforming such values and practices (Dryzek and Pickering, 2018). Ecological reflexivity may be operationalised by designing new or reforming existing institutional arrangements that enable: (i) the participation of a range of individuals and organisations; (ii) use of different types of knowledge to inform decision-making, and

(iii) promote collaboration and learning across governance networks. In the GBR, instances of these arrangements are associated with the development of partnerships with a range of stakeholders, integration of local knowledge in decision-making, and promotion of collective learning among various partners and stakeholders (Day and Dobbs, 2013). Nevertheless, addressing large-scale, complex issues, such as climate change remains challenging (see Chapter 3). This results, in part, from conflicting policies relating to the protection of the GBR and development based on fossil fuels. If only applied at a local/regional level, ecological reflexivity is a necessary but not sufficient condition for governing coral reefs in the Anthropocene. To be fully effective, ecological reflexivity will need to underpin decision-making at the national level and policies other than reef protection. For example, safeguarding coral reefs and dependent communities and industries should be one of the key priorities driving reef nations' emissions reduction policy. In this regard, transitioning to a low-carbon economy would greatly benefit from ecological reflexivity.

References

Anthony, K. R. N., Marshall, P. A., Abdulla, A., Beeden, R., Bergh, C., Black, R., Eakin, C. M., Game, E. T., Gooch, M., Graham, N. A. J., Green, A., Heron, S. F., van Hooidonk, R., Knowland, C., Mangubhai, S., Marshall, N., Maynard, J. A., Mcginnity, P., Mcleod, E., Mumby, P. J., Nyström, M., Obura, D., Oliver, J., Possingham, H. P., Pressey, R. L., Rowlands, G. P., Tamelander, J., Wachenfeld, D. & Wear, S. (2015). 'Operationalizing resilience for adaptive coral reef management under global environmental change'. *Global Change Biology*, 21, 48–61.

Australia & Queensland. (2018). *'Reef 2050 long-term sustainability plan'*. Australian Government; Queensland Government.

Blomquist, W. & DeLeon, P. (2011). 'The design and promise of the institutional analysis and development framework'. *Policy Studies Journal*, 39, 1–6.

Cinner, J. E., Pratchett, M. S., Graham, N. A. J., Messmer, V., Fuentes, M. M. P. B., Ainsworth, T., Ban, N., Bay, L. K., Blythe, J., Dissard, D., Dunn, S., Evans, L., Fabinyi, M., Fidelman, P., Figueiredo, J., Frisch, A. J., Fulton, C. J., Hicks, C. C., Lukoschek, V., Mallela, J., Moya, A., Penin, L., Rummer, J. L., Walker, S. & Williamson, D. H. (2016). 'A framework for understanding climate change impacts on coral reef social–ecological systems'. *Regional Environmental Change*, 16, 1133–1146.

Clement, S. & Standish, R. J. (2018). 'Novel ecosystems: Governance and conservation in the age of the Anthropocene'. *Journal of Environmental Management*, 208, 36–45.

Crowley, K. (2021). 'Fighting the future: The politics of climate policy failure in Australia (2015–2020)'. *WIREs Climate Change*, https://doi.org/10.1002/wcc.725.

Day, J. & Dobbs, K. (2013). 'Effective governance of a large and complex cross-jurisdictional marine protected area: Australia's Great Barrier Reef'. *Marine Policy*, 41, 14–24.

Dryzek, J. S. (2016). 'Institutions for the Anthropocene: Governance in a changing earth system institutions for the Anthropocene'. *British Journal of Political Science*, 46, 97–956.

Dryzek, J. S. & Pickering, J. (2017). 'Deliberation as a catalyst for reflexive environmental governance'. *Ecological Economics*, 131, 353–360.

Dryzek, J. S. & Pickering, J. (2018). *Politics of the Anthropocene*. Oxford: Oxford Academic Press.

Fidelman, P., Leitch, A. & Nelson, D. R. (2013). 'Unpacking multilevel adaptation in the great barrier reef'. *Global Environmental Change*, 23, 800–812.

Fidelman, P., McGrath, C., Newlands, M., Dobbs, K., Jago, B. & Hussey, K. (2019). 'Regulatory implications of coral reef restoration and adaptation under a changing climate'. *Environmental Science & Policy*, 100, 221–229.

Fidelman, P., Truong Van, T., Nong, K. & Nursey-Bray, M. (2017). 'The institutions-adaptive capacity nexus: Insights from coastal resources co-management in Cambodia and Vietnam'. *Environmental Science & Policy*, 76, 103–112.

GBRMPA. (2019a). *'Great barrier reef outlook report 2019'*. Townsville: Great Barrier Reef Marine Park Authority.

GBRMPA. (2019b). *'Reef advisory committees charter of operation'*. Townsville: Great Barrier Reef Marine Park Authority.

GBRMPA. (2021). '*Local marine advisory committee: terms of reference 2021–2024*'. Townsville: Great Barrier Reef Marine Park Authority.
Graham, N. A. J., Cinner, J. E., Norström, A. V. & Nyström, M. (2014). 'Coral reefs as novel ecosystems: embracing new futures'. *Current Opinion in Environmental Sustainability,* 7, 9–14.
Graham, N. A. J. & Hicks, C. C. (2015). 'Adaptive management for novel ecosystems'. In: Allen, C. R. & Garmestani, A. S. (eds.) *Adaptive management for Novel Ecosystems*. Dordrecht: Springer.
Heikkila, T. & Andersson, K. (2018). 'Policy design and the added-value of the institutional analysis and development framework'. *Policy & Politics,* 42, 309–324.
Hobbs, R. J., Hallett, L. M., Ehrlich, P. R. & Mooney, H. A. (2011). 'Intervention ecology: Applying ecological science in the twenty-first century'. *BioScience,* 61, 442–450.
Hobbs, R. J., Higgs, E. & Harris, J. A. (2009). 'Novel ecosystems: implications for conservation and restoration'. *Trends in Ecology & Evolution,* 24, 599–605.
Hollingsworth, J. R. (2000). 'Doing institutional analysis: Implications for the study of innovations'. *Review of International Political Economy,* 7, 595–644.
Hughes, T. P., Barnes, M. L., Bellwood, D. R., Cinner, J. E., Cumming, G. S., Jackson, J. B. C., Kleypas, J., Leemput, I. A. v. d., Lough, J. M., Morrison, T. H., Palumbi, S. R., Nes, E. H. V. & Scheffer, M. (2017a). 'Coral reefs in the Anthropocene'. *Nature,* 546, 82–90.
Hughes, T. P., Kerry, J., Álvarez-Noriega, M., Álvarez-Romero, J., Anderson, K., Baird, A., Babcock, R., Beger, M., Bellwood, D., Berkelmans, R., Bridge, T., Butler, I., Byrne, M., Cantin, N., Comeau, S., Connolly, S., Cumming, G., Dalton, S., Diaz-Pulido, G., Eakin, C. M., Figueira, W., Gilmour, J., Harrison, H., Heron, S., Hoey, A. S., Hobbs, J.-P., Hoogenboom, M., Kennedy, E., Kuo, C.-Y., Lough, J., Lowe, R., Liu, G., Malcolm McCulloch, H. M., McWilliam, M., Pandolfi, J., Pears, R., Pratchett, M., Schoepf, V., Simpson, T., Skirving, W., Sommer, B., Torda, G., Wachenfeld, D., Willis, B. & Wilson, S. (2017b). 'Global warming and recurrent mass bleaching of corals'. *Nature,* 543, 373–377.
Hughes, T. P., Kerry, J. T., Baird, A. H., Connolly, S. R., Dietzel, A., Eakin, C. M., Heron, S. F., Hoey, A. S., Hoogenboom, M. O., Liu, G., McWilliam, M. J., Pears, R. J., Pratchett, M. S., Skirving, W. J., Stella, J. S. & Torda, G. (2018). 'Global warming transforms coral reef assemblages'. *Nature,* 556, 492–496.
Hulme, M. (2020). 'One earth, many futures, no destination'. *One Earth,* 2, 309–311.
Imperial, M. T. (1999). 'Analyzing institutional arrangements for ecosystem-based management: Lessons from the Rhode Island salt ponds SAM plan'. *Coastal Management,* 27, 31–56.
Jordan, A. & Huitema, D. (2014). 'Policy innovation in a changing climate: Sources, patterns and effects'. *Global Environmental Change,* 29, 387–394.
Leverington, A., Hockings, M. & Leverington, F. (2019). *Reef 2050 Plan Insights Report*. Townsville: Great Barrier Reef Marine Park Authority.
Morrison, T. H. (2017). 'Evolving polycentric governance of the Great Barrier Reef'. *Proceedings of the National Academy of Science,* 114, 3013–3021.
Morrison, T. H., Adger, N., Barnett, J., Brown, K., Possingham, H. & Hughes, T. (2020). 'Advancing coral reef governance into the anthropocene'. *One Earth,* 2, 64–74.
Ostrom, E. (2005). *Understanding Institutional Diversity*. Princeton: Princeton University Press.
Ostrom, E. (2009). 'A general framework for analysing sustainability of social-ecological systems'. *Science*, 419–422.
Ostrom, E. (2011). 'Background on the institutional analysis and development framework'. *Policy Studies Journal,* 39, 7–27.
Pickering, J. (2018). 'Ecological reflexivity: Characterising an elusive virtue for governance in the Anthropocene'. doi: 10.1080/09644016.2018.1487148.
Seastedt, T. R., Hobbs, R. J. & Suding, K. N. (2008). 'Management of novel ecosystems: Are novel approaches required?'. *Frontiers in Ecology and the Environment,* 6, 547–553.
Sørensen, E. (2017). 'Political innovations: Innovations in political institutions, processes and outputs'. *Public Management Review,* 19, 1–19.
Steffen, W., Grinevald, J., Crutzen, P. & Mcneill, J. (2011). 'The anthropocene: Conceptual and historical perspectives'. *Philosophical Transactions of the Royal Society A: Mathematical, Physical and Engineering Sciences,* 369, 842–867.
Wooldridge, S. A. (2009). 'Water quality and coral bleaching thresholds: Formalising the linkage for the inshore reefs of the great barrier reef (Australia)'. *Marine Pollution Bulletin,* 58, 745–751.

16

FISHERIES AND AQUACULTURE IN SOUTHEAST ASIA

Managing the impacts of climate change

Achmad Poernomo and Anastasia Kuswardani

Located roughly between the latitudes of 10° South and 25° North, and between longitudes 94.5 East and 140° East, the region of Southeast Asia covers the mainland of the Asian continent between the basins of the Irrawaddy and Mekong River systems and two oceans: the Indian and the Pacific. Southeast Asian (SEA) countries are members of one regional organization, the Association of Southeast Asian Nations (ASEAN), which includes Brunei Darussalam, Cambodia, Indonesia, Laos, Malaysia, Myanmar, the Philippines, Singapore, Thailand, and Vietnam. (Timor-Leste is in the process of becoming a member.) The large marine ecosystems (LMEs), subregional marine and coastal areas, and watersheds of this region make it not only rich in marine and terrestrial resources but also one of the most socially, economically, and politically diverse regions of the world. Having about 9 million km^2 of waters, 3% of the world's ocean surface, the region covers the waters under Food and Agriculture Organisation (FAO) Major Fishing Areas 57 (Indian Ocean, Eastern), 61 (Pacific, Northwest), 71 (Pacific, Western Central), and 04 (Asia, Inland Waters), and provides significant resources of biodiversity including fish. Together with rice, fish is considered the staple food in the region especially in providing important animal protein and carbohydrate in the diet. Fish consumption in SEA was 35.1–38.4 kg per person per year in 2015, which is projected to increase to 61.5 kg per capita in 2050 (Chan et al., 2017; SEAFDEC, 2017). Fish account for at least 15% of animal protein intake for more than 100 million people and up to 50% for some communities (Gopal and Anbumozhi, 2019).

Fisheries and aquaculture production of the region had been increased considerably over the last decade (see Chapter 16). Currently, the region produces approximately 45 million tonnes per year, which is 20% of global production, involving more than 20 million workers. According to Funge-Smith and Bennet (2019), inland fisheries in the region employ about 11 million people, especially in fishing and postharvest activities. Southeast Asian fish production is projected to be 47.1 million tonnes in 2050 (Chan et al., 2017). Many ASEAN member states (AMSs) are among the highest producers or exporters of fish and fishery products in the world. Three AMSs (Indonesia, Vietnam, and the Philippines) are in the top ten global fishing nations, and four AMSs (Indonesia, Vietnam, Myanmar, and Thailand) are on top for aquaculture, while Vietnam and Thailand are among the top five exporting countries of fish and fish products in terms of value (FAO, 2020). In addition to their role in food security and direct livelihood, fisheries and aquaculture also generate forward and backward

DOI: 10.4324/9781315149745-20

linkage economy such as processing and trade, transportation, boat building, net making, and engine repair. According to World Bank data, these activities employ 10 million people (Gopal and Anbumozhi, 2019).

Although its contribution to global greenhouse gas emissions is small, at around 3%, Southeast Asia is a region that is vulnerable to climate change because the region is characterized by long coastlines and numerous freshwater bodies where people and their main economic activities are highly concentrated. According to a survey conducted by the ISEAS-Yusof Ishak Institute in Singapore, climate change was regarded as the third most pressing security concern in the region (Tang, 2020). A recent publication by German Watch indicated that Myanmar, the Philippines, and Thailand were on the list of ten countries most affected by climate disasters during 2000–2019 (Eckstein et al., 2021). Vietnam was previously included in the list. Climate change also poses the SEA region with an annual average temperature increase of 4.8°C and a sea-level rise of 70 cm by 2100, especially in Indonesia, the Philippines, Thailand, and Vietnam, and threatens major coastal cities in the region (ASEAN, 2021). Yusuf and Francisco (2009) mapped climate change vulnerability for Southeast Asia. They showed that almost all regions are very vulnerable, especially all regions in the Philippines, the Mekong River Delta in Vietnam, almost all regions in Cambodia, North and East Laos, the Bangkok region in Thailand, and West Sumatra, Western Java, and Eastern Java in Indonesia. If emissions are not controlled properly, agriculture, forestry, fisheries, and tourism will be affected and will likely result in greater economic losses than most other regions of the world. The Asian Development Bank reports that climate change could lead to an 11% decline in regional GDP by 2100 (Raitzer et al., 2015).

Impacts of climate change in Southeast Asian fisheries and aquaculture

The impact of climate change on fisheries and aquaculture was extensively discussed in a manuscript based on a series of workshops organized by FAO and edited by Barange et al. (2018). The text shows that the availability and trade of fish products could be affected, which may lead to geopolitical and economic consequences. The maximum potential for marine fish catches in the world's exclusive economic zones (EEZs) could decrease by 2.8–12.1% in 2050 (compared to 2010) depending on the greenhouse gas emission scenarios (i.e., RCP 2.6 = strong mitigation and RCP 8.5 = business-as-usual) of the Intergovernmental Panel on Climate Change (IPCC). These production changes are partly due to shifts in species distribution and can also result in conflicts between users in the region. Table 16.1 summarizes the predicted changes of potential catch in the EEZs of AMS under the two scenarios of greenhouse gas emissions (Cheung et al., 2018). Changes in fish migration due to climate change can cause problems with neighbouring countries, such as jurisdictional and fisheries management disputes, especially due to the entry of fishing vessels from other countries to gain access to fish that have migrated (Asche, 2018). Conflicts between countries that share freshwater and fishery resources, such as the Mekong River, are also very likely. Climate change will lead to changes in freshwater habitats and the fish stocks they support, and competition for scarce water resources combined with the effects of other anthropogenic stresses. For aquaculture, impacts can range from loss of production and infrastructure caused by extreme events to the emergence of disease, parasites, and harmful algal growth.

Blasiak et al. (2017) studied the impact of climate change on marine fisheries in 147 countries and calculated each country's vulnerability index as a function of exposure (E), sensitivity (S), and adaptive capacity (AC). Through a series of calculations taking into account

Table 16.1 Projected changes in catch potential (%) of ASEAN states' EEZs by 2050 and 2100 relative to 2000 under RCP2.6 and RCP8.5 based on outputs from the dynamic size-based food web model

	RCP2.6		*RCP8.5*	
AMS (or part thereof)	*2050*	*2100*	*2050*	*2100*
Brunei Darussalam	−1.62	−20.07	−5.53	−52.77
Cambodia	−9.66	−26.45	−11.85	−97.93
Indonesia	−12.70	−28.63	−18.35	−62.70
Indonesia (eastern)	−12.51	−31.60	−17.42	−64.43
Malaysia	−10.00	−25.17	−15.22	−64.94
Malaysia (Sabah)	−7.08	−20.80	−10.54	−55.24
Malaysia (Sarawak)	−15.35	−29.01	−17.45	−51.93
Myanmar	−0.12	1.37	1.76	−16.59
Philippines	−8.26	−23.72	−11.22	−59.21
Singapore	−35.13	−73.45	−41.47	−99.96
Thailand Andaman sea	0.66	−8.82	−1.62	−34.38
Thailand (Gulf of Thailand)	-9.87	−26.08	−12.21	−95.03
Vietnam	-6.77	−11.71	−7.49	−40.76

the many variables in E, S, and AC, they list countries by vulnerability index, from the most vulnerable to the least vulnerable. It was shown that seven small island developing countries dominated the top ten most vulnerable countries. For AMS, the rankings are Indonesia (26), Vietnam (43), Myanmar (68), Thailand (73), Philippines (88), Singapore (96), Brunei Darussalam (101), Malaysia (106), and Cambodia (114). Singapore's position in the ranking is 96, which is higher than Malaysia and Cambodia; this is questionable because even though Singapore is a small island country, its marine fisheries are very limited and are not a source of the people's economy. Overall, however, Singapore faces the threat of climate change in terms of rising sea levels that could flood its coastal zone.

Vietnam's economy is heavily dependent on agriculture and fishing. Fisheries and aquaculture contribute about 6–7% of GDP (The World Bank Group and The Asian Development Bank, 2020) It is currently the world's second largest exporter of seafood. Vietnam also has significant freshwater fish from the Mekong River and Red River deltas, in the southern and northern parts of the country, respectively. The Mekong River is considered to be the most biodiverse river system after the Amazon, with 781 scientifically described species, making it the largest freshwater fishery in the world (ICEM, 2010). Climate change impacts on these sectors could include river and coastal flooding, as well as saltwater intrusion, which are known to reduce the productivity of aquaculture operations (The World Bank Group and The Asian Development Bank, 2020). An increase of 0.5–1.5°C was recorded in the Mekong River region over the last 50 years (WWF, 2009). This will lead to changes in the environment and biodiversity. Thailand, Myanmar, Laos, and Cambodia, which also use the Mekong River, will be similarly affected. Evers and Pathirana (2018) described the impacts of climate change on the Mekong Basin, including increases in temperature and annual rainfall, deeper and longer flood durations in the Mekong Delta and Cambodian floodplains, runoff variability, prolonged agricultural drought, sea-level rise, and sea and salinity intrusion in the Mekong Delta. The construction of dams and hydropower plants along the

Mekong River has added to the problems, such as the loss of around 1 million tonnes of fish every year and rising tensions between countries (The Asean Post, 2021). Coastal flooding can have an impact on coastal infrastructure that is critical to the coastal economy, including from marine fisheries. Rising sea surface temperatures and acidification have become major concerns as these will lead to reduction in the maximum catch potential in marine fisheries (Table 16.1).

Thailand has more than 3,565 km of coastline, 2,700 km of the Gulf of Thailand, and 865 km of the Andaman Sea (Wikipedia, 2021). According to Sampantamit et al. (2019), in 2016, Thailand's fisheries production was around 2.4 million tonnes (63% came from capture fisheries), which accounted for about 0.8% of total GDP or 9.0% of agricultural GDP. The gradual rise in sea level was recorded at 3–6 mm per year from 1981 to 2006 (Panpeng and Ahmad, 2017: 2), which causes shoreline retreat, tidal flooding, and erosion.

With an annual marine fish production of more than 6 million tonnes, Indonesia ranks second after China as the world's largest fishing country (FAO, 2020). Fisheries and aquaculture contributed 2.6% to GDP, equivalent to US$26.9 billion in 2018 (Alifiaberizky, 2021). This sector provides 7 million jobs (California Environmental Associates, 2018: 12) and more than 50% of national animal protein consumption (Poernomo and Kuswardani, 2019: 107). As an archipelagic country with more than 95,000 kilometres of coastline and 17,000 islands, Indonesia is vulnerable to climate change (The World Bank Group and The Asian Development Bank, 2021). Huelsenbeck (2012) places Indonesia in the ninth rank of the ten countries most vulnerable to insecurity of fisheries for food as a result of climate change, although according to Allison et al. (2009) the impact of climate change on Indonesian fisheries will be moderate. According to Kulp and Strauss (2019), by 2050 rising sea levels and coastal flooding could affect 23 million Indonesians every year (see Poernomo and Kuswardani, 2019).

The Philippines produced 4.4 million metric tonnes of fish in 2019 (capture and aquaculture), which contributed about 1.2% of the country's GDP and provided employment to more than 1.9 million people (BFAR, 2020). Similar to Indonesia, as an archipelagic country, the Philippines is one of the countries that is most vulnerable to climate change (Badjeck et al, 2010). The Philippines is very sensitive to extreme weather events in terms of people affected and economic costs (Paun et al., 2018). Considering three scenarios, one baseline scenario and two climate change scenarios based on greenhouse gas concentration (RCP 2.6 and RCP 8.5), Suh and Pomerory (2020) projected that climate change would lead to a reduction of around 9–18% of fisheries GDP under the mitigation scenarios (RCP 2.6 and RCP 8.5) by 2060, compared to the baseline scenario (i.e., if conditions were not changed). However, they further conclude that there will be a 0.36% reduction in income for urban households and 0.38% for rural households in the Philippine economy. Although the impact is not large, it adds to the burden on these households because they are already in the poorest community group.

Similar to other AMS, climate change impacts were observed in Malaysia. According to Samah et al. (2019), the average temperature in Peninsular Malaysia has increased by 0.25°C per decade, Sabah by 0.20°C per decade, and Sarawak by 0.14°C per decade. Rising sea levels around Malaysia were noted by Izzati et al. (2018). The increase ranged from 3.27 ± 0.12 mm/yr in eastern Malaysia to 4.95 ± 0.15 mm/yr in western Malaysia. For fisheries, Shaffril et al. (2016) observed that the impact of climate change has reduced the number of fishing days, which can reach 90% a month during the northeast monsoon. In another study, Yaakob and Quah (2005) concluded that climate change resulted in a 32% reduction in fish catches in Malaysia.

Nationally determined contributions and ocean issues in Southeast Asia

All AMSs are signatories of United Nations Framework Convention on Climate Change (UNFCCC), and as of May 2021, all have submitted their Nationally Determined Contributions (NDCs), either first or second submissions (UNFCCC, 2021). The countries are required to update their NDCs every five years, and this was due in November 2021 before the 26th conference of the parties to the convention. Dedicatoria and Diomampo (2019) have studied the AMS' NDCs and concluded that their overall goal is to be 'climate resilient' to achieve sustainable development. They also noted that the top priorities for climate change adaptation are Water Resources, Health and Agriculture; and Coastal/Marine Resources. However, the latter is not prioritized by Singapore (for agriculture) and Laos (for coastal/ marine resources), as Singapore is not an agricultural country while Laos is a landlocked one. Gallo et al. (2017) have examined the submitted NDCs from 161 countries and concluded that 70% include marine issues. They categorize six marine sub-issues, namely ocean warming (in 76 NDCs; six from AMS), ocean acidification (in 14 NDCs), ocean deoxygenation (in one NDC), mangroves (in 45 NDCs; six from AMS), coral reefs (in 28 NDCs; one from AMS) and blue carbon (in 27 NDCs; three from AMS). They also found that the dominant concerns were coastal impacts, ocean warming, and fisheries (of which respectively 95, 77, and 72 NDCs cited them).

Recently, based on the adaptation measures in AMS' National Communications – reports that each Party to the Convention prepares periodically in accordance with the guidelines developed and adopted by the conferences of the parties – Pereira (2020: 19) summarizes the readiness of ASEAN member states to address projected climate change impacts for a particular region. There are seven projected impacts, and she further stated that except for rice, the level of preparedness to deal with these impacts is moderate. It is not surprising that Brunei Darussalam, Indonesia, Malaysia, the Philippines, Thailand, and Vietnam have placed adaptation measures to address the decline in rice yields as top priorities since rice is the main source of carbohydrates in these countries. In the marine and fisheries context, the impacts of climate change include sea-level rise, decreased marine biomass and fish stocks, decreased ocean net primary production, challenges to fisheries governance, and increased occurrence of ocean heat waves.

A regional workshop organized by the Centre for International Law at the National University of Singapore revealed that there are three challenges that AMSs face in implementing their NDCs (CIL, 2010). The challenges are the absence of a basic knowledge management system for collecting and storing data, the lack of funds and financial assistance, and the difficulty of coordination within government and between stakeholders. In their analysis of gaps and opportunities in climate change governance in ASEAN, Seah and Martinus (2021) also observe a similar problem in institutional gaps, and they recommend that concentrated efforts are needed among AMSs. After studying coverage of climate change in English-language ASEAN newspapers during 2002–2012, Freeman (2017) concludes that attention has grown substantially since 2006. However, such attention is usually in the form of general normative statements rather than solution-oriented actions. Other gaps observed by Seah and Martinus (2021) include the absence of a coordinated body, lack of data and information protocols, lack of ASEAN leadership, and lack of platforms to engage civil society. They recommend the establishment of a regional coordinating body on climate change to promote information sharing across ASEAN bodies, gaining support from dialogue partners and engaging civil society. Overland et al. (2017) recommend that ASEAN formulate a 'regionally determined contribution', taking into account the shared concerns of AMS as set out in their

NDCs to serve as a basis for cooperation or to create team spirit in ASEAN. This could be used to mobilize external funding to strengthen climate adaptation in the region.

The regional framework for climate change adaptation

There are three pillars in ASEAN, namely the Political-Security Community, the Economy Community, and the Socio-Cultural Community, under which the ASEAN Sectoral Ministerial Bodies are formed. Each body may have under its purview the relevant senior officials and subsidiaries to undertake its functions. The ASEAN (Sectoral) Working Group (AWG) is one example of these subsidiaries. There is no specific definition of 'sector' in the ASEAN Charter, but the term generally refers to ministerial mandates in AMS, such as transport, communications, forestry and agriculture, environment, energy and minerals, education, and so forth. The ASEAN Working Group on Climate Change is under the ASEAN Ministerial Meetings on the Environment and ASEAN Senior Officials on the Environment. This working group was formed in 2009 to enhance regional cooperation on climate change, promote collaboration between sectoral bodies, and articulate ASEAN concerns and priorities at the international level (Seah and Martinus, 2021). Climate change is a cross-cutting issue in the region, and it is reflected in the overlapping coverage of several AWGs on climate change and its impacts. Seah and Martinus (2021) point out that at least 13 programmes within ASEAN sectoral bodies have recognized climate threats.

In 2010, the ASEAN Climate Change Initiative (ACCI) was established as a regional consultative platform to further strengthen regional coordination and cooperation in addressing climate change, and to undertake concrete actions to respond to its adverse impacts. The scope of collaboration through the ACCI includes policy and strategy formulation, information sharing, capacity building, and technology transfer. The ASEAN Climate Change Initiative is coordinated by the ASEAN Working Group on Climate Change (AWGCC) (Letchumanan, 2010). However, according to Seah and Martinus (2021), since its establishment it has not been clear how ACCI operates. Earlier, Goron (2014: 108) indicated that, although the functions of ACCI are modest and limited to providing a 'consultative platform to exchange views on international climate negotiations', it has come short on delivering these limited functions.

The 'ASEAN Way' is a regional integration approach adopted by ASEAN (see Acharya, 2009). It is a way of implementing the 14 ASEAN Principles as set out in the ASEAN Charter, and it emphasizes member states adhering to the principles of sovereignty, non-intervention, and consensus decision-making. As such, ASEAN does not have a legal mandate to require AMS to adopt any specific policies at the national or regional level. Policies adopted at the ASEAN level must be approved by consensus through the appropriate ASEAN institutional organs, thus limiting the organization's ability to produce legally binding agreements on various issues, including on climate change. Regarding the climate change governance framework in ASEAN, Coen et al. (2020) consider it weak, and the ASEAN Way has a role in this weakness. The AWGCC is seen as failing as a consultative body that would support coordination and collaboration among AMS sectoral bodies and provides little means to align policies or lead climate action in the region. In the UNFCCC negotiations, Goron (2014) noted that ASEAN's voice was quite fragmented and lacked weight, and it limited the influence of ASEAN countries.

The ASEAN Sectoral Working Group on Fisheries (ASWGFi) is under the ASEAN Ministerial Meetings on Agriculture and Forestry (AMAF). Climate change in the fisheries is included in the issue of food security as indicated in the ASEAN Multi-Sectoral Framework for Climate Change: Agriculture and Forestry towards food security and nutrition, which was adopted by the 40th AMAF Meeting in 2018. Two years later, ASWGFi endorsed the Strategic Action Plan for ASEAN Cooperation in the Fisheries Sector 2021–2025. There are six strategic 'thrusts' in the Action Plan, one of which is in the field of fisheries, namely Thrust 4: Increasing the resilience of fisheries and aquaculture against climate change, natural disasters, and other shocks (see Table 16.2).

One of the partners to implement the Action Plan is the Southeast Asian Fisheries Development Centre (SEAFDEC) which has long collaborated with ASEAN, especially ASWGFi. SEAFDEC is considered the technical arm of ASWGFi and has been able to attract funding from international donors. Since 2000, SEAFDEC and AMAF have developed decades of Resolutions and Action Plans on Sustainable Fisheries for Food Security for the ASEAN Region, and for the period 2020–2030, they were endorsed by AMS agricultural ministers at the 42nd AMAF meeting in 2020. The issue of climate change has always been part of the

Table 16.2 Thrust 4 of Strategic Plan of Action on ASEAN Cooperation on Fisheries 2021–2025

Activities/sub-activities	*ASEAN bodies, lead countries*	*Timeline*	*Output indicators*
1 Information sharing on climate change impact and adaptation strategies for fisheries and aquaculture, early warning systems/ indicators of climate impacts, and climate-resilient aquaculture/fisheries technologies and approaches.	ASWGFi, AFCF	2021–2025	Climate change-related information on fisheries and aquaculture is compiled.
2 Stocktake on climate risk and mitigation and adaptation options across ASEAN fisheries and aquaculture sector.	ASWGFi, AFCF	2021–2025	Stocktaking report.
3 Develop regional guidelines on indicators for aquaculture and capture fisheries to facilitate ecosystem-based adaptation to the impact of climate change.	ASWGFi, AFCF Philippines and Vietnam	2021–2025	Regional guideline is endorsed by SOM-AMAF.
4 Identify and promote green technologies for adoption, such as low greenhouse gas emission, carbon sequestration from marine waters, technology for aquaculture, and Integrated Multi-trophic Aquaculture (IMTA).	ASWGFi, AFCF	2021–2025	List of green technologies identified and promoted.
5 Establish regional data and information on critical habitats such as mangrove, seagrass, and coral reef as well as linkage between relevant institutions in AMS.	ASWGFi	2021–2025	Regional information on critical habitats and linkage with institutions in AMS is established.

ASWGFi: ASEAN Sectoral Working Group on Fisheries.
AFCF: ASEAN Fisheries Consultative Forum.

SEAFDEC plan, which calls for 'Increasing the resilience of fishing communities in anticipating and adapting to changes in inland and coastal water environments, including those caused by climate change, which can adversely affect communities in fisheries and aquaculture operations' (SEAFDEC, 2000).

Indonesia, Malaysia, the Philippines, Papua New Guinea, the Solomon Islands, and Timor-Leste are located in an approximately 5.7 million km^2 area called the Coral Triangle, the boundaries of which are determined based on the diversity of corals and reef fish. In 2009, these countries established the Coral Triangle Initiative on Coral Reefs, Fisheries, and Food Security (CTI-CFF) to maintain outstanding marine and coastal resources by addressing critical issues such as food security, climate change, and marine biodiversity (see Chapter 15). This initiative is financially supported by international donors such as the Global Environment Facility, the United States Agency for International Development, the Australian government, and the consortium of international nongovernmental organizations. A working group on climate change adaptation was formed in 2018, and an Early Action Plan for climate change adaptation across the region was delivered in 2015. The plan requires, among other things, the implementation of effective adaptation measures for coastal communities and investment in the ability to conduct climate change vulnerability assessments and plan to increase the resilience of coastal communities. Fidelman (2019) has elaborated extensively the climate change impacts in the Coral Triangle and the adaptive capacity of the countries.

With support from development partners (Australia, Denmark, Luxembourg, Sweden, Finland, Germany, and the European Union), the Mekong River Commission (MRC) established a project called the Climate Change and Adaptation Initiative in 2009, with Cambodia, Laos, Thailand, and Vietnam as the main beneficiaries. This is a long-term initiative with at least three five-year phases starting in 2011 (MRC, 2009). The scope of this project is climate change impact assessment as well as adaptation planning and implementation integrated with development planning in the Lower Mekong Basin (LMB). The project is expected to address the transboundary impacts of climate change and also look at adaptation from a regional perspective. It aims to contribute to achieving Millennium Development Goals on poverty alleviation and increased food security.

Climate change adaptation by individual ASEAN member states

The AMS's responses to climate change vary. Overall, they acknowledge that the impacts are significant, but they have not yet taken the necessary actions to mitigate and adapt to these impacts. Issuing policies on climate change and creating or restructuring ministries and departments are among the actions that have been taken. Seah and Martinus (2021) have listed several new or modified government agencies in AMSs that deal with climate change, some even chaired by presidents and prime ministers. In terms of government organization, only Indonesia has its own ministry that handles fisheries, while other AMSs place fisheries under the ministry of agriculture or food. In addition, AMSs have issued legal frameworks on climate change in several forms, from acts or laws to strategies, as shown in Table 16.3. The Philippines is the only AMS that has issued a special act on climate change, while others have been in the form of policies, strategies, or action plans.

Similar to the regional framework, fisheries and aquaculture issues related to climate change in the AMS region are rarely discussed as separate strategies. The strategies usually fall under broad themes or sectors, such as agriculture, food security, oceans and coasts, management of water bodies such as rivers and lakes, or the environment. Countries such as Brunei Darussalam and Singapore do not even clearly mention fisheries and aquaculture in their frameworks,

Table 16.3 Legal frameworks on climate change in ASEAN member states

AMS	*Legal frameworks*
Brunei Darussalam	Brunei National Climate Change Policy (2020)
Cambodia	Cambodia Climate Change Strategic Plan (2014)
Indonesia	Indonesian National Action Plan on Climate Change Adaptation/RAN-API (2014)
Laos	Lao Climate Change Strategy (2010)
Malaysia	National Climate Change Policy 2009; National Agrofood Policy 2021–2030 or NAP 2.0
Myanmar	Myanmar Climate Change Strategy 2018–2030
Philippines	Climate Change Act of 2009; National Framework Strategy on Climate Change (2010–2022); National Climate Action Plan (2011–2028)
Singapore	National Climate Change Strategy 2012; Sustainable Singapore Blueprint 2015
Thailand	Master Plan on Climate Change (2015–2050)
Vietnam	National Strategy on Climate Change 2011; National Action Plan to Respond to CC 2012–2020

perhaps because fisheries and aquaculture do not play as big a role as other sectors. Brunei Darussalam focuses on forests and energy, while Singapore focuses on services. Indonesia, the largest fishing country in the region, places fisheries and aquaculture under the category of resilient coasts and small islands. Laos puts fisheries and climate issues under the management of the Mekong River. All of this indicates that fisheries and aquaculture are seen as integral parts of other sectors and, in terms of climate change adaptation, should be linked with them. Problems can arise in coordination between sectors, as priorities can sometimes differ according to the interests of the government or ministry. Sinaga (2020) notes that in Indonesia, the implementation of climate change policies is sometimes ineffective because of local governments and their autonomy. The Malaysian government has created a new ministry to deal with climate change, but more concerted efforts in this area are lacking given that there are ministries with overlapping jurisdictions (The Asean Post, 2019). Diomampo (2015: 203) notes that there are overlapping roles between national and sub-national sectors/stakeholders in AMS which cause confusion and lead to poor coordination and difficulty in mobilizing stakeholders.

Therefore, strong leadership is needed in each country so that efforts can be synergized and all sectors receive the right attention and priorities (on leadership, see Chapter 7).

While the impacts of climate change for fisheries and aquaculture have been addressed among the AMS, mitigation actions, such as emissions reductions and carbon sequestration, have not been considered. In the coastal and marine resources sector, the mitigation priority is to implement integrated coastal zone management plans, including mangrove conservation and plantation. Currently, there are more fossil-fuel-consuming fishing vessels operating than necessary to catch the available fish resources efficiently. It is assumed that reducing fleet excess capacity and switching from fuel-intensive techniques can substantially reduce carbon emissions. It is recommended to change techniques, such as dredging, bottom trawling, and beam trawling, to alternative techniques that use less fuel. (Ziegler and Hansson, 2003; Tyedmers, Watson and Pauly, 2005; Thrane, 2006). Improved governance, innovative technologies, and more responsible practices can generate increased and sustainable benefits from fisheries.

Conclusion

Southeast Asian countries are aware of the negative impacts of climate change and are responding to these impacts in various ways. At the regional level, ASEAN is the organization that coordinates action to address the impacts of climate change. Mutual understanding and cooperation between countries have been carried out through ASEAN organizations. Regional frameworks or mechanisms for addressing impacts have been issued by many technical working groups within the organization, or by the Secretariat in the form of policies or joint statements. For ocean issues and climate change, AMS concerns are ocean warming, coral bleaching, blue carbon systems, and mangroves. Fisheries and aquaculture are usually placed in the agricultural, food security or marine/coastal sectors. Regional initiatives to address the impacts of climate change on fisheries are contained in the Strategic Action Plan for ASEAN Cooperation in the Fisheries Sector 2021–2025. Overlaps between technical working groups within ASEAN and between ministries or departments within individual AMS have been identified. Gaps and challenges in implementing frameworks or mechanisms have been shown to hinder progress. Climate change adaptation is multi-sectoral, but unfortunately good coordination is sometimes lacking, while sectoral approaches are not uncommon. In this case, fisheries and aquaculture are often neglected. Representative coordinating bodies and strong leadership at regional and individual AMS levels, along with engagement of civil society, may pave the way for more effective adaptation to climate change in Southeast Asia.

References

Acharya, A. 2009, '*Constructing a security community in Southeast Asia: ASEAN and the problem of regional order*'. 2nd Edition. London and New York: Routledge, pp54–98.

Alifiaberizky. 2021, 'Garuda's efforts on increasing fisheries export: How is Indonesia's Fisheries Industry?' https://medium.com/with-bright-indonesia/garudas-efforts-on-increasing-fisheries-export-how-is-indonesia-s-fisheries-industry-2724d2716d9b

Allison, E. H., Perry, A. L., Badjeck, M.-C., Adger, W. N., Brown, K., Conway, D., Halls, A. S., Pilling, G. M., Reynolds, J. D., and Andrew, N. K. 2009, 'Vulnerability of national economies to the impacts of climate change on fisheries', *Fish and Fisheries* 10 (2):173–196.

Asche, F. 2018, 'Impacts of climate change on the production and trade of fish and fishery products', *The State of Agricultural Commodity Markets (SOCO) 2018- Background Paper*. Rome, FAO. 44p.

ASEAN. 2021, *ASEAN Cooperation on Climate Change*. https://environment.asean.org/asean-working-group-on-climate-change/

BFAR. 2016, 'Philippine fisheries profile 2015', *Quezon: Bureau of fisheries and aquatic resources*. 72p.

Barange, M., Merino, G., Blanchard, J.L., Scholtens, J., Harle, J., Allison, E.H., Allen, J.I, Holt, J. and Jennings, S. 2014, 'Impacts of climate change on marine ecosystem production in societies dependent on fisheries', *Nature Climate Change* 4: 211–216 (also available at https://doi.org/10.1038/nclimate2119).

Blasiak, R., Jessica Spijkers, J., Tokunaga, K., Pittman, J., Yagi, N, and Oesterblom, H. 2017, 'Climate change and marine fisheries: Least developed countries top global index of vulnerability', *PLoS One* 12(6): 1–15. https://doi.org/10.1371/journal.pone.0179632

California Environmental Associates. 2018, 'Trends in marine resources and fisheries management in Indonesia: A 2018 review, 145p.

Chan, C.Y., Tran, N., Dao, C.D., Sulser, T.B., Phillips, M.J., Batka, M., Wiebe, K. and Preston, N. 2017, 'Fish to 2050 in the ASEAN region', *Penang, Malaysia: WorldFish and Washington DC, USA: International Food Policy Research Institute (IFPRI). Working Paper*: 2017-01.

Cheung, W.W.L., Jorn Bruggeman, J., and Butenschön, M. 2018, Projected changes in global and national potential marine fisheries catch under climate change scenarios in the twenty-first century. In Barange, M., Bahri, T., Beveridge, M.C.M., Cochrane, K.L., Funge-Smith, S., and Poulain, F. (eds.) 2018, 'Impacts of climate change on fisheries and aquaculture: Synthesis of current knowledge, adaptation and mitigation options.' FAO Fisheries and Aquaculture Technical Paper No. 627 Rome, FAO. pp63–86.

CIL. 2019, 'Decoding the Paris rulebook for Southeast Asia: Implementation and its challenges', Singapore CIL-NUS, 44p. https://cil.nus.edu.sg/publication/decoding-the-paris-rulebook-for-southeast-asia-implementation-and-its-challenges/

Coen, D., Kreienkamp, J. and Pegram, T. 2020. 'Multilevel governance of global climate change - Problems, policies, and politics case studies of the EU, ASEAN, and National Climate Laws', Barcelona, Globe, 134p.

Dedicatoria, R.M.M. and Diomampo, C.B. 2019, 'Status of climate change adaptation in Southeast Asia Region' In Alam, M., Lee, J., and Sawhney, P., (eds.) *'Status of climate change adaptation in Asia and the Pacific'*, Springer, Cham. pp153–182. https://doi.org/10.1007/978-3-319-99347-8.

Diomampo, C.B. 2015, 'Monitoring and evaluation of climate change adaptation initiatives in South-East Asia', in Sawney, P. and Perkins, M. (eds) *Emerging climate change adaptation issues in the Asia-Pacific region*, Bangkok, IGES. pp181–207.

Eckstein, D., Kunzelm, V., and Shafer, L. 2021, 'Global Climate Index 2021', Bonn, Germanwatch e.V. 48p.

Evers. 2020, 'The State of World Fisheries and Aquaculture 2020', *Sustainability in action*. Rome. https://doi.org/10.4060/ca9229en

Evers, J., and Pathirana, A. 2018, 'Adaptation to climate change in the Mekong River Basin: Introduction to the special issue', *Climatic Change* 149: 1–11.

Fidelman, P. 2019, 'Climate change in the coral triangle: Enabling institutional adaptive capacity', In Harries, P.G. (ed.). *Climate change and ocean governance: Politic and policy for threatened seas*. Cambridge. Cambridge University Press, pp274–289.

Freeman, B.C. 2017, 'Claims, frames, and blame: Coverage of climate change in ASEAN's English-Language Newspapers, 2002–2012', *Sage Open* 7(1): 1–12.

Funge-Smith, S. and Bennet, A. 2019, 'A fresh look at inland fisheries and their role in food security and livelihoods', *Fish and Fisheries* 20: 1176–1195.

Gallo, N.D., Victor, D.G. and Lisa, L.A. 2017, 'Ocean commitments under the Paris Agreement', *Nature Climate Change*. 7p. DOI: 10.1038/nclimate3422

Gopal, T. and Anbumozhi, V. 2019, 'Effects of disasters and climate change on fisheries sectors and implications for ASEAN food security', in Anbumozhi, V., M. Breiling, and V. Reddy (eds.), *Towards a resilient ASEAN volume 1: Disasters, climate change, and food security: Supporting ASEAN resilience. Jakarta, Indonesia: Economic Research Institute for ASEAN and East Asia*, pp161–188.

Goron, C. 2014, 'EU-ASEAN relations in the Post-2015 climate regime: Exploring pathways for top-down and bottom-up climate governance' in Hofmeister, W and Rueppel, P. (eds,), *Climate change diplomacy. The way forward for Asia and Europe*. Singapore: Konrad Adenauer Stiftung, pp. 101–129.

Huelsenbeck, M. 2012, 'Ocean-based food security threatened in a high CO_2 world: A ranking of nations', *Vulnerability to Climate Change and Ocean Acidification*. Washington, DC, Oceana. 14p.

ICEM. 2010, 'Strategic environmental assessment of hydropower on the Mekong mainstream', *Final Report, International Centre for Environmental Management*.

Izzati, A., Hamid, A., Hassan, A., Hwang, C., Fadila, N., Tugi, A., and Mohd, K. 2018, 'Contemporary sea level rise rates around Malaysia: Altimeter data optimization for assessing coastal impact', *Journal of Asian Earth Sciences* 166: 247–259.

Kulp, S.A. and Strauss, B.H. 2019, 'New elevation data triple estimates of global vulnerability to sea-level rise and coastal flooding', *Nature Comunications*, 10(4844): 12. https://doi.org/10.1038/s41467-019-12808-z

Letchumanan, R. 2010, Climate change: Is Southeast Asia up to the challenge? Is there an ASEAN policy on climate change?, in Nicholas, K. (ed) *IDEAS reports - special reports*, Kitchen, Nicholas (ed.) (SR004).LSE IDEAS, London, pp 60–62.

MRC. 2009, 'Climate change and adaptation initiative framework', *Document for Implementation and Management*. 29p.

Overland, I. et al. 2017, 'Impact of climate change on ASEAN international affairs: Risk and opportunity multiplier', *Norwegian Institute of International Affairs and Myanmar Institute of International and Strategic Studies*, available at https://www.researchgate.net/publication/320622312_Impact_of_Climate_Change_on_ASEAN_International_Affairs_Risk_and_Opportunity_Multiplier)

Panpeng, J. and Ahmad, M.M. 2017, 'Vulnerability of fishing communities from sea-level change: A study of Laemsing district in Chanthaburi Province, Thailand', *Sustainability* 9: 1388; doi:10.3390/su9081388

Paun, A., Acton, L., and Chan, W. S. 2018, 'Fragile planet: Scoring climate risks around the world', London: HSBC Bank Plc, 64p.

Pereira, J.J. 2020, 'Climate Change: Building Resilience in the ASEAN Community', *The ASEAN*, :18–19.

Poernomo, A. and Kuswardani, A. 2019, 'Ocean policy perspectives: The case of Indonesia', In Harries, P.G. (ed.). *Climate change and ocean governance: Politic and policy for threatened seas*. Cambridge. Cambridge University Press, pp102–117.

Raitzer, D.D., Bosello, F., Tavoni, M., Orecchia, C., Marangoni, G. and Samson, J.N.G. 2015, 'Southeast Asia and the economics of global climate stabilization', Manila, Asian Developmet Bank, 166p.

Samah, A.A., Shaffril, H.A.M., Azimi Hamzah, A. and Samah, B.A. 2019, 'Factors affecting small-scale fishermen's adaptation toward the impacts of climate change: Reflections from Malaysian fishers', *SAGE Open*, July-September, 1–11. DOI: 10.1177/2158244019864204

Sampantamit, T., Noranarttragoon, P., Lachat, C. and Goethals, P. 2019, 'Evolution of fish and shell-fish supplies originating from wild fisheries in Thailand Between 1995 and 2015', *Sustainability* 11(24), 7198: 16. https://doi.org/10.3390/su11247198

SEAFDEC. 2017, 'Southeast Asian state of fisheries and aquaculture 2017', *Southeast Asian Fisheries Development Center*, Bangkok, Thailand. 167 pp.

SEAFDEC. 2020, 'Resolution and plan of action on sustainable fisheries for food security for the ASEAN region towards 2030', *Southeast Asian Fisheries Development Center*, Bangkok, Thailand; 23 pp.

Seah, S., and Martinus, M. 2021, 'Gaps and opportunities in ASEAN's climate governance', Singapore, ISEAS-Yusuf Ishak Institute, 29p.

Shaffril, H.A.M., Samah, A.A. and D'Silva, J.L. 2017, 'Adapting towards climate change impacts: Strategies for small-scale fishermen in Malaysia', *Marine Policy* 81: 196–201.

Sinaga, L.C. 2020, 'Assessing the commitment of Indonesian government towards climate change policy: The Yudhoyono presidency 2004–2014', *Politica* 11(2): 163–182. doi: 10.22212/jp.v11i2.1752

Suh, D and Pomeroy, R. 2020, 'Projected economic impact of climate change on marine capture fisheries in the Philippines', *Frontiers in Marine Science* 7, Article 232: 14p.

Tang, S.M., Hoang, T.H. and Ong, G. 2020, *The State of Southeast Asia: 2020 Survey Report*. Singapore: ISEAS-Yusof Ishak Institute. 56p.

The Asean Post. 2019, Global strike for climate change' https://theaseanpost.com/article/global-strike-climate-change

The Asean Post. 2021, 'What's at stake for the Mekong's fishery', https://theaseanpost.com/article/whats-stake-mekongs-fishery

The World Bank Group and The Asian Development Bank. 2020, 'Country Risk Profile: Viet Nam', Washington, The World Bank Group and The Asian Development Bank, 27p.

The World Bank Group and The Asian Development Bank. 2021, 'Country Risk Profile: Indonesia', Washington, The World Bank Group and The Asian Development Bank, 30p.

Thrane, M. 2006, 'LCA of Danish Fish Products: New Methods and Insights' *International of Journal Life Cycle Assessment* 11(1): 66–74

Tyedmers, P.H., Watson, R. and Pauly, D. 2005, 'Fuelling global fishing fleets' *Ambio* 34(8): 619–622.

UNFCCC. 2021, *The latest submission*. https://www4.unfccc.int/sites/ndcstaging/Pages/LatestSubmissions.aspx

Wikipedia. 2021. Fishing industry in Thailand. https://en.wikipedia.org/wiki/Fishing_industry_in_Thailand

WWF. 2009. 'The greater Mekong and climate change: Biodiversity, ecosystem services and development at risk', Bangkok, WWF. 31p.

Ziegler, F. and Hansson, P. 2003, 'Emissions from fuel combustion in Swedish cod fishery', *Journal Cleaner Production* 11: 303–314.

Yaakob, M. and Chau, Q.P. 2005, 'Weather downtime and its effect on fishing operation in Peninsular Malaysia', *Jurnal Teknologi* 42(A): 13–26. https://doi.org/10.11113/jt.v42.730

Yusuf, A.A., and Fransisco, H. 2009, 'Climate change vulnerability mapping for Southeast Asia', Singapore, EEPSEA. 26p.

17

THE BALTIC SEA AND GLOBAL ENVIRONMENTAL CHANGE

Best-in-class governance?

Marko Joas, Henrik Ringbom, and Nina Tynkkynen

The Baltic Sea provides a prime example of a multilevel and network-type system of environmental governance (Grönholm, 2020; Tynkkynen, 2013; VanDeveer, 2011; Haas, 1990). The Baltic can also be used as a time machine to study the consequences and mitigation of the effects of future global environmental changes because the region has already experienced many changes that are yet to be experienced in other sea areas (Reusch et al., 2018). Because of its small size and certain geophysical characteristics, the Baltic Sea is particularly vulnerable to environmental pressures from multiple interacting anthropogenic drivers at both local and global scales (Hyytiäinen et al., 2013). Its marine ecosystem suffers particularly from the problem of eutrophication (HELCOM, 2018a; 2018b; Fleming-Lehtinen et al., 2015; Wulff et al., 2007), which is an ongoing gradual increase in the concentration of nutrients (phosphorus and nitrogen) in the aquatic ecosystem of the Baltic Sea. That problem is further exacerbated by the effects of climate change (Korpinen et al., 2012).

The environmental governance of the Baltic Sea, which is achieved by the regulatory framework of the European Union, a regional convention for marine protection (the Helsinki Convention, 1974/1992), ambitious national environmental policies in many of the coastal countries, and active local actors, has enabled certain trend reversals, such as the return of top predators, recovering fish stocks and reduced input of harmful substances during the last couple of decades (Heiskanen et al., 2019; Elmgren et al., 2015). Yet, the nine countries surrounding the Baltic Sea have different historical trajectories, policy priorities, financial resources, industrial structures, and socio-economic development levels that impact how various policies, strategies, and plans are implemented across the region (Ollikainen et al., 2019; Tynkkynen et al., 2014; Tynkkynen, 2013). As the Cold War fault line between the Western liberal democracies and the Eastern state socialist countries, the region has experienced a number of (geo)political changes over time. This affects environmental governance in the region, particularly regarding the engagement of Russia in regional cooperation (Tynkkynen, 2018; Tynkkynen, 2015a; Korppoo et al., 2015). The expansion of EU-centred political transformation has also had a major impact.

There are two sets of challenges that make the Baltic Sea stand out when compared to the governance of other regional sea areas. The first challenge is global climate change, which is expected to have particularly significant effects in northern sea areas such as the Baltic. It is predicted to further exacerbate both eutrophication and biodiversity loss, making the

DOI: 10.4324/9781315149745-21

current efforts to combat these threats insufficient (Andersson et al., 2015; Jetoo et al., 2022). The second challenge is posed by the so-called ecosystem approach to management (EAM), which has become the guiding marine environmental policy principle in recent years (e.g., Waylen et al., 2014). This is particularly noticeable in EU marine legislation, where the Baltic Sea represents a model of collective action at the regional level, with the 2007 Baltic Sea Action Plan (BSAP, adopted by all Helsinki Convention HELCOM parties) at the heart of the process (Ringbom and Joas, 2018b). A key feature of the EAM is that it considers people as part of the social-ecological system (Berkes and Davidson-Hunt, 2012). This development necessitates a corresponding shift from sectoral management to multi-sectoral and holistic approaches (e.g., Ramirez-Monsalve et al., 2016), setting new requirements for political and administrative institutions.

This chapter describes the evolution of the multilevel governance system and regulation in the Baltic Sea Region and specifically asks how the key environmental problem in the region, eutrophication, is challenged by climate change and the shift towards an ecosystem-based governance approach. Given that the Baltic Sea Region is one of the most elaborate, but also complex, regions in terms of marine environmental governance, its example can provide valuable insight into the challenges that global environmental changes present for marine governance more generally, in practice as well as in research.

The Baltic Sea environment: interacting local and global changes

The Baltic Sea is a brackish body of water body encompassed by the Scandinavian peninsula and the mainland of northern Europe. It is bordered by Denmark, Estonia, Finland, Germany, Latvia, Lithuania, Poland, Russia, and Sweden. Belarus, the Czech Republic, Norway, Ukraine, and Slovakia have land on its drainage basin. There are significant differences in cultural, economic, geographical, and ecological features in the region, extending from the sparsely populated areas of the Northern Baltic Sea to the densely populated, heavily industrialised areas around the Baltic proper and in the Southern Baltic Sea (Henningsen et al., 2017). There is a large and densely populated catchment area covering nearly 20% of the European continent.

Whereas eutrophication of coastal waters is a global phenomenon, the entire sea area in the Baltic is affected by it (Fleming-Lehtinen et al., 2015). While there were already reports of algal blooms in the late 1800s (Murray et al., 2019), the change in the Baltic Sea ecosystem has been most tangible since the early twentieth century due to the increased human population, intensification of agriculture and forestry, and industrialisation (Gustafsson et al., 2012). The extensive nutrient input into the Baltic Sea originates from various sources, including natural outflow from land, up-welling of phosphorus-rich deep water (internal load), atmospheric deposition of nitrogen, and other anthropogenic sources, such as municipal and single household wastewater, agriculture, and industry (HELCOM, 2018b). Agriculture constitutes two-thirds of anthropogenic nutrient pressures. Nutrient emissions drift from estuaries and coastal waters to the Baltic proper. Russia and Poland are the major sources of waterborne nutrients, but they do not suffer from extensive eutrophication in their territorial waters as their nutrient loads drift to other parts of the Baltic Sea (Tynkkynen et al., 2014). This evidently decreases their incentives to solve the problem. HELCOM BSAP addresses the problem of eutrophication by setting emissions-reduction targets for each coastal country (HELCOM, 2007). By now, it is clear that these reductions were not reached in time to meet the ambitious goal of the BSAP to restore the good ecological status of the Baltic marine environment by 2021 (HELCOM, 2020).

Eutrophication has negative impacts on the quality of seawater, biodiversity (flora and fauna), recreational activities, and livelihoods, such as fishing and tourism (Andersen et al., 2017; Rönnberg and Bonsdorff, 2004). While a significant reduction of nutrient load from point sources (industry, urban wastewaters) has been achieved across the Baltic Sea's catchment area, nutrient input from diffuse sources, especially from agriculture, is still a major challenge (HELCOM, 2018b) and has been considered a 'wicked' governance problem, a problem with multiple sources, complex interdependencies and no easy solutions upon which key stakeholders on all societal levels could agree (Shortle and Horan, 2017; Patterson et al., 2013). As a key reason for the stubborn characteristic of the nutrient problem, recent research identifies the mismatch between the scales of societal and ecological problems, as well as the absence of suitable governance structures for the integration of agricultural and environmental policies (Westholm, 2021; Wiering et al., 2020; Tynkkynen, 2015b). Even if public awareness about the problem has risen (Lyytimäki, 2012), eutrophication lacks the definitional, ideological, and symbolic clarity typical of many other environmental issues (Tynkkynen et al., 2014).

Environmental problems threatening the Baltic Sea ecosystem cannot be treated separately from each other. The cumulative effects of different human pressures decrease the resilience of species and habitats, particularly in areas where several human pressures coincide and have additive and synergistic effects (Korpinen et al., 2012). According to predictions, global climate change – particularly changes in temperature, salinity, ice-cover, and hydrodynamics – will have a major impact on the Baltic Sea environment. The Baltic Sea is already one of the fastest warming large marine ecosystems on the globe (Rutgersson et al., 2014), with the sea surface temperature having increased by 1.35°C on average between 1982 and 2006 (Belkin, 2009). Climate-related changes are likely to affect biology in the Baltic Sea in several ways, from the individual organism and species level to that of population and community, as well as to the habitat and ecosystem levels (Andersson et al., 2015; Ojaveer et al., 2010).

Predicted changes in precipitation will affect runoff into the sea, with potential increases in mean annual river flow in the northernmost catchments and decreases in the southernmost catchments (Arneborg and Gustafsson, 2020; HELCOM, 2007). Furthermore, there is a growing risk of more frequent and more extreme flooding and more soil and riverbank erosion, which would increase particulate inputs to surface waters. Warming can induce new and modified agricultural and forestry practices and extend agricultural areas, potentially leading to higher nutrient loads (BACC II, 2015). The difficulty with climate change lies in the uncertainty regarding its cumulative effects and the limited means to predict future changes, chain reactions and interactions with existing major stressors such as eutrophication (see Arneborg and Gustafsson, 2020).

Baltic Sea environmental governance: some background

A stalemate in the Baltic Sea regional governance occurred as a result of division lines during the Cold War, but these gradually withered in the early 1970s. This period of détente between the Soviet Union and the United States and Western Europe created an atmosphere of higher levels of trust, leading to options to find regional solutions. The period of détente followed the environmental awakening that took place mostly in Western Europe but also within the Eastern bloc to some extent during the same period (see DeBardeleben, 1985). The early environmental movement highlighted the degrading environmental state of the Baltic Sea. Environmental issues were considered less politicised than security or human rights issues, thus opening the door for closer cooperation in the Baltic Sea Region (Darst, 2001). Additionally, bilateral discussions between the Soviet Union and the United States paved the way for intergovernmental multilateral solutions in environmental protection (Räsänen and Laakkonen, 2007).

The Convention on the Protection of the Marine Environment of the Baltic Sea Area (Helsinki Convention) was signed in 1974 and was ratified by all Baltic Sea coastal countries (which included seven states at the time). It entered into force in 1980, being the first regional seas agreement in the world and providing a model for subsequent regional seas agreements adopted within the framework of the United Nations Environment Program (UNEP). From the outset, the Helsinki Convention had a very broad scope aimed at protecting the Baltic Sea from all sorts of pollution from the air, land, and sea (Räsänen and Laakkonen, 2007; Hjort, 1994). The Convention aims to monitor changes and conserve habitats and biological diversity as well as foster sustainability. Today, the Convention has a permanent organisation through its governing body, the Helsinki Commission, with a rotating chair, regular high representative and heads of delegation meetings, a permanent secretariat, working- and expert-group structures, and stakeholder involvement (Backer Johnsen, 2020; Valman, 2014; Hammer et al., 2011; Kern, 2011).

The three Baltic states (Estonia, Latvia, and Lithuania) joined the Convention after regaining their independence following the dissolution of the Soviet Union in 1991. The Convention was comprehensively updated in 1992; the revised Convention entered into force in 2000, with ten contracting parties (including the European Union). After Finland and Sweden joined the EU in 1995, followed by Estonia, Latvia, Lithuania, and Poland in 2003, the Baltic Sea, with the exception of Russia, has become an internal sea of the EU. This has had a significant effect on regional cooperation. The EU has brought along a framework-based way of working with regulation, highlighting subsidiarity with common minimum environmental standards as the guiding principle for environmental regulation, as well as the need for multiple policy tools to approach complex environmental problems (Heiskanen et al., 2019; Ringbom and Joas, 2018a; 2018b; Hammer et al., 2011). While the EU's increased presence in the region has promoted environmental protection in many ways, it should be noted that its involvement also includes the risk of limiting the possibilities of its member states to take individual initiatives within HELCOM and other international fora due to the shift of regulatory competence from EU member states. Maintaining the largely positive impact of the European Union for the governance of the Baltic Sea therefore depends on a continued interest and proactive role of the EU institutions in the subject. Another, more obvious shortcoming of a strong reliance on the EU for marine governance in the Baltic Sea is the fact that one of the coastal states (Russia) remains outside the reach of EU measures and policies (Tynkkynen, 2018).

The regionalisation process, which has been facilitated by EU enlargement (Browning and Joenniemi, 2004), is a movement from national and international governance regimes towards a system that highlights multilevel and multi-actor governance as a way to approach complex problems. In the Baltic case, both horizontal governance (within one government layer, like regions within a nation state) and vertical governance (between government layers; for example, European Union and nation states) are at a high level (Gilek et al., 2016; Tynkkynen, 2013; Kern, 2011; Eckerberg and Joas, 2004). This development started in the early 1990s with the mushrooming of regional organisations (see Schymik 2012; Lehti, 2009), such as the Council of the Baltic Sea States (CBSS), that mostly cover policy preparation and the implementation of policies but also, to some extent, decision-making through network-based cooperation. Network governance (Grönholm, 2018; Tynkkynen, 2013) has been a key feature of the governance setting in the Baltic Sea Region since the early 1990s, with a system of policy institutions that include several governmental and nongovernmental actors, often from different societal levels (Schymik, 2012; Joas et al., 2007). Network governance enables a relatively well-balanced societal discussion on the needs and means of policy design and policy implementation, but it also highlights problems in institutional learning as many of the networks are mainly project-funded (Grönholm, 2020). Project-based activities and funding schemes can make a difference within a short time span, but many of the contemporary environmental problems, such as eutrophication, require solutions and monitoring over a long period of time. Furthermore, many of these network bodies have a weak legal foundation, which reduces their ability to enact or implement regulatory measures in a national context (Engström, 2018; Koivurova and Rosas, 2018).

Regulatory aspects of Baltic Sea environmental governance

As a result of the multilevel governance setting described above, the Baltic Sea can be labelled as the world's most regulated sea. In jurisdictional terms, since the widespread ratification of the 1982 UN Convention on the Law of the Sea (UNCLOS) (see Chapter 2), the entire sea area has been delimited and covered by coastal zones, meaning that all of it is subject to the environmental jurisdiction of its littoral states. There are no areas of high seas left in the Baltic Region (Franckx, 2018). Apart from this, the Baltic is subject to a broad range of regulations applied at different levels, or layers, through international law, including EU rules, national laws, subnational rules, and decisions by local or regional authorities. To some extent, all these regulatory layers are involved in the quest for an improved marine environment in the Baltic Sea, but the emphasis and centre of gravity between layers vary from one theme to another. There are no global treaties in place to address eutrophication. Instead, the key rules on this matter are to be found at the regional level, notably in EU law and in the Helsinki Convention. The setting is the opposite of climate change regulation, where global rules adopted under the UNFCCC framework dominate the regulatory landscape. Considering that the basic causes underlying eutrophication are well known, it is notable that only a handful of direct rules aimed at limiting the sources of eutrophication have been adopted.

In the Helsinki Convention framework in particular, a series of principles and guidance documents have been adopted, but the main rules of the convention and its annexes are held at a generic level, while the more concrete reduction targets have been mainly in the form of recommendations. At the EU level, some key rules, notably the nitrate directive (91/676/EEC) and the urban wastewater directive (91/271/EEC), place ceilings for the release of certain nutrients into the sea. These rules have been implemented and complemented by national rules, which extend, in part, beyond the scope of the regional requirements. Yet, while the low-hanging fruit of easily identifiable point sources has been picked over throughout past decades, the more difficult and diffuse sources of eutrophication remain largely unregulated. There are, for example, no rules placing limits on nutrient pollution from agriculture or aquaculture as such. The political and legal complexity involved in regulating diffuse sources of eutrophication has generated an interest in favour of employing marine geoengineering technology as a complementary method to reduce eutrophication in the Baltic Sea through sea-based measures, which in turn raises a series of other difficulties (see Ringbom, Bohman and Ilvessalo, 2019).

The main mechanism towards more holistic environmental governance for the Baltic Sea has been the shift towards ecosystem-based rules and management regimes. The implementation and effects of an ecosystem-based approach to management have been extensively studied in the Baltic Sea context (Ramirez-Monsalve et al., 2016; Hammer, 2015; Valman, 2015; Bohman and Langlet, 2015; Hassler et al., 2013). Through 'integrated maritime policy', all main aquatic legislation adopted in the EU since 2000 obliges states to achieve a certain ecological or socio-ecological goal and provide a certain procedural framework for that purpose. The Water Framework Directive (2000/60/EC), the Marine Strategy Framework Directive (2008/56/EC) and Directive 2014/89/EU establishing a framework for maritime spatial planning (MSPD) provide for generic goals, like 'good ecological/environmental status' of marine areas, to be met within a certain time frame (as stated in WFD art. 4 and MSFD art. 1). The rules focus on environmental goals and procedural obligations, including the adoption of cyclical and evolving plans and programmes, rather than prescriptive standards or substantive obligations, as well as the types of measures to be taken in order to achieve the objectives. Thus, the rules leave broad discretion for states to decide and prioritise the measures required to meet the goals.

Such goal-based rules have many advantages over traditional command-and-control rules because they are open-ended and flexible. In the context of marine environmental governance, they are particularly valuable as they can accommodate changing ecological features or new scientific knowledge and can permit varying requirements for regions with different ecological threats and regulatory needs. Thus, within this framework, the Baltic Sea states are free to – and are expected to – focus on the ecological challenges that dominate the region, notably eutrophication. A further key feature of goal-based rules is that they provide opportunities for states to use a multitude of legal and other tools to achieve them. These features allow for broader stakeholder involvement in the development and implementation phases, which enhances the legitimacy of regulation. Stakeholder engagement has become one of the dominant features of environmental governance in the Baltic Sea Region (Saunders et al., 2017; Linke and Jentoft, 2014; Linke et al., 2011). As noted by Evans et al. (2005), however, stakeholder engagement cannot replace institutional capacity as such, even if it makes the odds for success better in implementing new policies and regulations.

Yet, if the goal-based rules are not accompanied by the legal and actual powers necessary to meet them, the flexibility in terms of target, tools, and methods may create regulatory uncertainty compared to more detailed objectives and obligations. Measures adopted to date have largely been 'soft' governance measures, such as plans and action programmes of unclear legal status, which highlights the uncertainty as to the basic distribution of duties and responsibilities. This has resulted in variations as to how the rules are understood and implemented in different Baltic Sea states (Soininen et al., 2019; Ringbom and Joas, 2018b). All HELCOM measures can also be regarded as soft measures, including the BSAP. It has accordingly been acknowledged that greater coordination of different programmes and a clearer understanding of the roles, responsibilities and obligations of all parties are needed (see Soininen and Platjouw, 2019; ECA, 2016).

Lessons learned

Baltic Sea environmental governance stands out in terms of the robustness of governance institutions and its multilevel and multi-agency character, which may make it well-placed to combat climate change. The highly institutionalised inter-state cooperation adds an element of hierarchy for environmental governance, which gives the governance a certain institutional clarity. There are institutions that can quickly take action to combat new and emerging threats. That said, they are sometimes slow in employing new types of cross-sectoral and integrative management actions. The first lesson to be learned from five decades of Baltic

Sea governance is that different governance and regulatory solutions are required for dealing with different types of environmental challenges. Regarding eutrophication, a distinction should be made, on the one hand, between easily identifiable point sources of pollution, where traditional legislation targeting such as specific emissions standards may adequately address concerns and, on the other hand, diffuse sources of pollution, such as agriculture. By involving a much broader spectrum of polluters, substances, and scenarios, diffuse measures call for a more holistic regulatory approach that allows for flexibility and adaptability in both setting and meeting the requirements. What works in one location for a specific type of activity is not necessarily the best solution in another location.

A second lesson is that goal-oriented, ecosystem-based regulatory tools require a solid procedural and institutional backing to work. While the Baltic Sea governance system may be advanced, it has not yet resolved challenges relating to how to implement an ecosystem approach in terms of law, policy, and institutions (Kern and Söderström, 2018; Valman, 2014). In the EU and the Baltic Sea Region, much of the energy aimed at improving the environment of the Baltic Sea has been directed at meeting and operationalising the requirements emanating from the marine directives (Backer Johnsen, 2020; Soininen and Platjouw, 2019), whereas relatively little attention has been given to strengthening or broadening existing point source requirements (Ringbom and Joas, 2018a). Shifting away from traditional regulatory approaches involves a risk that potentially effective measures to address specific environmental concerns, including more precise standard-setting, clearer distribution of responsibilities and better enforcement mechanisms, are left unutilised. Legally precise and solid rules need not be in the form of command-and-control emissions standards, however. Law offers a broad spectrum of preventive, punitive, and economic tools to promote compliance. For example, in Baltic Sea governance, little creativity has been utilised to develop rules that involve financial consequences for persons who release nutrients into the sea and/or benefits for those who do not (Hautakangas and Ollikainen, 2011).

Conclusion

From an ecological point of view, it is difficult to characterise the environmental governance of the Baltic Sea over the past decades as a success. The main ecological threat to the sea – eutrophication – remains. There are no quick fixes to curb eutrophication. The emissions-reduction work that has been carried out over the past five decades through a variety of steering mechanisms is gradually beginning to show results. Patience and persistence accordingly appear to be essential when it comes to improving the sea from the threats it faces. Yet, both climate change and biodiversity loss are likely to exacerbate eutrophication so that they threaten to limit the effectiveness of current management schemes. Clearly, more could be done to reduce the outflow of nutrients in the sea area, not least in the regulatory sphere, for example by means of market-based regulation serving to link emissions of nutrients with financial consequences. Such efforts have not advanced in view of the practical and policy complexities they pose. The easy targets of regulation have largely been covered by regulation at national and regional levels.

In terms of governance, the Baltic Sea is a showcase of myriad actors, stakeholders, organisations, policies, and programmes that evolved over time in a particular geopolitical and historical context. This infrastructure can be used when preparing for global change. The Baltic Sea case offers a prism through which to analyse various multilevel governance arrangements and management practices. Such governance modes encompass uncertainties and knowledge gaps at different governance levels. Many questions remain regarding, for example, which modes work best at which governance level, and what the instances are when command-and-control governance and binding rules lead to better outcomes than other instruments. Future governance needs to be adaptive and flexible in order to meet the challenges of climate change, especially regarding scientific uncertainties, time scales, and disruptions to business as usual. The new strategies to deal with holistic, ecosystem-based approaches seem to require more time than expected to become operational and effective, and it is not yet entirely clear how these can be implemented in an encompassing way.

The ability to predict future environmental changes and chain reactions appears to be limited. While monitoring, modelling and other future research methodology can reduce the knowledge gaps, knowledge will remain incomplete and, to some degree, uncertain. It seems evident that global climate change will exacerbate other, more local-scale problems of marine ecosystems, such as eutrophication, and make them more difficult to manage. This also poses a challenge to research on the governance of these issues: there is an obvious need to move from knowing what is known to highlighting what is uncertain and how that uncertainty can be managed. This requires the recognition of experience and learning from the past.

Acknowledgements

This research has been conducted within the Strategic Research Profiling Area 'The Sea' at Åbo Akademi University, funded by the PROFI3–6 programmes by the Academy of Finland. Project grant 315715 by the Academy of Finland has supported the writing of this chapter.

References

Andersen, J. H., Carstensen, J., Conley, D. J., Dromph, K., Fleming-Lehtinen, V., Gustafsson, B. G., Josefson, A. B., Norkko, A., Villnäs, A. and Murray, C. (2017) 'Long-term temporal and spatial trends in eutrophication status of the Baltic Sea: Eutrophication in the Baltic Sea', *Biological Reviews*, vol 92, no 1, pp135–149.

Andersson, A., Meier, H. E. M., Ripszam, M. et al. (2015) 'Projected future climate change and Baltic Sea ecosystem management', *AMBIO* vol 44, suppl 3, pp345–356.

Arneborg, L. and Gustafsson B. (2020) *Framtidens Östersjön – påverkan av övergödning och klimatförändringar.* Nr 2/2020. Stockholms universitets Östersjöcentrum och SMHI, Stockholm.

BACC II Author Team. (2015) 'Second assessment of climate change for the Baltic Sea Basin', in *Regional Climate Studies*. Springer, Cham.

Backer Johnsen, H. (2021) 'The HELCOM Ecosystem Approach: Time for quantification, integration, and measures', PhD thesis, University of Helsinki, Helsinki.

Belkin, I. M. (2009) 'Rapid warming of large marine ecosystems', *Progress in Oceanography*, vol 81, pp207–213.

Berkes, F. and Davidson-Hunt, I. (2012) 'The cultural basis for an ecosystem approach. Sharing across systems of knowledge', in Waltner-Toews et al. (eds) *The Ecosystem Approach. Complexity, Uncertainty, and Managing for Sustainability. Complexity in Ecological Systems Series.* Columbia University Press, New York.

Bohman, B. and Langlet, D. (2015) 'Float or sinker for Europe's seas? The role of law in marine governance', in Gilek, M. and Kern, K. (eds) *Governing Europe's Marine Environment: Europeanization of Regional Seas or Regionalization of EU Policies?* Ashgate Publishing, Aldershot, pp53–73.

Browning, C. and Joenniemi, P. (2004) 'Regionality beyond security? The Baltic Sea Region after enlargement', *Cooperation and Conflict,* vol 39, no 3, pp233–253.

Darst, R. (2001) *Smokestack Diplomacy: Cooperation and Conflict in East-West Environmental Politics.* MIT Press, Cambridge, MA.

DeBardeleben, J. (1985) *The Environment and Marxism-Leninism. The Soviet and East German Experience.* Westview Press, Boulder, CO.

ECA. (2016) European Court of Auditors, Special Report, 'Combating eutrophication in the Baltic Sea: Further and more effective action needed', European Court of Auditors.
Eckerberg, K. and Joas, M. (2004) 'Multi-level environmental governance: A concept under stress?' *Local Environment*, vol 9, no 5, pp405–412.
Elmgren, R., Blenckner, T. and Andersson, A. (2015) 'Baltic Sea management: Successes and failures', *Ambio*, vol 44, suppl 3, pp335–344.
Engström, V. (2018) 'Complexities of the Baltic Sea regulatory framework', *Marine Policy*, vol 98, pp191–200.
Evans, B., Joas, M., Sundback, S. and Theobald, K. (2005) *Governing Sustainable Cities*. Earthscan, London.
Fleming-Lehtinen, V., Andersen, J. H., Carstensen, J., Łysiak-Pastuszak, E., Murray, C., Pyhälä, M. and Laamanen M. (2015) 'Recent developments in assessment methodology reveal that the Baltic Sea eutrophication problem is expanding', *Ecological Indicators* vol 48, no 1, pp380–388.
Franckx, E. (2018) 'Gaps in Baltic Sea maritime boundaries', in Ringbom, H. (ed) *Regulatory Gaps in Baltic Sea Governance*. Springer, Cham, pp7–20.
Gilek, M., Karlsson, M., Linke, S. and Smolarz, K. (eds) (2016) *Environmental Governance of the Baltic Sea*. Springer, Cham.
Grönholm, S. (2018) 'A tangled web: The Baltic Sea Region governance through networks', *Marine Policy*, vol 98, no 3, pp201–210.
Grönholm, S. (2020) 'Governing the Baltic Sea. A study of the functionality of contemporary environmental governance', PhD thesis, Åbo Akademi University, Turku.
Gustafsson, B. G., Schenk, F., Blenckner, T. et al. (2012) 'Reconstructing the development of Baltic Sea Eutrophication 1850–2006', *AMBIO*, vol 41, pp534–548.
Haas, P. M. (1990) *Saving the Mediterranean: Politics of International Environmental Cooperation*. Columbia University Press, New York.
Hammer, M. (2015) 'The ecosystem management approach: Implications for marine governance', in Gilek, M. and Kern, K. (eds) *Governing Europe's Marine Environment. Europeanization of Regional Seas or Regionalization of EU Policies?* Ashgate Publishing, Aldershot.
Hammer, M., Balfors, B., Mörtberg, U., Petersson, M. and Quin, A. (2011) 'Governance of water resources in the phase of change: A case study of the implementation of the EU Water Framework Directive in Sweden', *Ambio*, vol 40, pp210–220.
Hassler, B., Boström, M. and Grönholm, S. (2013) 'Towards an ecosystem approach to management in regional marine governance? The Baltic Sea context', *Journal of Environmental Policy & Planning*, vol 15, no 2, pp225–245.
Hautakangas, S. and Ollikainen, M. (2011) 'Making the Baltic Sea Action Plan workable: A nutrient trading scheme', in Pihlajamäki, M. and Tynkkynen, N. (eds) *Governing the Blue-Green Baltic Sea: Societal Challenges of Marine Eutrophication Prevention*. Finnish Institute of International Affairs, Helsinki, pp112–121.
Heiskanen, A.-S., Bonsdorff, E. and Joas, M. (2019) 'Baltic Sea: A recovering future from decades of eutrophication', in Wolanski, E., Day, J., Elliott, M. and Ramesh, R. (eds) *Coasts and Estuaries*, Elsevier, Amsterdam, pp343–362.
HELCOM. (2007) 'Baltic Sea Action Plan', Helsinki Commission, Helsinki.
HELCOM. (2018a) 'State of the Baltic Sea – Second HELCOM Holistic Assessment 2011–2016', *Baltic Sea Environment Proceedings* 155. Helsinki Commission, Helsinki.
HELCOM. (2018b) *The Sixth Pollution Load Compilation (PLC-6)*. Helsinki Commission, Helsinki.
HELCOM. (2020) 'Status of implementation of BSAP actions and recommendations. 2020 update'. Helsinki Commission, Helsinki, https://helcom.fi/wp-content/uploads/2020/10/Status-of-implementation-of-BSAP-actions-and-recommendations-2020-v9.pdf, accessed 23 May 2021.
Henningsen, B., Etzold, T. and Hanne, K. (2017) *The Baltic Sea Region: A Comprehensive Guide. History, Politics, Culture and Economy of a European Role Model*. Berliner Wissenshafts-Verlag, Berlin.
Hyytiäinen, K., Hasler, B., Ericsdotter, S., Nekoro, M., Blyh, K., Artell, J., Ahlvik, L. and Ahtiainen, H. (2013) 'Worth it: Benefits outweigh costs in reducing eutrophication in the Baltic', BalticSTERN Summary Report for HELCOM Ministerial Meeting 2013. Helsinki Commission, Helsinki.
Jetoo, S., Tynkkynen, N., Joas, M., Hellström, M., Sjöqvist, C. and Törnroos, A. (2022), 'Climate change and the governance of the Baltic Sea environment: Challenges and a research agenda', *Journal of Baltic Studies*, vol 53, no 1, pp65–84.

Joas, M., Kern, K. and Sandberg, S. (2007) 'Actors and arenas in hybrid networks: Implications for environmental policymaking in the Baltic Sea Region', *Ambio,* vol 36, nos 2–3, pp237–242.

Kern, K. (2011) 'Governance for sustainable development in the Baltic Sea Region', *Journal of Baltic Studies*, vol 42, no 1, pp21–35.

Kern, K. and Söderström, S. (2018) 'The ecosystem approach to management in the Baltic Sea Region: Analyzing regional environmental governance from a spatial perspective', *Marine Policy,* vol 98, pp271–277.

Koivurova, T. and Rosas, A. (2018) 'The CBSS as a vehicle for institutionalised governance in the Baltic Sea Area, in comparison with its two sister organisations in the North', *Marine Policy,* vol 98, pp211–219.

Korpinen, S., Meski, L., Andersen J. H. and Laamanen, M. (2012) 'Human pressures and their potential impact on the Baltic Sea ecosystem', *Ecological Indicators*, vol 15, pp105–114.

Korppoo, A. Tynkkynen, N. and Hønneland, G. (2015) *Russia and the Politics of International Environmental Regimes: Environmental Encounters or Foreign Policy?* Edward Elgar, Cheltenham.

Lehti, M. (2009) 'Baltic region in becoming: From the Council of the Baltic Sea States to the EU's Strategy for the Baltic Sea Area', *Lithuanian Foreign Policy Review*, vol 22, pp9–27.

Linke, S. and Jentoft, S. (2014) 'Exploring the phronetic dimension of stakeholders' knowledge in EU fisheries governance', *Marine Policy*, vol 47, pp153–161.

Linke, S., Dreyer, M. and Sellke, P. (2011) 'The regional advisory councils: What is their potential to incorporate stakeholder knowledge into fisheries governance?' *Ambio*, vol 40, pp133–143.

Lyytimäki, J. (2012) 'The environment in the headlines: Newspaper coverage of climate change and eutrophication in Finland', PhD thesis, University of Helsinki, Helsinki.

Murray, C., Muller-Karulis, B., Carstensen, J., Conley D., Gustafsson B. and Andersen J. (2019) 'Past, present and future eutrophication status of the Baltic Sea', *Frontiers in Marine Science*, vol 6, https://doi.org/10.3389/fmars.2019.00002.

Ojaveer, H., Jaanus, A., MacKenzie, B. R., Martin, G., Olenin, S., Radziejewska, T., et al. (2010) 'Status of biodiversity in the Baltic Sea', *PLoS One,* vol 5, no 9, pe12467 https://doi.org/10.1371/journal.pone.0012467

Ollikainen, M., Hasler, B., Elofsson, K. et al. (2019) 'Toward the Baltic Sea socioeconomic action plan', *Ambio*, vol 48, pp1377–1388.

Patterson, J. J., Smith, C. and Bellamy, J. (2013) 'Understanding enabling capacities for managing the 'wicked' problem of nonpoint source water pollution in catchments', *Journal of Environmental Management,* vol 128, pp441–452.

Ramírez-Monsalve, P., Raakjær, J., Nielsen, K. N., Laksá, U., Danielsen, R., Degnbol, D., Ballesteros, M. and Degnbol, P. (2016) 'Institutional challenges for policy-making and fisheries advice to move to a full EAFM approach within the current governance structures for marine policies', *Marine Policy*, vol 69, pp1–12.

Räsänen, T. and Laakkonen, S. (2007), 'Cold War and the environment: The role of Finland in international environmental politics in the Baltic Sea Region', *Ambio*, vol 36, nos 2–3, pp229–236.

Reusch, T. B. H. et al. (2018) 'The Baltic Sea as a time machine for the future coastal ocean', *Science Advances,* vol 4, no 5, eaar8195.

Ringbom, H. and Joas, M. (2018a) 'Introduction: Multi-level regulation in the Baltic Sea region', *Marine Policy,* vol 98, pp187–190.

Ringbom, H. and Joas, M. (2018b) 'Concluding article: The changing regulatory landscape of the Baltic Sea – An analysis', *Marine Policy*, vol 98, pp317–324.

Ringbom, H., Bohman, B. and Ilvessalo, S. (2019) 'Combatting eutrophication in the Baltic Sea: Legal aspects of sea-based engineering measures', *Legal Perspectives, The Law of the Sea*', Issue 2.4, Brill, Leiden.

Rönnberg, C. and Bonsdorff, E. (2004) 'Baltic Sea eutrophication: Area-specific ecological consequences', *Hydrobiologia,* vol 514, pp227–241.

Rutgersson, A., Jaagus, J., Schenk, F. and Stendel, M. (2014) 'Observed changes and variability of atmospheric parameters in the Baltic Sea Region during the last 200 years', *Climate Research*, vol 61, no 2, pp177–190.

Saunders, F. P., Gilek, M. and Linke, S. (2017) 'Knowledge for environmental governance: Probing science–policy theory in the cases of eutrophication and fisheries in the Baltic Sea', *Journal of Environmental Policy & Planning*, vol 19, no 6, pp769–782.

Schymik, C. (2012) 'The Baltic Sea region: Who cooperates with whom and why?', in Bellini, N. and Hilpert, U. (eds) *Europe's Changing Geography. The Impact of Inter-regional Networks*. Routledge, London, pp67–80.

Shortle, J. and Horan, R. D. (2017) 'Nutrient pollution: A wicked challenge for economic instruments', *Water Economics, Policy and Society*, vol 3, no 2, p1650033.

Soininen, N. and Platjouw, F. M. (2019) 'Resilience and adaptive capacity of aquatic environmental law in the EU: An evaluation and comparison of the WFD, MSFD, and MSPD', in Langlet, D. and Rayfuse, R. (eds) *The Ecosystem Approach in Ocean Planning and Governance Perspectives from Europe and Beyond*. Brill Nijhoff, pp17–79.

Soininen, N., Belinskij, A., Similä, J. and Kortet, R. (2019) 'Too important to fail? Evaluating legal adaptive capacity for increasing coastal and marine aquaculture production in EU-Finland', *Marine Policy*, vol 110, p103498.

Tynkkynen, N., Schönach, P., Pihlajamäki, M. and Nechiporuk, D. (2014) 'The governance of the mitigation of the Baltic Sea eutrophication: Exploring the challenges of the formal governing system', *Ambio*, vol 43, no 1, pp105–114.

Tynkkynen, N. (2013) 'The challenge of environmental governance in the network society: The case of the Baltic Sea', *Environmental Policy and Governance*, vol 23, no 6, pp395–406.

Tynkkynen, N. (2015a) 'Russia and the Baltic Sea. Frames and spaces of environmental problems', *Eurasian Geography and Economics*, vol 55, no 6, pp674–690.

Tynkkynen, N. (2015b) 'Baltic Sea environment, knowledge and the politics of scale', *Journal of Environmental Policy and Planning*, vol 17, no 2, pp201–216.

Tynkkynen, N. (2017) 'The Baltic Sea environment and the European Union: Analysis of governance barriers', *Marine Policy*, vol 81, pp124–131.

Tynkkynen, N. (2018) 'The "Russian Issue" in transnational governance of the Baltic Sea environment: Analysis of drivers and constraints of Russia's participation', *Marine Policy*, vol 98, no 4, pp220–226.

Valman, M. (2014) 'Three faces of HELCOM - Institution, organization, policy producer', PhD Thesis, Department of Political Science, Stockholm University, Stockholm.

VanDeveer, S. (2011) 'Networked Baltic environmental cooperation', *Journal of Baltic Studies*, vol 42, pp37–55.

Waylen, K. et al. (2014) 'The need to disentangle key concepts from ecosystem approach jargon', *Conservation Biology*, vol 28, no 5, pp1215–1224.

Westholm, A. (2021) 'Scaling marine water management', PhD thesis, School of Business, Economy and Law, University of Gothenburg.

Wiering, M., Liefferink, D., Boezeman, D., Kaufmann, M., Crabbé, A. and Kurstjens, N. (2020) 'The wicked problem the Water Framework Directive cannot solve', *Water*, vol 12, no 5, 1240.

Wulff, F., Savchuk, O. P., Sokolov, A. V., Humborg, C. and Mörth, M. (2007) 'Management options and effects on a marine ecosystem: Assessing the future of the Baltic', *Ambio*, vol 36, pp243–249.

18

GOVERNANCE OF THE BLACK SEA

Institutional arrangements for managing the impacts of global environmental change

Emma Avoyan and Jan van Tatenhove

Worldwide, the news is full of alarming headlines about global environmental change, ranging from the changing climate, degradation of marine ecosystems to the accelerated loss of biodiversity. For example, in 2021, a newspaper reported that 'Fish migrate opposite way in Black Sea for the first time due to global warming' (Daily Sabah, 2021). In conjunction with drastic impacts from other anthropogenic activities on oceans, marine ecosystems are affected by continuous and further projected climate change, resulting in increasing water temperatures, acidification, deoxygenation, and sea-level rise (Hoegh-Guldberg *et al.*, 2015; IPCC, 2019). The Black Sea's biodiversity and ecosystem functioning are especially vulnerable to climate change due to its geophysical features. The Black Sea is a semi-enclosed, largely land-locked sea having little water circulation with oceans as well as being limited by density stratification that limits interaction between surface and deep waters (Miladinova *et al.*, 2017). Regardless of its relatively large water volume, only a fine layer of the water mass, comprising about 10% of the total depth, supports marine life (Kideys, 2002). The deeper parts of the Black Sea lack oxygen, making it the world's largest anoxic marine basin (Hoegh-Guldberg *et al.*, 2015). Such anoxic conditions are aggravated by the very limited ability of organisms to migrate when ecosystem conditions change (Hoegh-Guldberg *et al.*, 2015). The Black Sea's peculiar geography also affects the ability of coastal sovereign states to mitigate and adapt to global environmental change.

The Black Sea is bordered by Romania, Bulgaria, Russia, Ukraine, Georgia, and Turkey (see Figure 18.1). These are countries with differing cultures and traditions, socio-economic and political realities, and often contradictory environmental regulations and management arrangements (Dimadama and Timotheou, 2010). As a crossroad between Europe and Asia for energy and transport flows, the Black Sea has strategic importance from both economic and geopolitical perspectives (O'Higgins *et al.*, 2014). Yet, foremost, it is a unique water body possessing a rich natural heritage, which has experienced drastic environmental decline (Coutto and Devlen, 2014). Overfishing, continuous discharges of organic substances and nutrients from agriculture and sewage treatment, oil spills from shipping accidents, the rapid development of mass tourism and, finally, climate change have all contributed to the degradation of the Black Sea marine ecosystem (Borysova *et al.*, 2005; Coutto and Devlen, 2014; Knudsen, 2015). Moreover, as the Black Sea is greatly impacted by non-climatic stressors,

 DOI: 10.4324/9781315149745-22

attribution of observed and confirmed deterioration trends to climate change is problematic (Hoegh-Guldberg *et al.*, 2015). In light of this uncertainty, regional cooperation to monitor, prevent, mitigate, and adapt to the challenges posed by transboundary impacts of climate change is essential. Yet, establishing adequate regional cooperation in the Black Sea region is a problematic and challenging task. First, the existing national institutional structures, the different power and political dynamics, and policy and legal frameworks lead to ambiguities in the perception and assessment of risks and challenges posed by global environmental change in general and climate change in particular. Second, the existing governance arrangements for regional cooperation, including the Black Sea Commission and Black Sea Economic Cooperation, struggle to function and adapt adequately to contemporary challenges, including that of climate change (Avoyan, Van Tatenhove and Toonen, 2017).

This chapter provides an overview of the governance arrangements in the Black Sea. To achieve climate change mitigation goals and to prepare for climate adaptation, the Black Sea countries will, despite their differences, need to work together. The main question posed in this chapter is whether the current Black Sea governance arrangement is able to address effective regional cooperation to deal with the challenges of climate change. The remainder of the chapter has the following structure: First, we provide a short overview of environmental problems in the Black Sea in the context of global environmental change. Second, through the lens of 'governance arrangement' (Van Tatenhove, 2013), we describe four dimensions of the Black Sea governance arrangement: discourses, key actors, rules, and resources. Third, we summarise and discuss the strengths and limitations of the Black Sea governance arrangement and its ability to achieve adequate cooperation in the region and to manage the impacts of climate change. We conclude our chapter with general recommendations for actors in the Black Sea region.

Figure 18.1 The Black Sea.

Source: Flanders Marine Institute (VLIZ), http://www.coastalwiki.org/wiki/Black_Sea_Convention.

Environmental problems and the consequences of climate change in the Black Sea

The Black Sea is one of the most degraded seas in the world (Knudsen, 2015). It has undergone severe environmental decline since the second half of the twentieth century, reaching the tipping point of deterioration in the early 1990s when its ecosystem was in nearly a catastrophic state (Kideys, 2002). Several factors have resulted in the decline of the marine ecosystem. For example, eutrophication and consequent hypoxia (oxygen depletion) caused by continuous and heavy discharges of organic substances from agriculture and sewage treatment were among the main pressures (Kideys, 2002; Borysova *et al.*, 2005). This ecological shift unfolded with the invasion of non-native species (e.g. the comb jelly Mnemiopsis) (Knudsen, 2015). The decline of the marine ecosystem was further exacerbated by over-exploitation of fish stocks. Overfishing and illegal fishing coupled with destructive practices caused decline of fish stocks and alterations to the trophic structure of many species (Coutto and Devlen, 2014). During the 1970s and 1980s, there were also serious oil spills from shipping accidents and pollution by a variety of harmful substances (Dimadama and Timotheou, 2010).

Although there is no consensus regarding the extent to which the Black Sea has recovered, scientists agree that, as from the mid-1990s, its marine ecosystem has shown signs of some recovery (BSC, 2007; Velikova and Oral, 2012). Improvements were especially observed in the level of eutrophication, with the total nutrient input from the Danube River stabilising and then decreasing (Kideys, 2002) due to reduced use of fertilisers (Mee, 2001)) and policy initiatives to tackle the problem among the coastal countries of the Black Sea and Danube regions (Kideys, 2002). However, the latest *State of the Environment of the Black Sea* report (2009–2014) concluded that only modest improvement in the water quality was observed, leaving a system change or ecosystem-regime shift as major concerns (BSC, 2019). Some even argue that the observed post-eutrophication state should not be considered as a recovery and is not even close to the 'pristine' state of the Black Sea in the 1960s (Oguz and Velikova, 2010).

Overall, research indicates that the recovery of the Black Sea is non-linear and unstable (Knudsen, 2015). Major environmental stressors are still eutrophication, chemical pollution, overfishing, and introduction of non-native species (BSC, 2019). There is growing uncertainty regarding the impact of climate change on the ecological instability of the Black Sea (Hoegh-Guldberg *et al.*, 2015). Yet, it is certain that the surface water temperature of the Black Sea has increased by approximately 1°C (according to the data from 2006) (Belkin, 2009). Furthermore, the frequency and strength of weather anomalies have increased in the wider Black Sea region due to climate change. As a consequence, some Mediterranean species (e.g., sardines) have invaded the Black Sea (BSC, 2019). It is expected that increasing air and seawater temperatures will increase the influx of alien species; the decline or even loss of native species is inevitable (Sezgin *et al.*, 2010). This makes the Black Sea very sensitive to climate change and further predicted seawater temperature rise. To summarise, a combination of climate change-induced oceanographic shifts (i.e., increased temperatures), eutrophication, and overfishing is causing major systemic changes in the ecosystem of the Black Sea. More effective marine environmental governance will be needed to achieve environmental rehabilitation of the Black Sea and adaptation to the challenges posed by climate change.

The governance arrangement of the Black Sea ecosystem

A *governance arrangement* is a temporary stabilisation of the substance and organisation of a (marine) policy domain (Arts, Leroy and van Tatenhove, 2006; Liefferink, 2006; Van Tatenhove, 2013).

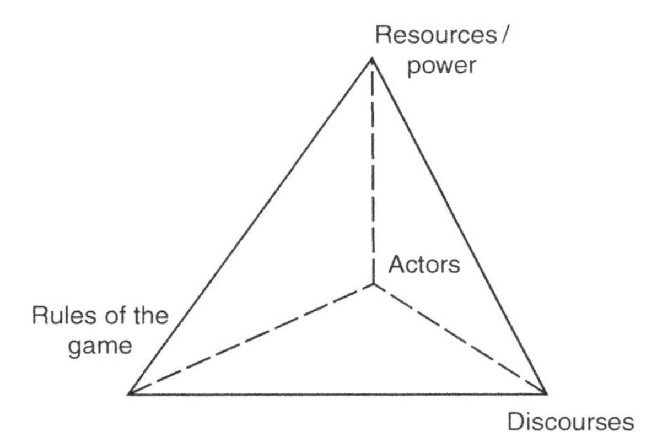

Figure 18.2 Dimensions of a policy arrangement.
Source: After Liefferink (2006) and Arts et al. (2006).

Substance refers to discourses resulting in distinct policy and regulatory goals, whereas organisation refers to the types of actors involved, the rules of the game (instruments, procedures, division of tasks) and the available resources (Van Tatenhove, Arts and Leroy, 2000). In a marine governance arrangement, coalitions of public and private actors try to influence activities within a policy domain and to design legitimate solutions, based on shared discourses (ideas and categorisations) and to manage resources (budgets, expertise, information, knowledge, permits, etc.) and existing formal and informal rules at different levels (Van Tatenhove, 2013; van Tatenhove *et al.*, 2020). The four dimensions of a governance arrangement are inextricably interwoven; change in one of the dimensions may induce change in the other dimensions, resulting in a change in the marine governance arrangements (see Figure 18.2).

The changes in the dimensions of a governance arrangement could be the result of processes of political modernisation, such as the interplay of contextual processes of structural political and social change, or problem-oriented renewal of policy making and decision-making by agents in day-to-day (policy) practices (Arts and Van Tatenhove, 2006).

Dominant discourses

Within the framework of marine governance, Black Sea regional cooperation unfolds around two dominant discourses: economic development and environmental protection. Both discourses distinctly emerged at the beginning of the 1990s. After the collapse of the Soviet Union, many countries in the wider Black Sea region gained sovereignty and, consequently, free-market economies. The shock event of the end of the Soviet era, coupled with the near non-existence of national economic structures, broken economic ties among the countries, reduction in revenues, closure of large industries, and unemployment, led to an economic crisis in the region (Rudneva and Petzold-Bradley, 2001; Manoli, 2004). The discourse on economic development in the region by means of regional cooperation was put forward as a way to overcome the crisis and consolidate the post-Soviet countries. However, economic growth returning to the region has brought environmental pressures resulting from extensive industrial development and large-scale infrastructure construction (Tavitian *et al.*, 2008). With the entry of the European Union (EU) into the Black Sea region – with the accession of Romania and Bulgaria in 2007 – the discourse of economic development shifted towards the discourse of sustainable economic development in sectors, such as aquaculture, tourism, and energy. The idea that environmental protection could coexist next to sustainable economic development of the Black Sea was increasingly receiving attention (Velikova and Oral, 2012). Also, the discourse on environmental protection called for regional cooperation as a mechanism for dealing with the catastrophic state of the environment in the Black Sea in the 1990s. The discourse on environmental protection was centred around the sense of urgency among the Black Sea countries to rehabilitate and protect the marine ecosystem (Avoyan, Van Tatenhove and Toonen, 2017). Over time, an integrative perspective on prioritising the Black Sea ecosystem, along with sustainable ecosystem use, has been suggested, particularly by the EU.

Key regional actors

The central actor in the economic development discourse coalition is the Black Sea Economic Cooperation (BSEC). The BSEC, established in 1992, is a regional organisation to encourage trade and economic development in the broader Black Sea region and has the environmental protection as one of the fields for cooperation (Velikova and Oral, 2012). The BSEC includes 12 member states of the wider Black Sea region. Although a Working Group on Environmental Protection was nominated from the outset, it was not until 2002, after the agreement on cooperation between BSEC and UNEP was signed, that marine governance as a priority received more attention (Oral, 2013). Overall, the ministerial meetings and working group activities are organised in accordance with relatively specific guidelines, which include implementation and monitoring recommendations (Celac, 2011). The Working Group meets once in two years or whenever necessary. It has developed an environmental policy and action plan, defining environmental protection quite broadly to include environmental issues beyond the marine environment. As a number of BSEC member states have no direct access to the Black Sea, it covers a wide range of environmental issues, with particular attention to the areas not covered by the Black Sea Commission (Velikova and Oral, 2012).

The signing of the Convention for the Protection of the Black Sea Against Pollution, known as the Bucharest Convention, in 1992 resulted in the establishment of the Black Sea Convention (BSC), with clearly defined responsibilities to execute and promote the implementation of the Convention, its protocols and action plans, and to cooperate with competent international organisations (Avoyan, Van Tatenhove and Toonen, 2017). The BSC consists of one representative from each of the six coastal countries. It meets once a year and adopts a work programme or plan for the coming year (Velikova and Oral, 2012). Since 2000, the Black Sea Permanent Secretariat, based in Istanbul, has supported the implementation of annual work programmes. Furthermore, seven advisory groups for specific areas of environmental protection in the Black Sea consult the BSC Commission and Secretariat (DiMento and Hickman, 2012). Each of the Advisory Groups, consisting of two representatives (an expert and a representative of the relevant state ministry), has corresponding Activity Centres located in all six coastal countries.

Rules in place

To enable regional cooperation, the economic and environmental discourse coalitions, represented by BSEC and BSC are supported by formal, legally defined rules acknowledging the need for enhanced cooperation. The main legal framework for BSEC's operation – the Summit Declaration on Black Sea Economic Cooperation, adopted in 1992 – requests that BSEC member states cooperate and take appropriate steps for the sustained growth of their national economies, the social well-being of their peoples, as well as the protection of the environment of the Black Sea (BSEC charter). To this end, the BSEC's Member States have elaborated a general cooperation framework, the BSEC Action Plan for Cooperation in the Field of Environmental Protection, adopted in 2006. Furthermore, the Joint Declarations on Combating Climate Change in the wider Black Sea (2010) and Climate Change and Green Economy (2012) focus on different aspects of climate change policy and highlight the member states' intention to address climate change and its impacts in the region. Based on these declarations, BSEC's Working Group on Environmental Protection elaborated the 'BSEC Climate Change Adaptation Strategy for the Black Sea', which was adopted by the Council of Ministers of Foreign Affairs in 2017. The strategy aligns with one of the goals of the BSEC Economic Agenda, which is related to developing plans for mitigation and adaptation measures to address the impact of climate change. Yet an action plan for the implementation of this strategy has not been developed so far. Over the years, BSEC's commitment to promote cooperation in favour of the environment of the region, and recently to address climate change related issues, has been confirmed at the political level. However, the rules to enable and facilitate implementation actions have not been forthcoming.

The BSC's legal framework for cooperation is the Bucharest Convention, which includes action-based protocols (Mee, 2001). Four Ministerial Declarations (of 1993, 2002, 2007, and 2009) and the Black Sea Strategic Action Plan (BS SAP, first adopted in 1996 and updated in 2009) supplement the Convention. The 1993 Odessa Ministerial Declaration expanded the scope of the Bucharest Convention beyond pollution and more precisely elaborated environmental goals to protect, preserve, and rehabilitate the Black Sea marine environment by means of regional cooperation and sustainable management (DiMento, 2001). This declaration was notably ambitious as it was the first policy paper on regional seas with implementation deadlines and inclusion of public participation, the precautionary principle, accountability, and the need for cooperation and coordination of regional activities (DiMento, 2001; Mee, 2002). A drawback of the declaration was the lack of a visible enforcement mechanism for regional cooperation and fulfilment of commitments (Avoyan, Van Tatenhove and Toonen, 2017), thus the need for the BS SAP. Unlike the BSEC's action plan, the latter contained ambitious targets and a timetable for the implementation of the Bucharest Convention. However, it had to be updated in 2009 due to lack of implementation.

Parallel to the rules that BSEC and BSC have to promote regional cooperation, there are also rules that externally support and promote states' engagement in regional initiatives for addressing different environmental problems. One notable example is the Danube-Black Sea Task Force (DABLAS), initiated by the EU as a platform for cooperation for the protection of water-related ecosystems in the Black Sea and Danube regions. The DABLAS members include not only the Black Sea countries but also a number of European countries from the Danube River basin, different international organisations, NGOs, and regional arrangements, including BSC and the International Commission for the Protection of the Danube River (ICPDR). The initiative provided coordinating rules and financial mechanisms for the implementation of various pollution-reduction and ecosystem-rehabilitation joint projects (Stribis, 2009). Overall, DABLAS was a major success during its operation (2001–2011), bringing together BSC, ICPDR, the EU, and countries of the wider Black Sea region (Knudsen, 2015) to address the problem of municipal and industrial discharges by successfully implementing 354 investment projects and eventually having positive effects on the reduction of nutrient discharges into the Black Sea (Knudsen, 2015). Another example of a rule for cooperation initiated by the EU is the Common Maritime Agenda for the Black Sea (CMA), which was adopted in 2019. This sea-basin initiative aims at supporting regional cooperation for a more sustainable blue economy in the Black Sea while fostering economic growth (on the European blue economy, see Chapter 6). Notably, CMA prioritises the protection and sustainability of the marine ecosystem and calls for further research on climate change impacts on the Black Sea. Both BSEC and BSC are consultative bodies of the CMA Steering Group.

Resources

The BSEC and BSC discourse coalitions that have enabled regional cooperation have been largely dependent on the availability of resources among the member states. We refer here to financial resources, human capital, expertise, and knowledge. For example, although BSC formally has six Activity Centres located in all six member states, only two (in Bulgaria and Ukraine) are currently functioning and supporting the BSC while having financial assistance from corresponding governments (e.g., to fund the acquisition of necessary labs, equipment, or staff). The Activity Centres could be consequential for enabling effective regional cooperation as they are meant to be national focal points for technical support, accumulation of knowledge and expertise for the implementation of BS SAP and the facilitation of regional cooperation (on expertise, see Chapter 8). The lack of willingness among the Black Sea countries to allocate funds also affects the operation of the BSC. The BSC's Permanent Secretariat is not able to expand the number of employees due to lack of funding. Currently staffed by only three people, the secretariat facilitates the work of BSC's 16 subsidiary bodies (seven advisory groups, formally six activity centres, and three ad hoc working groups), implementation of the Annual Work Programme, coordination of BSC's regional projects and activities, and communication with international organisations while also having the additional task to elaborate proposals for new regional projects. Evidently, the responsibilities of the secretariat do not match its human and financial resources (Avoyan, Van Tatenhove and Toonen, 2017). The number of employees at the BSEC's secretariat is bigger, but only one person is handling the work of the working group on Environmental protection. At the same time, because BSEC promotes several other areas of cooperation (e.g., institutional renewal, good governance, and tourism), the turnover of expertise within the organisation is possible. This places BSEC in a relatively favourable position in terms of acquired human resources and expertise.

Struggles with funding within the Black Sea governance arrangement in general are clearly seen also in a well-rooted joint project-implementation tradition. Starting in the 1990s, the Black Sea countries, being developing states with transition economies, requested financial assistance and aid from different international organisations to stimulate both economic and environmental cooperation in the region and to develop a long-term action plan for the implementation of the Bucharest Convention (Oral, 2013). As a result, the Black Sea Environmental Programme (BSEP) has been launched with the financial contribution of the Global Environmental Facility and the EU. For over 15 years, until BSEP was suspended in 2008, substantial financial resources were invested in the implementation of national and regional projects along with developing the BS-SAP – the major achievement of BSEP (Knudsen, 2015). Similarly, following the accession of Romania and Bulgaria to the EU in 2007, the EU adopted the Black Sea Synergy as an initiative to develop cooperation within the Black Sea region (EC, 2015). Considerable funds have flowed from this initiative to the Black Sea region. This has extended the prevailing tradition of enacting regional cooperation through project implementation. As a consequence, existing institutional structures (e.g., BSEC, BSC) and implementation of the rules (e.g., Bucharest Convention, ministerial declarations, actions plans, etc.), have been heavily reliant on financial assistance from donor organisations for project implementation. Yet, this should not undermine the value of these large-scale multinational projects. These projects have resulted in many positive outcomes, particularly in terms of generating a substantial volume of scientific knowledge, setting up a network of scientists active across the Black Sea countries, and establishing collaboration among scientific institutions and public, private, and non-profit organisations (Avoyan, Tatenhove and Toonen, 2017). These outcomes could be of use if new environmental agendas like climate change are prioritised in the region.

Discussion: strengths and limitations of the Black Sea governance arrangement

The main advantage of the Black Sea governance arrangement is that it accommodates two distinct discourse coalitions with institutionalised structures, rules, strategies, and resources that could be complementary for advancing a climate change agenda if integrated together. However, collaboration between BSEC and BSC has never advanced beyond attending each other's meetings on an ad hoc basis. In general, it might be questionable whether BSEC, which was initiated primarily to facilitate economic development and cooperation, has the capacity to deal with climate change and other environmental policies. Climate change adaptation measures might be viewed as hampering economic development, particularly in the Black Sea region where not all national authorities prioritise environmental protection (Dimadama and Timotheou, 2010). As noted by a BSC representative, 'we are dealing with environmental problems of the Black Sea, the BSEC is dealing with the need to coordinate the efforts of the countries in many directions. We are not staying until midnight [to] fight and make political statements like BSEC does' (Avoyan, 2016: 29). In this light, the co-existence of the BSC and the BSEC also poses some questions regarding duplication and overlap of measures and responsibilities in marine governance, particularly in the implementation of regional norms (DiMento and Hickman, 2012). Both organisations, while having formal requirements to cooperate with each other (BSC in BS SAP, BSEC in the charter), decided upon a mutual observer status in 2002, thus making it formally possible to participate in each other's meetings (Stribis, 2009). However, coordination of efforts and cooperation between BSC and BSEC has not gone further to concrete, action-based results (Knudsen, 2015).

Furthermore, BSC and BSEC independently have struggled to achieve major outcomes through effective regional cooperation. Despite the existence of rules, legal frameworks, instruments, and procedures, the implementation of BS SAP concerning regional governance structures of the Black Sea has been limited. The countries failed to address the jointly agreed targets (e.g., to adopt a regulative mechanism to control the transboundary movement of hazardous wastes and illegal traffic, agree to a Fisheries Convention, or develop regional Black Sea strategy for integrated coastal zone management), or to agree on substantial regional strategies and enforcement/regulatory mechanisms. The lack of progress of BS SAP was a result of inadequate resources within the BSC and across its member countries, top-down decision-making traditions in the countries, outdated rules and regulations (e.g., the Bucharest Convention has never been amended to address recent developments in marine governance, including climate change) and, to some extent, political instability (Avoyan, Tatenhove and Toonen, 2017). The major achievements of BSC are seen in facilitating scientific cooperation and organising technical country reporting largely through project execution. Similarly, within the BSEC, environmental issues are dealt with in a project-implementation manner (Velikova and Oral, 2012).

Regional integrative management based on a common set of rules seems unrealistic given the diversity of BSEC member countries, particularly in terms of their commitments under various international treaties (e.g., not all have signed the Energy Charter Treaty) (Celac,

2011). Furthermore, being a platform for 'Blue' (sea-oriented) growth and recently also climate change concerns, the organisation is left out of discussions around major economic developments of the region that have direct impact on the ecology of the Black Sea. Examples include Turkish President Erdogan's 'crazy project' of the Istanbul canal: a proposed waterway between the Black Sea and the Sea of Marmara parallel to the Bosporus Strait (Daniel and Lindenstrauss, 2020: 2). Overall, the discourse coalition of BSEC lacks common perspectives and priorities mostly because it attempts to cover a range of different areas, ranging from the environment and climate change to agriculture, crime, culture, education, energy, and so forth.

Finally, the political situation and power struggles in the region affect the work of both discourse coalitions in their efforts to advance marine governance of the Black Sea. The region is a major oil and gas producer where three powerful actors (the EU, Russia, and Turkey) meet to compete for the political and economic 'control', particularly related to energy security (Emerson, 2008; Knudsen, 2015). Although the geopolitical situation in the Black Sea region represents a challenge to regional cooperation and implementation of marine policies (O'Higgins et al. 2014), it should be noted that regional cooperation between the Black Sea countries has existed even during times of strong political tensions, economic instability, and armed conflicts between riparian countries (Avoyan, Tatenhove and Toonen, 2017). Even the Russian-Georgian war in 2008 and the Russian-Ukrainian conflict around Crimea in 2014 did not cause those countries to reject completely the existence of cooperation on environmental issues, although those conflicts did slow down the process in number of respects. For example, those political conflicts limited the choice of organising the Black Sea Commission's events based on the principle of rotation. Georgian and Ukrainian governmental and nongovernmental representatives do not principally accept the invitations to participate in the events concerning Black Sea environmental management organised in the territory of the Russian Federation and vice versa (Avoyan, Tatenhove and Toonen, 2017). Most joint meetings are now being organised in Istanbul (Turkey), the 'neutral zone' for all participant countries and hosting both BSEC's and BSC's headquarters. In addition, because of the Crimea crisis (i.e., Russia's annexation), EU-funded environmental projects in Crimea have been suspended (European Commission, 2015) and the political status of a number of highly knowledgeable experts and scientific institutions from this region is not clear. Thus, an atmosphere of political instability coupled with power struggles in the region has negatively affected cooperative efforts of the Black Sea countries.

Conclusion

The main question we posed in this chapter was whether the current Black Sea governance arrangement is able to address the challenges of climate change through effective regional cooperation. This is a critical question in the light of formally arranged commitments of the Black Sea countries to promote regional cooperation, introduce policies on climate change mitigation, devote resources to greening the regional economy, increase public awareness of climate change and so forth under BSEC's joint Declaration on combating climate change in the wider Black Sea area (2010) and BSEC's climate change adaptation strategy (2017). However, in the past three decades, regional cooperation among the Black Sea countries has been challenging for both economic development and environmental protection. The BSEC and the BSC have had some achievements (particularly the latter in catalysing scientific research) that have been supported by necessary rules (e.g., declarations, strategies, and action plans) and sufficient resources (e.g., secretariats, employees, and project reports) within the Black Sea governance arrangement. Despite these achievements, we believe that the geopolitical situation, the power imbalances between the countries, the lack of coordination (both between the countries and dominant actors/institutions), and a lack of willingness to move towards integrative management of marine ecosystem will inevitably hamper efforts to deal with climate change impacts. Our main conclusion is that the governance of climate change in the Black Sea will require substantially more political will among the countries, socio-economic incentives, and major institutional reforms.

Regional cooperation among Black Sea countries has been possible even during times of strong political tensions, economic instabilities, and armed conflicts. Yet, to be prepared for the climate change agenda, the mere presence of cooperation will not suffice. To be fully prepared for its future tasks, the Black Sea governance arrangement needs institutional reforms. Therefore, to effectively develop and implement climate change measures, we recommend several major legal, institutional, and procedural reforms of the Black Sea governance arrangement. Firstly, the Bucharest Convention should be amended to meet the standards of best legal practices to ensure the application of current environmental principles and approaches in marine governance, including as it relates to climate change. Secondly, the Black Sea governance arrangements should stimulate the accession of international organisations (e.g., the EU) to open formal opportunities for financial resources, technical assistance, and best practices to develop and implement climate change measures. Finally, a revision of internal BSEC and BSC regulations is needed to ensure a better accountability between the subsidiary bodies and enhanced coordination between the two organisations. If these and similar reforms are implemented, it may be possible in the future to achieve more effective marine governance in the region even as global environmental changes become more pronounced.

References

Arts, B. and Van Tatenhove, J. (2006) 'Political modernisation', in *Institutional dynamics in environmental governance*. Dordrecht: Springer Netherlands, pp. 21–43. doi: 10.1007/1-4020-5079-8_2.

Arts, B., Leroy, P. and van Tatenhove, J. (2006) 'Political modernisation and policy arrangements: A framework for understanding environmental policy change', *Public Organization Review*, 6(2), pp. 93–106. doi: 10.1007/s11115-006-0001-4.

Avoyan, E. (2016) *Marine collaborative governance in the Black Sea: assessing the performance of Black Sea Commission*. Wageningen: Wageningen University.

Avoyan, E., Tatenhove, J. van and Toonen, H. (2017) 'The performance of the Black Sea Commission as a collaborative governance regime', *Marine Policy*. doi: 10.1016/j.marpol.2017.04.006.

Belkin, I. M. (2009) 'Rapid warming of large marine ecosystems', *Progress in Oceanography*, 81(1–4), pp. 207–213. doi: 10.1016/j.pocean.2009.04.011.

Borysova, O. *et al.* (2005) *Eutrophication in the Black Sea region; Impact assessment and Causal chain analysis*. Kalmar, Sweden: University of Kalmar.

BSC (2007) *Black Sea transboundary diagnostic analysis: BS TDA*. Istanbul.

BSC (2019) *The Black Sea State of Environment report 2009–2014(5)*. Istanbul.

Celac, S. (2011) *The Role and Potential of the Organization of the BSEC, Neighbourhood Policy Paper no. 01*.

Coutto, T. and Devlen, B. (2014) 'Environmental concerns in EU-Black Sea affairs', *Journal of Balkan and Near Eastern Studies*. doi: 10.1080/19448953.2014.928537.

Daily Sabah (2021) 'Fish migrate opposite way in Black Sea for 1st time due to global warming', *13.01.2021*.

Daniel, R. and Lindenstrauss, G. (2020) 'The Istanbul Canal: Erdogan's "crazy project" raises concerns | INSS', *Institute for National Security Studies; INSS Insight No. 1269*.

Dimadama, Z. and Timotheou, A. (2010) 'Greening the Black Sea : Overcoming inefficiency and fragmentation through environmental governance', *ICBSS POLICY BRIEF no. 21*, 21.

DiMento, J. F. (2001) 'Black Sea environmental management: Prospects for new paradigms in transitional contexts', in *Reflections on water : New approaches to transboundary conflicts and cooperation*. Cambridge, MA: MIT Press, pp. 239–266.

DiMento, J. F. and Hickman, A. J. (2012). *"The Black Sea". In environmental governance of the great seas: law and effect.* Cheltenham: Edward Elgar Publishing. doi: 10.4337/9781781006153.00009

Emerson, M. (2008) 'The EU ' s New Black Sea policy what kind of regionalism is this ?', *SSRN Electronic Journal*, 297(June), pp. 1–18. doi: 10.2139/ssrn.1337499.

European Commission. (2015) *Joint staff working document black sea synergy: review of a regional cooperation initiative.* doi: 10.1017/CBO9781107415324.004.

Hoegh-Guldberg, O. *et al.* (2015) *The ocean. In climate change 2014 – Impacts, adaptation and vulnerability: Part B: Regional aspects: Working group II contribution to the IPCC fifth assessment report.* Cambridge: Cambridge University Press.

IPCC. (2019) *IPCC Special Report on the Ocean and Cryosphere in a Changing Climate* [H.-O. Pörtner, D.C. Roberts, V. Masson-Delmotte, P. Zhai, M. Tignor, E. Poloczanska, K. Mintenbeck, A. Alegría, M. Nicolai, A. Okem, J. Petzold, B. Rama, N.M. Weyer (eds.)]. In press.

Kideys, A. E. (2002) 'Ecology: Fall and rise of the Black Sea ecosystem', *Science.* doi: 10.1126/science.1073002.

Knudsen, S. (2015) 'Marine Governance in the Black Sea', in *Governing Europe's marine environment: Europeanization of regional seas or regionalization of EU policies?* Ashgate, pp. 225–247. doi: 10.1111/reel.12133.

Liefferink, D. (2006) 'The dynamics of policy arrangements: turning round the tetrahedron', in Arts, B. and Leroy (eds) *Institutional dynamics in environmental governance.* Springer, pp. 45–68. doi: 10.1007/1-4020-5079-8_3.

Manoli, P. (2004) *The formation of Black sea economic cooperation a case study of subregionalism.* University of Warwick.

Mee, L. (2001) 'Can the marine and coastal environment of the black sea be protected?', in Aybak, T. (eds) *Politics of the Black Sea: dynamics of cooperation and conflict*, New York: I.B. Tauris, pp. 133–161. doi: 10.2307/2697174

Mee, L. (2002) 'Protecting the Black Sea environment: A challenge for cooperation and sustainable development in Europe', in *Europe's Black Sea dimension.* Centre for European Policy Studies, pp. 79–140.

Miladinova, S. *et al.* (2017) 'Black Sea thermohaline properties: Long-term trends and variations', *Journal of Geophysical Research: Oceans.* doi: 10.1002/2016JC012644.

O'Higgins, T. *et al.* (2014) 'Achieving good environmental status in the Black Sea: scale mismatches in environmental management', *Ecology and Society*, 19(3). doi: 10.5751/ES-06707-190354.

Oguz, T. and Velikova, V. (2010) 'Abrupt transition of the northwestern Black Sea shelf ecosystem from a eutrophic to an alternative pristine state', *Marine Ecology Progress Series*, 405, pp. 231–242. doi: 10.3354/meps08538.

Oral, N. (2013) *Regional Co-operation and protection of the marine environment under international law.* Brill. doi: 10.1163/9789004250864.

Rudneva, I. and Petzold-Bradley, E. (2001) 'Environment and security challenges in the Black Sea region', in *Responding to environmental conflicts: Implications for theory and practice.* Springer Netherlands, pp. 189–207. doi: 10.1007/978-94-010-0395-7_12.

Sezgin, M. *et al.* (2010) 'Likely effects of global climate change on the black SEA benthic ecosystem', *Journal of Environmental Protection and Ecology*, 11(1), pp. 238–246.

Stribis, I. (2009) 'Pooling forces in protecting the Black Sea marine environment : Actors and Actions', *ICBSS POLICY BRIEF*, 17.

Tavitian, N. *et al.* (2008) *Greening the Black Sea Synergy, "Greening the Neighbourhood Policy" project.* Brussels.

Van Tatenhove, J., Arts, B. and Leroy, P. (2000) *Political modernisation and the environment. The renewal of environmental policy arrangements.* Edited by J. Van Tatenhove, B. Arts, and P. Leroy. Dordrecht/Boston/London: Kluwer Academic Publishers.

Van Tatenhove, J. P. M. (2013) 'How to turn the tide: Developing legitimate marine governance arrangements at the level of the regional seas', *Ocean & coastal management.* Elsevier, 71, pp. 296–304. doi: 10.1016/j.ocecoaman.2012.11.004.

Van Tatenhove, J. P. M. *et al.* (2020) 'The governance of marine restoration: insights from three cases in two European seas', *Restoration Ecology.* doi: 10.1111/rec.13288.

Velikova, V. and Oral, N. (2012) 'Governance of the protection of the Black Sea: A model for regional cooperation', in *Environmental security in watersheds: The Sea of Azov*, pp. 159–171. doi: 10.1007/s13398-014-0173-7.2.

19

POLAR SEAS

Governing extreme change in the Arctic and Southern Oceans

Nicole Wienrich, Hugh McDonald, Arne Riedel, Tim Packeiser, and Janos Hennicke

Globally unique marine ecosystems exist in the Arctic and Antarctic. These ecosystems are well-adapted to the extreme environmental conditions prevalent at the Poles, including sea ice, extreme weather events, and strong seasonal changes. But these polar marine ecosystems are increasingly under threat from pollution, resource exploitation and, most significantly, the changes brought about by climate change. Especially in the Arctic, warming sea temperatures, acidification, and decreasing sea ice are leading to major shifts in the marine ecosystems, with species moving northwards and ice-dependent species increasingly facing a loss of habitat. In comparison to the Arctic, climate change impacts in the Antarctic are less pronounced, but it is expected that major changes will also occur in the Southern Ocean soon. Against this background, effective governance and management that is able to cope with rapid change in these polar waters becomes an urgent need. This chapter provides an overview of the impacts of climate change on marine ecosystems in the Arctic and Antarctic and of the governance frameworks in place in the two regions. It also highlights challenges and opportunities for improving ocean governance to cope with extreme climate-induced changes in the Arctic and Antarctic.

Impacts of climate change on Arctic and Antarctic marine ecosystems

Climate change has a profound impact on the ecosystems of the Arctic Ocean and the Southern Ocean. Recent review reports, including the 2019 IPCC Special Report on Ocean and Cryosphere in a Changing Climate (Meredith et al., 2019), as well as the Snow, Water, Ice, and Permafrost in the Arctic (SWIPA) assessment series of the Arctic Monitoring and Assessment Programme (first published in 2011, updated in 2017 and 2019) and the Antarctic Climate Change and the Environment Report of the Scientific Committee on Antarctic Research (first published in 2009, updated in 2019) clearly identify a range of physical changes. As a first stark indicator of rapid change, the Arctic has warmed at three times the global average temperature rate in recent decades (AMAP, 2021: 2), resulting in warmer water temperatures and a strong decline in the coverage, thickness, and age of Arctic sea ice (Meredith et al., 2019: 213, 216). In the Antarctic, some evidence was found that waters are warming, but no discernible trend in Antarctic sea ice change could be observed (SCAR, 2019: 3, 5). This is illustrated by Figures 19.1 and 19.2, which show the monthly sea

DOI: 10.4324/9781315149745-23

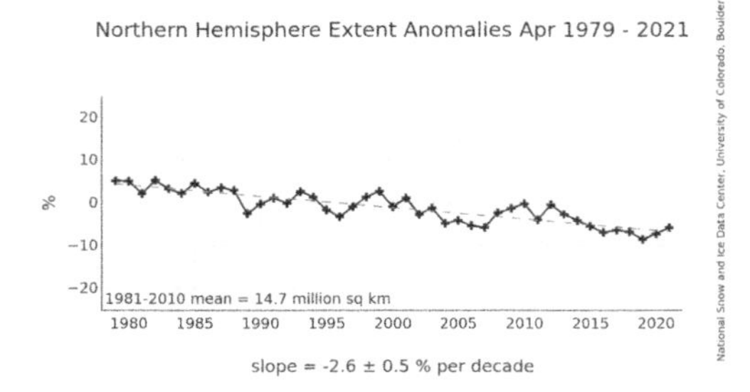

Figure 19.1 Arctic sea ice extent anomalies, 1979–2021.

Source: Fetterer et al. (2017). Available at https://nsidc.org/data/seaice_index

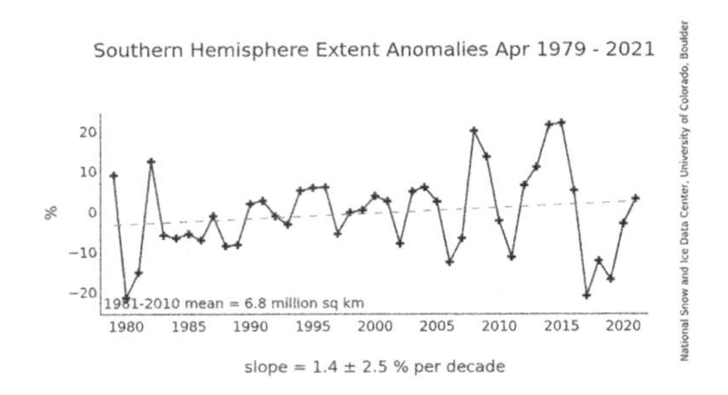

Figure 19.2 Antarctic sea ice extent anomalies, 1979–2021.

Source: Fetterer et al. (2017). Available at https://nsidc.org/data/seaice_index

ice extent anomaly in the Arctic and Antarctic, plotted as a percentage difference between the specific month and the mean sea ice extent for that month (1981–2010). In both regions, sea water is also becoming increasingly acidic, causing conditions which are corrosive for calcium carbonate shell-producing organisms such as corals, oysters, clams, and mussels (AMAP, 2018: 49; Meredith et al., 2019: 207).

These physical changes are resulting in variable and somewhat uncertain biological changes in both oceans. In the Arctic, observed changes include an increase in net primary production of up to 30%, as well as shifts in timing and location of primary production (Meredith et al., 2019: 226). Benthic species, including economically valuable snow crabs, are subject to changes in productivity and distribution, and there is evidence that fish species have expanded their ranges northward by hundreds of kilometres (Meredith et al., 2019: 228). Similar impacts are expected in the Southern Ocean, though less data means that uncertainty is generally higher. For example, krill are expected to shift southwards, with flow-on effects on other species (Meredith et al., 2019: 232, 234). In addition, there is increasing evidence that Antarctic birds, mammals, and fish are being affected by climate change (SCAR, 2019: 6). The physical and biological changes brought about by climate change are expected to further accelerate and become more extreme in the future, even under relatively optimistic climate scenarios (Meredith et al., 2019: 205).

In the Arctic, the decline in sea ice has created increasing interest in intensifying human activities in the region, including shipping, cruise tourism, as well as oil and gas exploration and exploitation (Spohr et al., 2021: 70, 119). In addition, effects on commercial and subsistence fisheries are expected to be profound because climate change is projected to alter the spatial distribution and productivity of target species (Meredith et al., 2019: 234). While uncertainty remains about the future impacts of climate change and intensified human activities on the Arctic Ocean (O'Rourke et al., 2020: 99), it is expected that intensifying human activities will further aggravate climate change impacts by multiplying pressures on the marine ecosystems (IPBES, 2018: 368). In the Antarctic, tourism has been rapidly increasing in the past years (Carey, 2020: 1) and interest in current and future commercial exploitation of toothfish, krill, and other fish species is sustained by several states, potentially increasing the need to more strictly protect the vulnerable ecosystems of the Southern Ocean in the face of increased human activity (Teschke et al., 2021: 8).

Governance of marine ecosystems in the Arctic and Antarctic under climate change

Against the background of the rapid and still somewhat uncertain changes at the Poles and the risk posed by intensifying human activities, ocean governance in both the Arctic and Antarctic is challenged to understand and react to shifts in the polar ecosystems and to manage potential conflicts between harmful human activities and environmental conservation objectives. While the challenges are comparable in both regions, the governance approaches in place in the regions differ widely due to historical and political reasons.

The Antarctic consists of a continent that is covered by ice and surrounded by the Southern Ocean. Territorial claims made in Antarctica by several countries were set aside in 1959 through the establishment of the Antarctic Treaty. While seven states still sustain sovereign claims to Antarctica, the continent and its surrounding waters are currently governed as a 'de facto international space' (Scott and Vanderzwaag, 2020: 3). No permanent residents live in Antarctica; the population is mainly comprised of scientists and support staff. In contrast, the Arctic is an ocean surrounded by land that has been inhabited for millennia. Large parts of the Arctic Ocean are located within the national jurisdiction of the Arctic coastal states of Canada, the United States (US), the Russia, and Greenland (Denmark). This means that national rules and regulations established by these countries govern Arctic waters, in addition to global and regional governance frameworks. Apart from the national waters, several pockets of international waters (high seas) exist in the Arctic, the largest being the Central Arctic Ocean (CAO).

International governance framework

The overarching international legal framework applicable to ocean governance at both Poles is provided by the United Nations Convention on the Law of the Sea (UNCLOS) (see Chapter 2). Relevant aspects addressed by UNCLOS include the management of living and non-living resources, scientific research, as well as the protection and preservation of the marine environment. According to UNCLOS, national governance frameworks determine the use of the waters and their protection in coastal areas and their exclusive economic zones of up to 200 nautical miles offshore. A special provision for ice-covered areas is included in Article 234 UNCLOS, which gives coastal states the right to also adopt and enforce national measures to prevent, reduce, and control marine pollution from vessels. This competence has been used by Canada and Russia for regulation of the waters in the Northwest Passage and the Northern Sea Route, respectively (Solski, 2021: 2). It is noteworthy that the United States is the only Arctic coastal state that has not acceded to UNCLOS, though it accepts most of its provisions in practice.

An important sectoral governance agreement applicable to the international waters of the Southern Ocean as well as the high seas of the Arctic Ocean is the UN Fish Stocks Agreement (UNFSA), which provides rules on the conservation and management of straddling and highly migratory fish stocks. All activities related to the exploration or exploitation of marine mineral resources in the international seabed area are organised, regulated, and controlled by the International Seabed Authority (ISA). As of yet, there has been no mining for mineral resources from the deep seabed beyond national jurisdiction, as the ISA has not yet agreed upon the requisite regulatory framework (the so-called 'Mining Code').

Regarding shipping activities, the treaties and governance instruments established under the International Maritime Organization (IMO) provide almost universal geographic coverage (see Chapter 23). The International Convention for the Safety of Life at Sea's (SOLAS), existing rules on navigation and safety, as well as the International Convention for the Prevention of Pollution from Ships' (MARPOL) regulations on avoiding pollution by oil and other hazardous substances (including sulphur dioxide and nitrous dioxide emissions) have been amended to include specific standards and protocols for ships navigating in Arctic and Antarctic waters. The amendments were introduced through the so-called Polar Code, which entered into force in 2017. In addition, the Antarctic area was designated as MARPOL Special Area, resulting in stricter regulations for the discharge of oil, noxious liquid substances, and garbage, as well as a ban on heavy fuel oil both as cargo and fuel (see Chapter 20).

Regional ocean governance in the Arctic

The main regional governance framework in the Arctic is provided by the Arctic Council. The Arctic Council, which was founded through the Ottawa Declaration in 1996, has as permanent members the eight Arctic states (Canada, Greenland/Denmark, Finland, Iceland, Norway, Russia, Sweden, and the US) and six Indigenous Peoples organisations. The permanent members cooperate with an increasing number of observer states and organisations on knowledge creation and coordination. Work within the Arctic Council is mainly conducted by six working groups and several task forces, which gather and generate up-to-date knowledge at the circumpolar scale to facilitate collaboration and joint action by the Arctic states. Work on marine governance and marine conservation has accelerated in the past years with ecosystem-based management as the conceptual basis (Balton and Zagorski, 2020: 4). This work is mainly conducted by working groups on Conservation of Arctic Flora and Fauna (CAFF), Protection of the Arctic Marine Environment (PAME), Arctic Monitoring and Assessment Programme (AMAP), and Emergency Preparedness, Prevention and Response (EPPR). Responsibilities of these working groups include, *inter alia*, the elaboration of assessments and guidelines related to marine biodiversity monitoring, marine litter, environmental protection, and shipping and traffic data. Important assessments by the working groups include the 2013 Arctic Marine Shipping Assessment Report, which identified ecologically and culturally sensitive marine areas, the 2015 Framework for a Pan-Arctic Network of Marine Protected Areas and the 2017 State of the Arctic Marine Biodiversity Report.

Apart from the Arctic Council working groups, a wide range of actors carry out research on the Arctic and monitoring of the Arctic marine environment, including national research institutes, Indigenous communities, and regional scientific bodies such as the International Council for the Exploration of the Sea (ICES) and the North Pacific Marine Science Organization (PICES). However, these efforts remain largely uncoordinated, as there is no central scientific body in place (Balton and Zagorski, 2020: 18). While the Arctic Council does not have the competence to regulate human activities and issue legally binding decisions, it has served as a basis for exchange by the Arctic states on the following legally binding regional governance instruments: Search and Rescue Agreement (2011), Marine Oil Preparedness and Response Agreement (2013) and Scientific Cooperation Agreement (2017) (Spohr et al., 2021: 299). No additional agreements are currently under development.

Although the Arctic Council and its working groups have succeeded in promoting regional cooperation on marine issues, the non-binding nature of commitments made by Arctic Council members leaves open room for a lack of compliance and accountability (Balton and Zagorski, 2020: 18). This is, for example, evidenced by the discussions around marine protected areas (MPAs), where efforts under the Arctic Council have spurred discussions on the need for increased protection of ecosystems by establishing a network of MPAs, but it is ultimately up to the Arctic states to put measures in place to protect the waters within their jurisdiction. A regional assessment conducted by PAME and CAFF indicated that in 2016, only 4.7% of marine waters in the Arctic were protected as opposed to 7.4% globally, with efforts being concentrated in a few very large areas, while the high seas and several other areas were lacking protection (PAME and CAFF, 2017: 11).

Another regional framework with competences in environmental governance in the Arctic Ocean is the OSPAR Convention for the Protection of the Marine Environment of the North-East Atlantic, which also covers parts of the Arctic Ocean within its 'Region I'. Under the Convention's Annex V and its Biodiversity and Ecosystems Strategy, the OSPAR Commission has a mandate for the protection of the marine environment, including through the creation of a coherent and well-managed network of MPAs. In addition, OSPAR's efforts in monitoring and assessing the state of the environment of the North-East Atlantic provide important foundations for marine governance. The OSPAR Commission became an observer to the Arctic Council in 2017 to better exchange knowledge and best practices.

Fisheries in the Arctic are managed by, in addition to national and bilaterally agreed rules and regulations, several Regional Fisheries Management Organizations (RFMOs), such as the North-East Atlantic Fisheries Commission (NEAFC), the Northwest Atlantic Fisheries Organization (NAFO) and the International Commission for the Conservation of Atlantic Tunas (ICCAT), all of which cover limited parts of Arctic and subarctic waters. In late 2018, Canada, Iceland, Denmark, Norway, the US, and Russia, as well as China, Japan, South Korea, and the European Union, signed an agreement to prevent unregulated high seas fisheries in the CAO. The agreement is expected to be ratified by 2021, and it provides a de facto moratorium on commercial fishing within its boundaries for at least 16 years. During this period, it aims to foster scientific cooperation with a monitoring programme on fish stocks and fisheries management. If the scientific work under the agreement concludes that commercial high seas fisheries in the CAO are reasonable, the agreement may serve as a basis for establishing a CAO-specific RFMO (Hoel, 2020: 4).

Discussions on options to strengthen ocean governance in the Arctic have been ongoing in recent years and have led to the decision of the Arctic Council to establish in 2015 a Task Force on Arctic Marine Cooperation (TFAMC) to assess options for improving institutional arrangements, including the option to establish a new subsidiary body or a regional seas programme. The mandate of the task force was limited in 2018 due to a lack of political support among Arctic states. In September 2020, a new 'Marine Mechanism' was launched by the Senior Arctic Officials of the Arctic states to promote regional cooperation on marine issues and address some of the needs identified by the TFAMC. The effectiveness of this body and their way forward remains to be seen (Balton and Zagorski, 2020).

Regional ocean governance in the Antarctic

The Antarctic Treaty provides the overarching regional governance framework for Antarctica and the surrounding Southern Ocean. It endorses peaceful use and scientific cooperation in the region and has 54 parties to date, 29 of which have consultative (voting) status and participate in decision-making. Apart from the Antarctic Treaty, the Antarctic Treaty System comprises the 1972 Convention on the Conservation of Antarctic Seals, the 1980 Convention on the Conservation of Antarctic Marine Living Resources (CAMLR Convention), the 1991 Protocol on Environmental Protection to the Antarctic Treaty (Madrid Protocol), and around 200 additional measures, resolutions, and decisions. Precaution plays an important role in the governance of the Southern Ocean and has been an underlying principle to the CAMLR Convention and the approach of the Madrid Protocol (Scott and Vanderzwaag, 2020: 11). The Antarctic Treaty System regulates all human activities in the Antarctic related to research, tourism, fishing, minerals exploration, and area-based protection. Shipping is the only activity in the Antarctic which is not regulated through the Antarctic Treaty System but by the IMO. In the Antarctic, many research initiatives with a focus on environmental change and its impacts in the region have been and are still being carried out, inter alia by the Scientific Committee on Antarctic Research (SCAR) and the Scientific Committee on Oceanic Research (SCOR).

When it comes to marine conservation and sustainable use of living resources in the Antarctic, the Commission for the Conservation of Antarctic Marine Living Resources (CCAMLR) established by the CAMLR Convention has a central role. CCAMLR oversees fisheries management as well as the designation of MPAs in the Southern Ocean, following a precautionary, ecosystem-based, and science-led management approach (Teschke et al., 2021: 1). CCAMLR engages in adaptive management by regularly reviewing the opening and closing of fishery areas and by setting catch limits at the annual Commission meetings. In 2009, CCAMLR Contracting Parties committed to establishing a representative system of MPAs in the Convention Area by 2012. Since then, only two MPAs have been designated by CCAMLR: the Southern Shelf Area of the South Orkney Islands MPA and the Ross Sea region MPA. In recent years, CCAMLR has reviewed three additional MPA proposals for the East Antarctic, Weddell Sea, and western Antarctic Peninsula but has not reached the needed consensus for approval for any of them. In the case of the Weddell Sea MPA proposal, interest in exploiting the Antarctic toothfish population in the area appears to be the main reason for some countries' lack of support for the proposal (Teschke et al., 2021: 8).

In addition to the work on MPAs, CCAMLR has also created a registry to recognise and report vulnerable marine ecosystems (VMEs) and allows for the precautionary measure of designating areas newly exposed by ice retreat as Special Areas for Scientific Study (Scott and Vanderzwaag, 2020: 11). In these areas, significant limitations on fishing apply for a certain period ranging up to ten years. So far, two such areas have been designated: the Larsen C Special Area for Scientific Study and the Pine Island Glacier Special Area for Scientific Study.

The way ahead: future ocean governance in the Arctic and Antarctic

To be effective, ocean governance in the Arctic and the Antarctic must be capable of understanding changes in marine ecosystems brought about by climate change and of designing swift and appropriate actions to cope with such changes. In addition, strong coordination of governance approaches and marine conservation measures among competent bodies across jurisdictions is necessary in the areas where management borders do not align with marine ecosystem dynamics, as is the case in many parts of the Arctic (Balton and Zagorski, 2020: 10). Increasing evidence exists that conservation measures such as MPAs can make marine ecosystems more resilient to climate change impacts by supporting the biological processes which underlie resilience (Roberts et al., 2017: 6168) and that networks of MPAs can play a key role in promoting resilience to climate change as part of an ecosystem-based approach to management (PAME, 2021: 16). In addition to static approaches, the polar regions need dynamic and adaptive/anticipatory management approaches set up to offer protection to those habitats or species which react to climate change impacts by shifting their ranges (Frazão Santos et al., 2020: 511). As part of such dynamic approaches, new information should be used to iteratively update management plans, goals, and methods, including through rezoning or re-delineating MPA network boundaries (Wilson et al., 2020: 3260).

Challenges in addressing climate change impacts in the Arctic and Southern Oceans

While the existing governance approaches in place at both Poles provide a basis for addressing current and future environmental change in the Arctic and Antarctic marine environment, challenges remain when it comes to implementing effective and dynamic governance approaches that can address rapid and significant impacts of climate change. One persistent challenge is economic interests of countries which can create barriers to conservation efforts at both Poles (Wenzel et al., 2016: 79). In the Arctic, Arctic and non-Arctic states have significant interests related to exploiting the energy, fishery, and mineral resource of the region and increasing the use of the area for shipping (Wenzel et al., 2016: 79). In the Antarctic, several attempts under CCAMLR to designate large parts of the Southern Oceans as MPAs were frustrated because contracting parties defended their right to exploit marine resources and are concerned that MPAs may interfere with current or future fishing activity (Wenzel et al., 2016: 77).

In the Arctic, another challenge for the implementation of integrated, ecosystem-based management arrangements is the division of the marine waters into areas under different national jurisdictions as well as areas beyond national jurisdiction, regulated through an array of rules and regulations set at the international, regional, national, and sub-national levels. A central institution or multilateral Arctic governance initiative which would be able to coordinate and harmonise the existing rules and regulations among the Arctic states and stakeholders is currently not in place, making it difficult to address the profound changes in the marine ecosystem brought about by climate change in an integrated and regionally co-ordinated manner. This coordination challenge is less prevalent in the 'de facto international space' of the Antarctic, where all human activities other than shipping are regulated through the Antarctic Treaty System (Scott and Vanderzwaag, 2020: 3).

Finally, more scientific information about the effects of climate change on marine ecosystems and dependent human uses is needed. In the Arctic, knowledge of the marine environment is still limited in many areas (Balton and Zagorski, 2020: 10) and much uncertainty remains about the impacts of climate change and intensified human activities (O'Rourke et al., 2020: 99). In a similar manner, additional information is needed to understand the impacts of climate change on marine ecosystems in the Southern Ocean and to attain the robust scientific knowledge needed to inform protection measures (Gutt et al., 2020: 2).

International marine biodiversity policy with relevance to the Arctic and Southern Oceans

When considering options for strengthening ocean governance frameworks in the Arctic and Antarctic, it is crucial to consider a range of ongoing processes at the global, regional, and national level that are expected to contribute to and shape future global and regional frameworks and approaches. Two important ongoing multilateral processes are the development of the 'post-2020 Global Biodiversity Framework' (post-2020 GBF) within the context of the United Nations Convention on Biological Diversity (CBD) and the development of an international legally binding instrument (ILBI) under the United Nations Convention on the Law of Sea (UNCLOS) on the conservation and sustainable use of marine biological diversity of areas beyond national jurisdiction (BBNJ). The post-2020 GBF, which is to be adopted by the 15th Conference of the Parties to the CBD, will include goals and targets for the conservation and sustainable use of biodiversity until 2030. Among the targets currently being negotiated is the target to 'by 2030, protect and conserve through a well-connected and effective system of protected areas and other effective area-based conservation measures at least 30% of the planet, with the focus on areas particularly important for biodiversity'. (CBD, 2020: 5). If adopted, this target would substantially raise the ambition and the need to take further action, including in the polar regions, to pursue science-based approaches for the identification and designation of protected areas.

The envisioned UN BBNJ ILBI will address existing gaps in high seas governance by providing legal provisions on: (i) marine genetic resources, (ii) measures such as area-based management tools, (iii) environmental impact assessments, and (iv) capacity building and transfer of marine technology, (UNGA, 2017: 1,2). It remains difficult at present to predict the scope and strength of the final ILBI, as there are still divergent and to some extent conflicting views among states regarding several key issues (Humphries and Harden-Davies, 2020: 3,4,5). Also, it remains to be seen how the provisions of the ILBI can and will interact with the well-established governance arrangements in place under the Antarctic Treaty System, including CCAMLR. In the Arctic, on the other hand, the ILBI is expected to provide an overarching legal framework and a strong impetus for further developing governance approaches for marine areas beyond national jurisdiction in the CAO (Spohr et al., 2021: 299).

Towards integrated, ecosystem-based management in the Arctic

At the regional level, the 'Senior Arctic Officials Marine Mechanism' recently launched under the Arctic Council may provide an opportunity to fulfil some of the unmet needs identified by the TFAMC. To allow for the Marine Mechanism to strengthen ocean governance in the Arctic, it should receive a strong mandate to generate concrete and actionable measures aimed at institutionalising the strategic planning and implementation of ecosystem-based management of the Arctic Ocean, including the high seas.

Apart from strengthening structures and processes within the Arctic Council, the Arctic Council and its newly established Marine Mechanism need to closely cooperate and coordinate with a broad set of actors, including the Observer States and Observer Organisations, industry, officials responsible for making decisions and for implementing policy on marine issues in each of the Arctic Council member states, as well as with other existing regional instruments, such the CAO Fisheries Agreement, the OSPAR Convention, and competent sectoral organisations such as the IMO and the ISA in order to develop a well-integrated Arctic marine governance system. In addition, the involvement of Indigenous rights holders is crucial, as it allows for an improved understanding of climate change impacts and is necessary to ensure economic sustainability of Indigenous coastal communities, for example by introducing co-management approaches.

In the longer term, the option of building a stronger architecture under the umbrella of the Arctic Council or outside the Arctic Council should be considered. Such an architecture could include a marine management body with a clear mandate, capable of receiving scientific advice and recommendations and of acting decisively on such advice by adopting management measures to achieve effective ecosystem-based management in the Arctic Ocean (Balton and Zagorsky, 2020: 22).

Towards stronger conservation in the Antarctic

In contrast to the Arctic, the Antarctic is subject to a comprehensive, long-established governance instrument: the Antarctic Treaty System and its related agreements, including the CAMLR Convention. In theory, the Antarctic Treaty System provides a good basis for ecosystem-based management of the Southern Ocean and its biodiversity. However, adjustments seem necessary for addressing the current and emerging challenges. Currently, any CCAMLR Contracting Party can block decisions and thus inhibit, for example, the designation of MPAs and their associated management and research plans. This hinders timely and effective decision-making in the face of significant changes occurring in the region, including decisions on the implementation of ambitious conservation measures. A way out of the deadlock generated by veto rights for every Contracting Party would be to allow for decision-making through majority votes. Depending on the nature and scope of the decision to be taken, a simple or two-thirds majority could be set as required. Furthermore, the introduction of the possibility to 'opt-out', which would allow willing countries to take measures even if not all Contracting Parties would do so, could be considered. While such far-reaching adjustments to the current governance structures could be difficult to pursue, the increasing pressure to act in light of rapid climate change could gather a 'coalition of the willing' to move ahead with proposing such amendments to the Antarctic Treaty System, including CCAMLR.

The potential role of competent international organisations

Given the broad scope of environmental changes faced by both polar regions, the establishment of protected areas through tools such as MPAs is but one part of a broader ecosystem approach to management that is essential to protect ecosystems and their biodiversity. In addition, an essential building block for any future ocean governance approach in the Arctic are the competent international organisations capable of addressing the effects of specific human activities, such as the IMO, ISA and RFMOs. Options exist, for example, to designate Particularly Sensitive Sea Areas and Special Areas (PSSAs) under the IMO, Areas of Particular Environmental Interest (APEIs) and Preservation Reference Zones (PRZs) under the ISA and Vulnerable Marine Ecosystems (VMEs) under RFMOs. These opportunities for area-based management measures applicable to specific human activities could play an important role within an overarching governance approach.

Conclusion

While climate change will lead to changes to marine ecosystems at both Poles, many differences exist with regard to the speed of change, the expected impacts, and the approaches taken in both regions to address and adapt to the changes. In the 'de facto international space' of the Antarctic, climate change impacts are currently less pronounced than in the Arctic and the Antarctic Treaty Consultative Meeting has been less outspoken and proactive with regards to identifying climate change impacts and adaptation strategies when compared to the Arctic Council (Scott and Vanderzwaag, 2020: 11). In contrast, the Arctic Ocean is already seeing significant physical and biological changes as a result of rapid climate change in the region. The impacts of climate change on the marine ecosystems in the Arctic have been studied for decades and have partly positive and partly negative implications for local communities and economies that depend on the natural resources provided by the Arctic Ocean.

Governance frameworks established at the Poles actively aim to ensure marine conservation in the wake of the challenges brought about by climate change. In both the Arctic and the Antarctic, an ecosystem-based approach to management is being promoted and precautionary actions have been adopted, for example, by protecting areas newly exposed due to ice loss from fishing. Arctic governance structures have shown that they are capable of responding proactively to the rapid change and uncertainty brought about by climate change, as illustrated by the establishment of the CAO Fisheries Agreement. In addition, efforts to identify and establish networks of MPAs are being undertaken, and scientific research and cooperation with Indigenous knowledge holders are being promoted to better understand the impacts of climate change in the regions.

Notwithstanding these achievements, the existing governance schemes in the Arctic and Antarctic have shortcomings and gaps with regard to ensuring effective marine conservation. Strong interest in the exploitation of natural resources, insufficient cooperation, and collaboration of competent bodies, and a lack of scientific information create obstacles to effective conservation efforts at the Poles. This results in slow overall progress, especially when compared to the urgent action required by rapid climate change induced impacts on marine ecosystems. To improve management of polar marine ecosystems, cooperative and collaborative processes must be advanced, and new management approaches and tools, including dynamic tools, must be developed and piloted. Political will and momentum will be crucial to achieve progress. Strong international commitments within the framework of the BBNJ negotiations as well as the CBD post-2020 GBF offer unique opportunities in this regard.

Acknowledgements

We appreciate the comments of the reviewers Nina Döring, Sebastian Unger, Lina Röschel, and Olga Lukyanova, which led to considerable improvements.

References

AMAP. (2017) *Snow, water, ice and permafrost in the arctic (SWIPA) 2017.* Arctic Monitoring and Assessment Programme (AMAP), Oslo, Norway.

AMAP. (2018) *AMAP Assessment 2018: Arctic ocean acidification.* Arctic Monitoring and Assessment Programme (AMAP), Tromsø, Norway.

Balton, D., and Zagorski, A. (2020) *Implementing marine management in the Arctic Ocean.*

Carey, P.W. (2020) Is it time for a paradigm shift in how Antarctic tourism is controlled? *Polar Perspectives,* no 1.

CBD. (2020) *Update of the zero draft of the Post-2020 Global Biodiversity Framework.* 17 August 2020. www.cbd.int/doc/c/3064/749a/0f65ac7f9def86707f4eaefa/post2020-prep-02-01-en.pdf, accessed 28 May 2021

Fetterer, F., Knowles, K., Meier, W. N., Savoie, M., and Windnagel, A. K. (2017) Sea Ice Index, Version 3. Updated daily. Monthly Antarctic and Arctic Sea Ice Extent Anomaly. Boulder, Colorado USA, NSIDC: National Snow and Ice Data Center, https://nsidc.org/data/seaice_index/, accessed 28 May 2021, https://doi.org/10.7265/N5K072F8

Frazão Santos, C., Agardy, T., Andrade, F., Calado, H., Crowder, L. B., Ehler, C. N., García-Morales, S., Gissi, E., Halpern, B. S., Orbach, M. K., Pörtner, H. O., and Rosa, R. (2020) Integrating climate change in ocean planning. *Nature Sustainability,* vol 3, no 7, pp 505–516, https://doi.org/10.1038/s41893-020-0513-x

Gutt, J., Isla, E., Xavier, J. C., Adams, B. J., Ahn, I. Y., Cheng, C. H. C., Colesie, C., Cummings, V. J., di Prisco, G., Griffiths, H., Hawes, I., Hogg, I., McIntyre, T., Meiners, K. M., Pearce, D. A., Peck, L., Piepenburg, D., Reisinger, R. R., Saba, G. K., Schloss, I. R., Signori, C. N., Smith, C. R., Vacchi, M., Verde, C., Wall, D. H. (2020) Antarctic ecosystems in transition – life between stresses and opportunities. *Biological Reviews,* vol 96, no 3, pp 798–821, https://doi.org/10.1111/brv.12679

Hoel, A. H. (2020) *The geopolitics of fish in the arctic.* Policy Brief, 11, Norwegian Institute of International Affairs, https://doi.org/10.13140/RG.2.2.22156.13448

Humphries, F., and Harden-Davies, H. (2020) Practical policy solutions for the final stage of BBNJ treaty negotiations. *Marine Policy,* vol 122, 104214, pp 1–7, https://doi.org/10.1016/j.marpol.2020.104214

IPBES. (2018) *The IPBES regional assessment report on biodiversity and ecosystem services for the Americas.* Rice, J., Seixas, C. S., Zaccagnini, M. E., Bedoya-Gaitán, M., and Valderrama N. (eds). *Secretariat of the intergovernmental science-policy platform on biodiversity and ecosystem services*, Bonn, Germany, https://doi.org/10.5281/zenodo.3236252

Meredith, M., M. Sommerkorn, S. Cassotta, C. Derksen, A. Ekaykin, A. Hollowed, G. Kofinas, A. Mackintosh, J. Melbourne-Thomas, M.M.C. Muelbert, G. Ottersen, H. Pritchard, and E.A.G. Schuur (2019) *Polar Regions.* Pörtner, H.-O., Roberts, D.C., Masson-Delmotte, V., Zhai, P., Tignor, M., Poloczanska, E., Mintenbeck, K., Alegría, A., Nicolai, M., Okem, A., Petzold, J., Rama, B., Weyer, N.M. (eds). *IPCC Special Report on the Ocean and Cryosphere in a Changing Climate.* https://www.ipcc.ch/site/assets/uploads/sites/3/2019/11/07_SROCC_Ch03_FINAL.pdf

O'Rourke, R., Comay, L. B., Folger, P., Frittelli, J., Humphries, M., Leggett, J. A., Ramseur, J. L., Sheikh, P. A., and Upton, H. F. (2021) *Changes in the arctic: Background and issues for congress.* Congressional Research Service.

PAME. (2021) *Marine protected areas in a changing Arctic.* Information Brief, Protection of the Arctic Marine Environment.

PAME and CAFF. (2017) *Arctic protected areas: Indicator report 2017.* Conservation of Arctic Flora and Fauna and Protection of the Arctic Marine Environment, Akureyri, Iceland.

Roberts, C. M., O'Leary, B. C., Mccauley, D. J., Cury, P. M., Duarte, C. M., Lubchenco, J., Pauly, D., Sáenz-Arroyo, A., Sumaila, U. R., Wilson, R. W., Worm, B., and Castilla, J. C. (2017) Marine reserves can mitigate and promote adaptation to climate change. *Proceedings of the National Academy of Sciences of the United States of America*, vol 114, no 24, pp 6167–6175, https://doi.org/10.1073/pnas.1701262114

SCAR. (2019) *Antarctic climate change and the environment – 2019 update.* Information Paper presented at the 42nd Antarctic Treaty Consultative Meeting, 2019, Prague, Czech Republic. www.scar.org/library/policy/antarctic-treaty/atcm-xlii-and-cep-xxii-2019-prague-czech-republic/5328-atcm42-ip136/, accessed 28 May 2021

Scott, K. N., and Vanderzwaag, D. L. (eds) (2020). *Research handbook on polar law.* Edward Elgar Publishing.

Solski, J. J. (2021) The Genesis of Article 234 of the UNCLOS. *Ocean Development & International Law*, vol 52, no 1, pp 1–19, https://doi.org/10.1080/00908320.2020.1835026

Spohr, K., Hamilton, D. S. and Moyer, J. C. (eds) (2021) *The arctic and world order.* Brookings Institution Press.

Teschke, K., Brtnik, P., Hain, S., Herata, H., Liebschner, A., Pehlke, H., and Brey, T. (2021). Planning marine protected areas under the CCAMLR regime – The case of the Weddell Sea (Antarctica). *Marine Policy*, vol 124, p 104370, https://doi.org/10.1016/j.marpol.2020.104370

UNGA. (2017) *Resolution 72/249.* Adopted by the General Assembly on 24 December 2017. International legally binding instrument under the United Nations Convention on the Law of the Sea on the conservation and sustainable use of marine biological diversity of areas beyond national jurisdiction, https://undocs.org/en/a/res/72/249, accessed 28 May 2021

Wenzel, L., Gilbert, N., Goldsworthy, L., Tesar, C., Mcconnell, M., and Okter, M. (2016). Polar opposites? Marine conservation tools and experiences in the changing Arctic and Antarctic. *Aquatic Conservation: Marine and Freshwater Ecosystems*, vol 26, pp 61–84. https://doi.org/10.1002/aqc.2649

Wilson, K. L., Tittensor, D. P., Worm, B., and Lotze, H. K. (2020). Incorporating climate change adaptation into marine protected area planning. *Global Change Biology*, vol 26, no 6, pp 3251–3267, https://doi.org/10.1111/gcb.15094

20

OIL POLLUTION AND BLACK CARBON IN THE ARCTIC

Dynamic shipping governance in a rapidly warming region

Benjamin Hofmann

Climate change has multiple adverse effects on ocean environments (Henson *et al.* 2017; Harris 2019). These effects are particularly pronounced in the Arctic, which, between 1971 and 2019, warmed three times faster than the global average (AMAP 2021: 2). Rapid warming directly affects Arctic ecosystems and enables increasing regional economic activity (see Chapter 19). Higher economic activity comes with new environmental externalities that put additional pressure on vulnerable Arctic marine ecosystems. An important growth sector in a warming Arctic is shipping. Shipping routes through the Arctic Ocean become more accessible as sea ice retreats and more demanded as regional economic activity rises. Ship traffic in the Arctic has been growing accordingly (Dawson *et al.* 2018; Gunnarsson 2021), with new cruise ships, tankers, bulk carriers, and container ships venturing into the area. A matter of environmental concern is that many ships are operating on heavy fuel oil (HFO) (Prior and Walsh 2018; Chircop 2020). HFO entails high oil pollution risks in case of accidental spillage and contributes to climate-forcing emissions of black carbon. Finding adequate governance responses to HFO use in growing Arctic shipping is important for protecting the Arctic marine environment in an era of environmental change.

The effects of global environmental change can be governed through traditional, state-based arrangements or more innovative, transnational arrangements (Hale 2020). Governance encompasses all kinds of regulatory, programmatic, and knowledge-generating activities. Traditional governance responses are given by individual states on the domestic level or by states cooperating with each other on regional or global levels (DeSombre 2000). By contrast, innovative governance responses are developed by civil society organisations and/or business, sometimes also together with state actors (Abbott and Snidal 2009; Andonova, Betsill and Bulkeley 2009). The scholarly literature has recently begun to integrate these two research streams by examining interactions between state-based and transnational governance (Cashore *et al.* 2021). This chapter studies governance responses of states and nonstate actors in an integrated way to generate insights into flexible and adaptive societal responses to global environmental change. The Arctic is an informative case as rapid warming has transformed this region into a laboratory for governing change. Lessons from this laboratory can inform future governance efforts in other sea regions for which major environmental changes are projected, too.

DOI: 10.4324/9781315149745-24

This case study examines the sequence and interplay of traditional and innovative governance arrangements regarding HFO use in Arctic shipping in five steps. First, I outline the distinction between traditional, state-based governance and innovative, transnational forms of governance. Second, I explain that Arctic warming is enabling shipping growth and that HFO use by ships entails oil pollution risks and climate effects. Third, I analyse the gaps that traditional governance responses have left in addressing these threats. Fourth, I show that innovative governance responses have partly filled these gaps and facilitated the development of traditional regulatory measures. Fifth, I conclude that nonstate leadership is an encouraging lesson for ocean protection well beyond the Arctic but is contingent on public concern about environmental changes and their impacts.

The interaction of traditional and innovative governance responses

Researchers have studied how states as well as transnational initiatives involving nonstate actors have governed environmental problems. The study of state-based governance includes the making, design, and effectiveness of domestic policies (Harrison and Sundstrom 2010; Schaffrin, Sewerin and Seubert 2014) and of international environmental agreements and regimes (Young 2014; Mitchell *et al.* 2020). The study of transnational governance is a growing field exploring, amongst others, the growth and impacts of private environmental standards (Green 2013; Auld 2014). One way to bring these research streams together is the concept of institutional complexity (Zelli, Möller and Asselt 2017). For instance, the static notion of a regime complex captures the observation that an issue is governed by various bodies rather than a single international regime (Abbott 2012). Another way to capture governance diversity is to study interactions of state-based and transnational arrangements (Cashore *et al.* 2021) and their dynamic development over time. Environmental change promises to be a fruitful context for identifying sequences of different governance responses and interactions between them.

For this case study, I make a simple distinction between traditional and innovative forms of governance. Traditional governance refers to state actions on different geographical levels. First, coastal states take regulatory and other policy measures on the domestic level. This is useful for addressing local marine problems and industries not subject to global competition. Second, states cooperate with other states through regional organisations like the Arctic Council, organisations established under the UN regional seas programme, regional fisheries organisations, and other regional arrangements (cf. Mahon and Fanning 2019). This form of cooperation seems appropriate for targeting problems in a specific sea region and regional economic context. Third, states cooperate with each other on the global level through international organisations with broad membership. This matches globalised industrial sectors and processes of environmental change extending across several sea regions (cf. Mendenhall 2019). Examples include the UN Environment Programme (UNEP) for general ocean challenges, the International Maritime Organization (IMO) for shipping, and the Food and Agriculture Organization (FAO) for fisheries. States thus govern on domestic, regional, and global levels.

Innovative, transnational governance is different from state-led responses as it relies on nonstate actors like business and civil society organisations. The governance literature has identified initiatives with different actor combinations (Abbott and Snidal 2009). Some initiatives are led by civil society or business actors alone, while other initiatives are led jointly by civil society and business (see Chapters 9–11). Finally, civil society and/or business cooperate with states in hybrid governance arrangements (Andonova, Betsill and Bulkeley 2009). Like state-based responses, these transnational arrangements may develop at different geographical scales. Importantly, the simple distinction between traditional and innovative governance is not meant to replace more fine-grained governance typologies (e.g., Abbott and Snidal 2009). Rather, it is a tool broadly capturing how state-based and transnational responses to ocean changes differ and interact.

Shipping in a warming Arctic: oil pollution risks and emissions of black carbon

The dynamic development and interaction of traditional and innovative governance forms are studied here for environmental threats posed by shipping in a warming Arctic. Warming has led to a rapid decrease in Arctic sea ice extent and thickness (Kwok 2018). Retreating sea ice makes Arctic shipping lanes more accessible. In the long term, trans-Arctic shipping may become a commercially viable alternative to existing shipping routes (Theocharis *et al.* 2018). For instance, the Northern Sea Route along Russia's north coast is the shortest connection between ports in East Asia and Northwest Europe. In the short to medium term, better accessibility of the Arctic region will enable more economic activity in sectors like resource extraction and tourism. This will increase shipping to destinations in the Canadian and Russian Arctic, and perhaps also Greenland (Lasserre 2019; Gunnarsson 2021). These developments are likely to continue even under ambitious international climate mitigation efforts. The vast majority of latest climate models predicts a practically ice-free Arctic Ocean in September before 2050 (Notz and SIMIP Community 2020). As sea ice retreats further, shipping from, to, and across the Arctic will grow. Growing Arctic shipping entails many new environmental threats for this vulnerable ocean area (Hofmann 2019).

Two major threats related to Arctic shipping are oil pollution and emissions of black carbon due to the use of HFO as fuel. HFO consists mainly of residues from refining processes, is cheaper than distillate fuel, and is the most widely used fuel in global shipping (Comer *et al.* 2017: 2). An accidental spill of HFO would threaten seabirds, marine mammals, and other organisms living close to the water surface and on beaches. An HFO spill would be difficult to clean up in Arctic conditions and biodegradation be reduced in cold climate (DNV 2011; Fritt-Rasmussen *et al.* 2018). Spill risks in the Arctic are generally high, given navigational challenges and harsh weather. Oil pollution risks could be reduced by using lighter distillate fuel instead of HFO (DNV 2011: 42) or eliminated through alternative fuels, like liquefied natural gas. Moreover, incomplete combustion of HFO in ship engines leads to emissions of black carbon. Black carbon is a type of particulate matter and a potent climate-forcing agent (cf. Brewer 2019). It accelerates Arctic warming when deposited on ice and snow by increasing sunlight absorption and reducing the albedo of these surfaces (IPCC 2019: 247). Black carbon also adversely affects human health (Janssen *et al.* 2012). A phaseout of HFO is one among many potential control measures for black carbon. However, it would be feasible in a relatively short timeframe, reduce fleet-wide emissions of black carbon by around one third, and enable the use of even more effective control measures like diesel particulate filters (Zhang *et al.* 2019). Hence, the phaseout of HFO as a fuel in Arctic shipping would reduce both oil pollution impacts and climate-forcing emissions.

The use of HFO in Arctic shipping is a revelatory case for studying governance dynamics. First, pre-existing institutions enabled state actions at different levels. This includes domestic regulation by Arctic coastal states, regional cooperation in the Arctic Council, and measures by the global shipping regulator IMO (Stokke 2013; Rayfuse 2014; see Chapter 23). Second, the warming Arctic, with its high symbolic value, has mobilised governments, scientists, environmental activists, and companies alike to think about sustainable shipping (Keil 2019). This is a favourable condition for innovative governance responses. Third, Arctic warming and shipping growth exemplify a process of rapid change with environmental impacts that were initially little regulated (Hofmann 2019). This should be conducive to sequences of innovative and traditional governance responses within a relatively short time span. The case study draws on rich primary and secondary sources: documents and records of IMO meetings; participatory observation in IMO meetings and events organised by nonstate actors; scientific literature; media sources; and background interviews with state, NGO, and business representatives. The analysis ranges from the issue's emergence on the international agenda in 2009/10 to the IMO's agreement on an HFO ban in 2020/21.

Traditional governance responses by states

States responded to the oil pollution risks and climate effects of HFO use in Arctic shipping in four ways. The first response was domestic regulation in Norway. Well before the issue appeared on the international agenda, Norway had introduced a local HFO ban in protected areas around Svalbard in 2007 (Norwegian Ministry of Justice and Public Security 2017: 60). Internationally, within the IMO, HFO use by ships in Arctic waters received attention in connection with an HFO ban for the Antarctic area in 2010. In contrast to Antarctica, however, the Arctic has a permanent population and much more economic activity. Arctic warming has fuelled aspirations to expand activities in oil and gas, mining, transport, tourism, and fisheries. As all these activities usually involve shipping, a regulation of HFO use in the Arctic would affect a much broader range of interests than in the Antarctic case. Accordingly, the international community did not take any swift action. Even Norway extended the exemptions from its partial HFO ban around Svalbard until 2015 (Norwegian Ministry of Justice and Public Security 2017: 60). Because of this and its local scale, this measure did not have any larger, direct impact on Arctic-wide shipping.

The second response of states to threats from HFO use was to generate more knowledge on the regional level (cf. Bai and Chircop 2020: 267–269). Based on a Norwegian initiative, the Arctic Council launched a series of studies through its working group on the Protection of the Marine Environment (PAME 2010: annex II). This approach promised to enable international cooperation in two ways. First, more knowledge could help the Arctic states better understand their own interests. The studies assessed how much HFO was used, by which ship types, and on which routes (DNV 2011, 2013). Second, the idea was to build regional consensus within the Arctic Council before going to the IMO. Regional consensus was considered important as Canada and Russia with their long coastlines play key roles in implementing IMO regulations for the Arctic. The hopes for a knowledge-based, regional consensus around an HFO phaseout did not materialise, though. Pre-defined interests guided the knowledge gathering and not vice versa, for instance, when the policy option of an HFO ban was excluded from further studies (PAME 2012: 4). Besides knowledge generation, the Arctic Council served to conclude the Agreement on Cooperation on Marine Oil Pollution Preparedness and Response in the Arctic in 2013. This agreement provides a basis for coordinated oil spill response but does not address HFO use as a major risk factor.

The third state-led response was the adoption of a nonbinding recommendation not to use HFO as a fuel in Arctic shipping by the global regulator IMO. Disagreements among the Arctic states prevented a binding provision on HFO in the IMO Polar Code. The Polar Code regulates safety and environmental aspects of shipping in the polar regions. Amongst others, it prescribes the installation of fuel tank protection for certain categories of ships to prevent accidental oil spills (IMO 2015: II-A, §1.2.1). During the Code's development, environmental NGOs also proposed a mandatory ban on HFO use by ships in the Arctic (FOEI *et al.* 2013). Negotiations in the IMO globalised the issue, with non-Arctic countries like

Germany, France, and Spain supporting a ban. Among the Arctic states, however, Canada and Russia were firmly opposed to banning HFO (cf. Bognar 2018: 38–39). Given the lack of regional consensus, even Arctic states generally in favour of a ban, like Norway and the US, did not support its incorporation into the Code. As a softer alternative, in 2015, the IMO adopted a nonbinding provision as part of the Polar Code recommending all ships to refrain from using HFO in Arctic waters (IMO 2015: II-B, §1.1). While this provision has not had any legal teeth to effectively reduce HFO use, it flagged the issue for further consideration.

The fourth response by states was the decision to replace the IMO recommendation with a mandatory ban on HFO use as fuel in Arctic shipping. Two years after the Polar Code's adoption, IMO decided to reopen negotiations about risk mitigation measures (IMO 2017: 61). In 2019, most Arctic coastal states presented impact assessments confirming the benefits of an HFO ban, while Canada and Russia took sceptical positions (IMO 2020: 42–44). They expected that a ban would increase consumer prices and adversely affect the supply of northern settlements. Moreover, higher shipping prices for raw materials would undermine the competitiveness of Arctic resource extraction. Given these objections, IMO weakened the envisaged ban (IMO 2020: 44–47). The text agreed on in 2020 and formally adopted in June 2021 delays the ban's entry into force to 2024 and allows for exemptions until 2029 (IMO Secretariat 2021). Observers consider these to be major loopholes in reducing oil spill risks and black carbon emissions (Comer *et al.* 2020). Parallel IMO work on other black carbon control measures will not close these gaps either because, at the time of writing in June 2021, it aimed for nonbinding guidelines only (IMO 2021: 11–12). In light of slow global progress, Norway announced plans to extend its local HFO ban to all areas around Svalbard (*Euractiv* 2020). Yet, despite the observed shortcomings, IMO regulation is set to force changes in fuel choice within the next decade.

Overall, state-based responses to the new environmental challenges of HFO use in growing Arctic shipping evolved over time on the domestic, regional, and global level. The partial HFO ban by Norway around Svalbard did not target Arctic-wide shipping. Regional cooperation in the Arctic Council generated new knowledge but did not attempt to reduce HFO use. The nonbinding recommendation not to use HFO in the Arctic by the global agency IMO was primarily symbolic. Its replacement with a mandatory ban marks an improvement with loopholes. Thus, while traditional governance responses to HFO use by ships in a warming Arctic have become stronger, they have also left important gaps (see Table 20.1).

Table 20.1 Traditional governance responses to HFO use in Arctic shipping

Level	*State or international organisation*	*Year*	*Governance response*
Domestic	Norway	2007	Partial HFO ban with exemptions in protected areas around Svalbard
		2015	Exemptions of partial HFO ban around Svalbard expired
		2020	Plans to extend HFO ban to all areas around Svalbard
Regional	Arctic Council	2011-	Studies on HFO use in Arctic shipping
		2013	Agreement on Cooperation on Marine Oil Pollution Preparedness and Response in the Arctic
Global	IMO	2011-	Knowledge collection on black carbon and potential control measures
		2015	Polar Code: recommendation to refrain from using HFO, provision on fuel tank protection
		2021	Mandatory ban of HFO with exemptions and delayed implementation

Source: Compiled by the author.

Innovative governance responses by nonstate actors

The gaps left by traditional, state-led governance have partly been filled by innovative governance arrangements involving nonstate actors. Among civil society organisations, environmental NGOs were early proponents of an Arctic HFO ban by the IMO (FOEI *et al.* 2010). In 2016, a group of environmental NGOs stepped up their efforts by forming the Clean Arctic Alliance (CAA; 22 members in 2021). This coalition launched the 'HFO-Free Arctic' campaign to build support among IMO member states for a rapid and comprehensive HFO ban. One of its main achievements has been the mobilisation, collection, and dissemination of scientific and indigenous knowledge about risks posed by HFO to Arctic ecosystems and inhabitants (e.g., FOEI *et al.* 2018, 2019; FOEI, WWF, and Pacific Environment 2019). In addition to environmentalists, research organisations have become active. For instance, the US-based International Council on Clean Transportation (ICCT) gathered extensive knowledge about HFO and black carbon control measures (e.g., Comer *et al.* 2017; ICCT 2018). By injecting new knowledge into IMO negotiations, different civil society organisations complemented the studies carried out under the auspices of the Arctic Council.

In the business sector, some shipping actors decided to transition away from HFO in their Arctic operations. The Norwegian expedition cruise operator Hurtigruten was the first notable company to phase out HFO (Hurtigruten 2018: 14). Its decision reflected the need to comply with local HFO bans around Svalbard and concerns about public reputation in case of accidental fuel oil spillage. Other expedition cruise companies advertising a pristine Arctic to their customers, like Ponant (2020) from France, have followed Hurtigruten's example. The Association of Arctic Expedition Cruise Operators (AECO) also issued a recommendation for its members to use distillate fuel in Arctic operations (AECO 2013: 4). In 2019, AECO turned this recommendation into a mandatory provision (AECO 2019). The expedition cruise sector was the first to demonstrate the feasibility of an HFO phaseout.

Business and civil society also joined forces. In 2018, the Clean Arctic Alliance together with Hurtigruten initiated the Arctic Commitment, which calls for an HFO phaseout in Arctic shipping (CAA 2018). The Arctic Commitment targets both the practices of shipping companies and international regulation by the IMO. Its signatories mainly come from environmental activism, academia, and the expedition cruise sector. Prominent supporters are the Icelandic ports and IKEA Supply. However, the Arctic Commitment has not found as much business support as its initiators may have hoped for. For instance, even the rather progressive shipowners' associations of Denmark and Norway have not signed it. Nevertheless, the initiative further increased public awareness of HFO risks. The Arctic Commitment also gave the small expedition cruise sector a visible platform for its use of less polluting fuels in contrast to the much larger overseas cruise sector.

A more far-reaching initiative of civil society and business is the Arctic Corporate Shipping Pledge. Since 2019, this initiative has committed member firms to refrain from using trans-Arctic shipping routes (Ocean Conservancy 2020). This limits future growth of HFO use and other adverse impacts of shipping on the Arctic environment. The Arctic Corporate Shipping Pledge was initiated by the environmental NGO Ocean Conservancy and the company Nike. Signatory companies include major shippers (e.g., Bestseller, Gap, H&M, and Li & Fung) and some of the world's largest container shipping companies (e.g., CMA CGM, MSC, Evergreen Line, and Hapag Lloyd). While this has been a setback for trans-Arctic shipping visions in the short to medium term, it appears questionable whether shipping companies can afford this pledge in the long run. Trans-Arctic routes may become more accessible and competitive in the future and other shipping companies, like the Chinese COSCO (*High North News* 2019), may be willing to exploit their advantages. This governance arrangement may thus not endure under global competition and continued environmental change. For now, however, the Arctic Corporate Shipping Pledge reduces the number of potential trans-Arctic shipping actors with an interest in blocking an HFO ban.

Pressure by environmental NGOs, mounting public attention, and new business opportunities have led more, but far from all, companies to end HFO use in Arctic operations. Targeted by the HFO-Free Arctic campaign, Crystal Cruises, Carnival Corporation, and Maersk declared to use alternatives to HFO in Arctic operations (*The Guardian* 2016; *World Maritime News* 2018; *Marine Insight* 2019). Other companies that have been moving into cleaner fuels include Hapag Lloyd Cruises (2019) and the Finnish ice-breaking company Arctia (2016). This development has not extended to all shipping segments, though. Resupply shipping in Canada and Russia still uses HFO. The Russian Federation (2019: 30–31) noted that techno-economic barriers were preventing its relatively old Northern Supply fleet from switching to lighter fuels. Canada (2019: 7–11) stressed socio-economic reasons for the continued use of HFO in its sealift operations. Similarly, bulk shipping of Arctic mining products relies on HFO to remain competitive on the world market. Both resupply and bulk shipping play special roles in the Arctic as fuel price increases may lead to higher consumer prices and loss of mining jobs and revenues. Environmental threats appear to be politically more accepted when originating from shipping segments with high strategic priority.

Collaborations of civil society, business, and states have not played a major role in governing HFO use. One initiative that includes representatives from all three types of actors is the Arctic Shipping Best Practice Information Forum established by the Arctic Council. The forum primarily contributes to the implementation of the IMO Polar Code (Engtrø, Gudmestad and Njå 2020: 59). Furthermore, it makes progressive business actors like AECO visible. While not specifically focused on HFO, the forum is a channel for communicating green shipping advances.

The innovative governance responses of nonstate actors have interacted with international regulatory negotiations in the IMO. Knowledge provision by environmental NGOs drew non-Arctic states like Germany, France, and the Netherlands into the negotiations (cf. Denmark *et al.* 2019). Change in business practices played a crucial role in reducing potential opposition to a ban. Countries in which major shipping companies had gone green, including Finland, Norway, and the US, clearly supported a ban (cf. Finland *et al.* 2018). By contrast, the hesitance of resupply and bulk shipping to switch to cleaner fuels was mirrored in sceptical positions of Canada, Russia, and major non-Arctic flag states like the Marshall Islands (cf. Canada and Marshall Islands 2018; Russian Federation 2018). Hence, innovative governance responses by nonstate actors and their limitations foreshadowed the political compromises in the realm of traditional governance (see Table 20.2).

In sum, innovative governance responses by nonstate actors have fulfilled two functions. First, they have filled gaps left by traditional, state-led governance. Environmental NGOs put pressure on companies to reveal and justify their fuel choices. Commitments of business to phase out HFO or refrain from using Arctic shipping routes made shipping somewhat greener even in the absence of public regulation. Second, nonstate actors also injected new momentum into international regulatory negotiations. Knowledge gathering and

Table 20.2 Innovative governance responses to HFO use in Arctic shipping

Type of actors	*Organisation or initiative*	*Year*	*Governance response*
Civil society	Environmental NGOs, Clean Arctic Alliance, research institutes (ICCT)	2010–	Knowledge generation, gathering, and dissemination on HFO use and risks
Business	Various companies (e.g., Arctia, Carnival Corp., Hurtigruten, Hapag Lloyd, Ponant, Maersk)	2007–	Voluntary phaseout of HFO from Arctic operations (sometimes in response to NGO or public pressure)
	Association of Arctic Expedition Cruise Operators (AECO)	≤2013	Voluntary operational guideline for member firms to use distillate fuel in the Arctic
		2019	No-HFO policy made mandatory for members
Civil society and business	Clean Arctic Alliance and Hurtigruten; other companies, AECO	2018–	Arctic Commitment: signatories call for phaseout of HFO
	Ocean Conservancy and Nike; other shippers and major shipping companies	2019–	Arctic Corporate Shipping Pledge: member firms do not use trans-Arctic shipping routes
States, business, civil society	Arctic Shipping Best Practice Information Forum of Arctic Council	2017–	Exchange about experiences and challenges in implementation of Polar Code, incl. safety

Source: Compiled by the author.

dissemination by civil society organisations complemented the study activities of the Arctic Council. Moreover, green commitments of major companies reduced potential opposition to an HFO ban among IMO member states. Innovative, transnational governance initiatives were thus an important precursor to progress in traditional state-based regulation.

Conclusion

Observations on the governance of HFO use in Arctic shipping can inform the search for suitable governance models in other sea regions and marine issues. The Arctic is subject to rapid climate change, which enables more economic activity and brings new environmental threats to the region's ecosystems and inhabitants. I used Arctic shipping growth as a case that involves increasing risks of accidental oil pollution and emissions of climate-forcing black carbon. Both threats could be mitigated by phasing out HFO as a ship fuel in the region. The case study examined how traditional governance responses by states and innovative governance responses involving nonstate actors have evolved and interacted. Without doubt, many specific elements of this case are unique to the Arctic and its climatic, economic, and navigational conditions. More generally, however, rapid warming has turned the Arctic into a laboratory for governing global environmental changes that may unfold in different forms and over longer time spans in other sea regions, too (see Chapters 17, 18 and 25).

So, what can researchers and practitioners learn from the Arctic laboratory for governing environmental change? I propose three lessons that also add to ideas on how to use the United Nations Decade of Ocean Science for Sustainable Development (2021–2030) (Claudet *et al.* 2020). First, do not wait for states to respond. Traditional governance responses by states have emerged only slowly and left important policy gaps. Domestic responses have not been comprehensive enough to govern changes extending beyond national borders. Regional and global responses have been hampered by the many diverging interests in the cumbersome search for political compromise. One should therefore not overestimate the capacity of states to quickly respond to new problems in marine ecosystems. Other actors need to fill the gaps left by traditional governance.

Second, mobilise nonstate actors for rapid responses to new ocean challenges. Innovative governance responses involving civil society and business actors have evolved in a flexible way and filled policy gaps left by states. Environmental NGOs and research institutes have facilitated decision-making of states by gathering and disseminating knowledge. Even more importantly, they have reached out to business in search for pragmatic ways to improve environmental performance. Several business actors have assumed the role of green leaders that go beyond regulatory compliance. Green leadership by companies, in turn, has reduced barriers to an agreement among states on regulatory responses. Innovative, transnational governance arrangements of nonstate actors can facilitate traditional governance responses by states.

Third, develop systematic linkages between innovative and traditional forms of governance. The sequence identified in this case study only worked because of sufficient links between green business practices and regulatory negotiations of states. The consultative status of civil society organisations and business actors in international organisations like the IMO is an important institutional element in this respect. Forums fostering exchange between practice and policy may be another useful building block. Yet, these institutional links do not appear sufficient for kick-starting and maintaining dynamic governance sequences. Public attention has strongly contributed to company decisions to phase out HFO in Arctic shipping. Likewise, it has created pressures for states to address the issue using available expertise from civil society and business. Hence, both the public and different types of political actors need to be mobilised to successfully govern the challenges of global change in marine environments.

Acknowledgements

The chapter builds on Hofmann (2021). The Institute of Political Science at the University of St.Gallen, Switzerland, generously funded field research. I thank the Swiss Maritime Navigation Office for the opportunity to participate as an observer in IMO MEPC meetings (2017–19) and also thank all of my interlocutors.

References

Abbott, K. W. (2012) 'The transnational regime complex for climate change', *Environment and Planning C: Government and Policy*, 30(4), pp. 571–590.

Abbott, K. W. and Snidal, D. (2009) 'The governance triangle: Regulatory standards institutions and the shadow of the state', in Mattli, W. and Woods, N. (eds) *The politics of global regulation*. Princeton, NJ: Princeton University Press, pp. 44–88.

AECO. (2013) 'AECO's guidelines for expedition cruise operations in the Arctic'.

AECO. (2019) *Expedition cruise industry charts course for sustainable Arctic tourism, Press releases*. Available at: https://www.aeco.no/2019/11/expedition-cruise-industry-charts-course-for-sustainable-arctic-tourism/ (Accessed: 31 May 2021).

AMAP. (2021) *Arctic climate change update 2021: Key trends and impacts. Summary for policymakers*. Oslo: Arctic Monitoring and Assessment Programme (AMAP). Available at: https://www.amap.no/documents/doc/arctic-climate-change-update-2021-key-trends-and-impacts.-summary-for-policy-makers/3508 (Accessed: 24 May 2021).

Andonova, L. B., Betsill, M. M. and Bulkeley, H. (2009) 'Transnational climate governance', *Global Environmental Politics*, 9(2), pp. 52–73.

Arctia. (2016) *Next-generation icebreaker Polaris ready for action*, *News*. Available at: http://arctia.fi/en/2016/10/31/next-generation-icebreaker-polaris-ready-for-action/ (Accessed: 2 February 2020).

Auld, G. (2014) *Constructing private governance: The rise and evolution of forest, coffee, and fisheries certification*. New Haven, CT: Yale University Press.

Bai, J. and Chircop, A. (2020) 'The regulation of heavy fuel oil in arctic shipping: Interests, measures, and impacts', in Chircop, A. et al. (eds.) *Governance of Arctic shipping: Rethinking risk, human impacts and regulation*. Cham: Springer, pp. 265–284.

Bognar, D. (2018) 'Russia and the polar marine environment: The negotiation of the environmental protection measures of the mandatory Polar Code', *Review of European, Comparative & International Environmental Law*, 27(1), pp. 35–44.

Brewer, T. L. (2019) 'Black carbon emissions and regulatory policies in transportation', *Energy Policy*, 129, pp. 1047–1055.

CAA. (2018) 'The arctic commitment'. Available at: https://www.hfofreearctic.org/wp-content/uploads/2018/01/The-Arctic-Commitment-text-EN.pdf (Accessed: 31 May 2021).

Canada. (2019) 'Assessment of the benefits and impacts associated with a ban on the use and carriage of heavy fuel oil as fuel by ships operating in the Arctic'. PPR 7/INF.16.

Canada and Marshall Islands. (2018) 'Comments on document MEPC 72/11/1 on measures to reduce risks of use and carriage of heavy fuel oil as fuel by ships in Arctic waters'. MEPC 72/11/4.

Cashore, B. *et al.* (2021) 'Private authority and public policy interactions in global context: Governance spheres for problem solving', *Regulation & Governance*, Early view online. Available at: https://onlinelibrary.wiley.com/doi/abs/10.1111/rego.12395 (Accessed: 25 May 2021).

Chircop, A. (2020) 'The polar code and the arctic marine environment: Assessing the regulation of the environmental risks of shipping', *The International Journal of Marine and Coastal Law*, 35(3), pp. 533–569.

Claudet, J. *et al.* (2020) 'A roadmap for using the UN decade of ocean science for sustainable development in support of science, policy, and action', *One Earth*, 2(1), pp. 34–42.

Comer, B. *et al.* (2017) *Prevalence of heavy fuel oil and black carbon in Arctic shipping, 2015 to 2025*. Washington, DC: ICCT. Available at: https://theicct.org/sites/default/files/publications/HFO-Arctic_ICCT_Report_01052017_vF.pdf (Accessed: 18 August 2019).

Comer, B. *et al.* (2020) *The international maritime organization's proposed Arctic heavy fuel oil ban: Likely impacts and opportunities for improvement*. White Paper. Washington, DC: ICCT, p. 46. Available at: https://theicct.org/sites/default/files/publications/Arctic-HFO-ban-sept2020.pdf (Accessed: 31 May 2021).

Dawson, J. *et al.* (2018) 'Temporal and spatial patterns of ship traffic in the Canadian Arctic from 1990 to 2015', *Arctic*, 71(1), pp. 15–26.

Denmark *et al.* (2019) 'Draft language for a ban of use and carriage of heavy fuel oil as fuel by ships in Arctic waters'. PPR 7/14/4.

DeSombre, E. R. (2000) *Domestic sources of international environmental policy: Industry, environmentalists, and U.S. power*. Cambridge, MA: MIT Press.

DNV. (2011) *Heavy fuel in the Arctic (Phase I)*. Akureyri: PAME.

DNV. (2013) *Heavy fuel in the Arctic - Phase II*. 2013-1542-16G8ZQC-5/1. Høvik.

Engtrø, E., Gudmestad, O. T. and Njå, O. (2020) 'Implementation of the polar code: Functional requirements regulating ship operations in polar waters', *Arctic Review*, 11, pp. 47–69.

Euractiv. (2020) 'Norway plans heavy oil ban around Svalbard', 9 November. Available at: https://www.euractiv.com/section/energy/news/norway-plans-heavy-oil-ban-around-svalbard/ (Accessed: 28 May 2021).

Finland *et al.* (2018) 'Proposal to ban heavy fuel oil use and carriage as fuel by ships in Arctic waters'. MEPC 72/11/1.

FOEI *et al.* (2010) 'Additional MARPOL provisions for the Polar Code'. DE 54/13/8.

FOEI *et al.* (2013) 'Heavy fuel oil use by vessels in Arctic waters'. DE 57/11/11.

FOEI *et al.* (2018) 'Residuals bunker fuel ban in the IMO Arctic waters – an assessment of costs and benefits'. PPR 6/INF.25.

FOEI *et al.* (2019) 'Infographic on reducing Black Carbon emissions from international shipping'. MEPC74/INF.31.

FOEI, WWF, and Pacific Environment (2019) 'Arctic Indigenous Support for the Ban of Heavy Fuel Oil in the Arctic'. PPR 7/14/1.

Fritt-Rasmussen, J. *et al.* (2018) *Heavy Fuel Oil (HFO): A review of fate and behaviour of HFO spills in cold seawater, including biodegradation, environmental effects and oil spill response*. 549. Copenhagen: Nordic Council of Ministers. Available at: http://urn.kb.se/resolve?urn=urn:nbn:se:norden:org:diva-5380 (Accessed: 24 May 2021).

Green, J. F. (2013) *Rethinking private authority: Agents and entrepreneurs in global environmental governance*. Princeton, NJ: Princeton University Press.

Gunnarsson, B. (2021) 'Recent ship traffic and developing shipping trends on the Northern Sea Route—Policy implications for future arctic shipping', *Marine Policy*, 124, 104369.

Hale, T. (2020) 'Transnational actors and transnational governance in global environmental politics', *Annual Review of Political Science*, 23, pp. 203–220.

Hapag-Lloyd Cruises. (2019) *The Hapag-Lloyd Cruises' fleet is set to operate without heavy fuel oil - Hapag-Lloyd Cruises, Press Releases*. Available at: https://www.hl-cruises.de/presse/presse-meldungen/detail/the-hapag-lloyd-cruises-fleet-is-set-to-operate-without-heavy-fuel-oil (Accessed: 13 January 2020).

Harris, P. G. (2019) 'Climate change at sea: Interactions, impacts, and governance', in Harris, P. G. (ed.) *Climate change and ocean governance: Politics and policy for threatened seas*. Cambridge: Cambridge University Press, pp. 3–26.

Harrison, K. and Sundstrom, L. M. (2010) *Global commons, domestic decisions: The comparative politics of climate change*. Cambridge, MA: MIT Press.

Henson, S. A. *et al.* (2017) 'Rapid emergence of climate change in environmental drivers of marine ecosystems', *Nature Communications*, 8(1), 14682.

High North News. (2019) 'Chinese shipping company COSCO to send record number of ships through arctic', 13 June. Available at: https://www.highnorthnews.com/en/chinese-shipping-company-cosco-send-record-number-ships-through-arctic (Accessed: 31 May 2021).

Hofmann, B. (2019) 'Policy responses to new ocean threats: Arctic warming, maritime industries and international environmental regulation', in Harris, P. G. (ed.) *Climate change and ocean governance: Politics and policy for threatened seas*. Cambridge: Cambridge University Press, pp. 215–235.

Hofmann, B. (2021) *Arguing over technology: The coproduction of business practices and international maritime environmental regulation*. Dissertation, University of St. Gallen.

Hurtigruten. (2018) 'Environmental Report 2017'. Available at: https://secure.viewer.zmags.com/publication/72ebfbed#/72ebfbed/1 (Accessed: 2 February 2020).

ICCT. (2018) *Workshop summary: Fifth ICCT workshop on marine black carbon emissions: Appropriate black carbon control measures*. San Francisco: International Council on Clean Transportation. Available at: https://theicct.org/sites/default/files/Workshop%20summary_5th%20ICCT%20BC%20workshop_vf_rev4.pdf (Accessed: 18 August 2019).

IMO. (2015) 'Resolution MEPC.264(68): International Code for ships operating in polar waters (polar code)'.

IMO. (2017) 'Report of the Marine Environment Protection Committee on its Seventy-First Session'. MEPC 71/17.

IMO. (2020) 'Report to the Marine Environment Protection Committee'. PPR7/22.

IMO. (2021) 'Report to the Marine Environment Protection Committee'. PPR8/13.

IMO Secretariat. (2021) 'Draft amendments to MARPOL Annex I'. MEPC76/3/1.

IPCC. (2019) *IPCC Special report on the ocean and cryosphere in a changing climate*. Pörtner, H.-O. *et al.* (eds.). Intergovernmental Panel on Climate Change.

Janssen, N. A. H. *et al.* (2012) *Health effects of black carbon*. Bonn: WHO European Centre for Environment and Health.

Keil, K. (2019) 'Sustainability understandings of Arctic shipping', in Gad, U. P. and Strandsbjerg, J. (eds) *The politics of sustainability in the Arctic: Reconfiguring identity, space, and time*. London: Routledge, pp. 34–51.

Kwok, R. (2018) 'Arctic sea ice thickness, volume, and multiyear ice coverage: Losses and coupled variability (1958–2018)', *Environmental Research Letters*, 13(10), 105005.

Lasserre, F. (2019) 'Arctic shipping: A contrasted expansion of a largely destinational market', in Finger, M. and Heininen, L. (eds) *The globalarctic handbook*. Cham: Springer, pp. 83–100.

Mahon, R. and Fanning, L. (2019) 'Regional ocean governance: Polycentric arrangements and their role in global ocean governance', *Marine Policy*, 107, 103590.

Marine Insight. (2019) 'Carnival makes bombshell claim to only use cleaner fuel on cruise ships in arctic', 12 March. Available at: https://www.marineinsight.com/shipping-news/carnival-makes-bombshell-claim-to-only-use-cleaner-fuel-on-cruise-ships-in-arctic/ (Accessed: 29 August 2019).

Mendenhall, E. (2019) 'The ocean governance regime: International conventions and institutions', in Harris, P. G. (ed.) *Climate change and ocean governance: Politics and policy for threatened seas*. Cambridge: Cambridge University Press, pp. 27–42.

Mitchell, R. B. *et al*. (2020) 'What we know (and could know) about international environmental agreements', *Global Environmental Politics*, 20(1), pp. 103–121.

Norwegian Ministry of Justice and Public Security. (2017) 'Svalbard. Meld. St. 32 (2015–2016) Report to the Storting (white paper)'.

Notz, D. and SIMIP Community. (2020) 'Arctic sea ice in CMIP6', *Geophysical Research Letters*, 47(10), e2019GL086749.

Ocean Conservancy. (2020) *Take the arctic corporate shipping pledge, protecting the arctic*. Available at: https://oceanconservancy.org/protecting-the-arctic/take-the-pledge/ (Accessed: 11 June 2020).

PAME. (2010) 'Working Group Meeting Report PAME I-2010'.

PAME. (2012) 'Working Group Meeting Report PAME I-2012'.

Ponant. (2020) *Environment, Ponant*. Available at: https://au.ponant.com/why-ponant/environment/ (Accessed: 13 January 2020).

Prior, S. and Walsh, D. (2018) 'A vision for a heavy fuel oil-free arctic', *Environment: Science and Policy for Sustainable Development*, 60(6), pp. 4–11.

Rayfuse, R. (2014) 'Coastal state jurisdiction and the polar code: A test case for Arctic Ocean governance?', in Stephens, T. and VanderZwaag, D. L. (eds) *Polar oceans governance in an era of environmental change*. Cheltenham: Edward Elgar, pp. 235–252.

Russian Federation. (2018) 'Comments on the document with the proposal to ban heavy fuel oil use and carriage as fuel by ships in Arctic waters (MEPC 72/11/1)'. MEPC 72/11/3.

Russian Federation. (2019) 'Impact Assessment Report'. PPR7/INF.13.

Schaffrin, A., Sewerin, S. and Seubert, S. (2014) 'The innovativeness of national policy portfolios – climate policy change in Austria, Germany, and the UK', *Environmental Politics*, 23(5), pp. 860–883.

Stokke, O. S. (2013) 'Regime interplay in Arctic shipping governance: explaining regional niche selection', *International Environmental Agreements: Politics, Law and Economics*, 13(1), pp. 65–85.

The Guardian. (2016) 'Large cruise ship voyage through Arctic ice rekindles rows', 13 August. Available at: https://www.theguardian.com/environment/2016/aug/13/large-cruise-ship-voyage-arctic-ice-crystal-cruises (Accessed: 7 February 2020).

Theocharis, D. *et al*. (2018) 'Arctic shipping: A systematic literature review of comparative studies', *Journal of Transport Geography*, 69, pp. 112–128.

World Maritime News. (2018) 'Venta Maersk starts historic arctic voyage using ultra-low sulphur fuel', 23 August. Available at: https://worldmaritimenews.com/archives/259311/venta-maersk-starts-historic-arctic-voyage-using-ultra-low-sulphur-fuel/ (Accessed: 9 February 2020).

Young, O. R. (2014) 'The effectiveness of international environmental regimes: Existing knowledge, cutting-edge themes, and research strategies', in Betsill, M. M., Hochstetler, K., and Stevis, D. (eds) *Advances in international environmental politics*. 2nd edn. Basingstoke: Palgrave Macmillan, pp. 273–299.

Zelli, F., Möller, I. and Asselt, H. van (2017) 'Institutional complexity and private authority in global climate governance: The cases of climate engineering, REDD+ and short-lived climate pollutants', *Environmental Politics*, 26(4), pp. 669–693.

Zhang, Q. *et al*. (2019) 'Reducing black carbon emissions from arctic shipping: Solutions and policy implications', *Journal of Cleaner Production*, 241, 118261.

21
THE HIGH SEAS
Adapting to changes in pelagic ecosystems

Mitchell Lennan

The pelagic marine environment is the water column extending from the surface of the ocean down to the floor of the deep sea. It is the largest space on earth and represents 99% of the physically habitable space for biodiversity (Angel, 1993; A/72/112, 2015). It is characterised by its sheer vastness and dynamism. There are several direct impacts on pelagic ecosystems from global environmental change, including climate change, overfishing, and pollution. These impacts are pervasive, interconnected, and require collective action for adaptation. Consider climate change, for example. The inertia of the global climate system, in conjunction with historical greenhouse gas emissions, means that all adverse changes to the climate cannot be prevented by emission reductions alone, and the Intergovernmental Panel on Climate Change (IPCC) has stressed that 'there are some impacts for which adaptation is the only available and appropriate response' (IPCC, 2007: 198). Adaptation to global change is necessary, and the governance structures applicable to pelagic ecosystems must facilitate this. The vast and open nature of pelagic ecosystems, in conjunction with the fact that the high seas are a 'commons' and do not fall under any single State's jurisdiction, makes this unique area of the ocean a classic collective action problem requiring interstate cooperation to manage marine living resources (MLRs) and address issues brought about by environmental change. However, the fragmented nature of the existing governance framework, and its inability to simultaneously provide synergistic solutions to these impacts, makes adapting to change a challenge across scales, including on the high seas. The fact that pelagic ecosystems are far from shore and often very remote makes them difficult to manage both in terms of access and regulatory enforcement (Game *et al.*, 2009).

The term 'pelagic' is not found in any of the key instruments that make up the applicable governance framework for the high seas. Pelagic ecosystems and the high seas are not synonymous, as the latter is a legal term and the former an oceanographic one. The two terms are not mutually exclusive; the ocean space constituting the high seas can be said to be a pelagic ecosystem, while not all pelagic ecosystems are high seas; around 64% of pelagic ecosystems are outside of national jurisdiction and therefore constitute the high seas (Norse, 2006: 35). The high seas are understood, in legal terms, as the water column beyond the exclusive economic zone (EEZ) or territorial sea where no EEZ has been declared. 'The Area' (UNCLOS, Art. 1(1)) is the legal term for the seabed beyond the limits of the continental shelf, according to Article 76 of the 1982 United Nations

DOI: 10.4324/9781315149745-25

Convention on the Law of the Sea (UNCLOS). Areas beyond the limits of national jurisdiction (ABNJ) include both the high seas and the area. This chapter focuses on the water column, not the seabed.

This chapter first briefly describes pelagic ecosystems, their characteristics and importance, and their vulnerability to global environmental change. Second, it explains the governance framework of the high seas applicable to MLRs. Third, it illustrates that framework's ability to adapt to environmental change through using intergovernmental bodies for fisheries management on the high seas, known as regional fisheries bodies (RFBs).

Pelagic ecosystems

This section defines pelagic ecosystems, outlines their characteristics, and discusses the effects that global environmental change is having on them.

Definition and characteristics

The word pelagic originates from the Greek word *pelagikos* meaning 'open sea'. In oceanographical terms, pelagic or oceanic habitats are anything seaward of the continental shelf. With roughly 71% of the surface of the Earth, an average depth of 3,682 m, and a volume of 1.33 trillion km^3 (Beaugrand, 2017: 141), the vastness of pelagic ecosystems is hard to fathom. Pelagic ecosystems are characterised by plankton, tiny plants, and animals that live suspended in the water column at the mercy of ocean currents. Phytoplankton (microscopic photosynthetic algae) are at the base of the marine food web and contribute to around half of global primary production (Beaugrand, 2017). They also uptake carbon from photosynthesis, which travels up the food chain and is eventually stored in the deep sea as organic matter (Sabine *et al.,* 2004). Pelagic ecosystems provide many ecosystem goods and services, including regulation of climate and ocean circulation, recycling and balance of carbon, and production of living and genetic resources (Dickey-Collas *et al.*, 2017). For example, they produce over 80% of MLRs consumed by humans and support nearly all marine biodiversity directly or indirectly (Game *et al.*, 2009: 360; Pauly, 2002: 691). Productivity and vulnerability in these open-ocean ecosystems vary over temporal and spatial scales.

Pelagic ecosystems and global environmental change

The vastness and interconnectivity of the open ocean present several challenges. It is a great challenge to both study and comprehend environmental impacts on the pelagic zone simply due to its physical size. Because of this, '[w]hen compared with other components of the marine ecosystem, the impact of human activities on pelagic systems may appear minimal' (Dickey-Collas *et al.*, 2017: 2334). Despite appearing minimal and being hard to both measure and comprehend, there are several large-scale impacts on pelagic ecosystems from global environmental change, including changes in the physical and chemical makeup of the ocean through warming seas, ocean acidification and deoxygenation from climate change, as well as overharvesting/overexploitation. Ocean warming has a profound impact on pelagic ecosystems. An example is warming that causes loss of sea ice at the poles, which also negatively affects pelagic primary production and CO_2 uptake (IPCC, 2019b). Loss of sea ice may also create new pelagic ecosystems, and open-up fishing opportunities for States which, if not managed adequately, could result in overexploitation of fish stocks. A further issue which is gaining attention is temperature-driven shifts in the distribution of marine biodiversity. This causes shifts in the distribution of pelagic species, changes in their abundance, reduction in body size, and changes in migration patterns (IPCC, 2019b).

Plankton, invertebrates, and most fish cannot regulate their own body temperature apart from through behavioural means, such as moving to preferred environmental conditions. Although this is not possible for planktonic organisms since they are at the mercy of ocean currents. Since their body temperature and biological performance are linked with their external environment, they are especially vulnerable to ocean warming (Pörtner and Peck, 2010). Warming alters the rate of an array of metabolic processes which influence vital rates such as growth, reproductive success, and energy demand (Pörtner and Peck, 2010). Warming oceans can also affect the mortality, abundance, and phenology (seasonality or rhythm of life cycle) of marine species (Cheung *et al.*, 2013). The main example of behavioural temperature regulation is that of mobile marine species moving to an area with a preferred temperature for optimum metabolism. The direction of these shifts is either poleward or into deeper waters, and this phenomenon has been observed across all regions of the ocean. This has knock-on ecological effects invertebrates and fish shift their distribution, and seabirds, reptiles, and marine mammals shift as they follow their food source (Poloczanksa *et al.*, 2016). According to the IPCC in its *Special Report on Oceans and the Cryosphere,*

> [r]ates of poleward shifts in distributions across different marine species since the 1950s are 52 ± 33 km per decade ... for organisms in the epipelagic.... The direction of the majority of shifts of epipelagic organisms are consistent with a response to warming.
>
> *(IPCC, 2019b: 451)*

Ocean acidification and deoxygenation limit the ability of the ocean to uptake and store carbon through phytoplankton, as described above (IPBES, 2019; for the legal implications of ocean acidification, see Chapter 4). Warming exacerbates the effects of ocean acidification on photosynthetic plankton, particularly those planktonic organisms that are calcified (IPCC, 2019b). Deoxygenation, a pervasive and overlooked issue, can create oxygen minimum zones (OMZs) in the oceans. The expansion of OMZs combined with ocean acidification and warming can weaken pelagic ecosystem structure and function, which highlights the interconnectivity of the impacts of global environmental change (IPCC, 2019b: 481).

Overexploitation is a key challenge facing pelagic ecosystems. The state of global fish stocks is troubling and has been in general decline since the 1970s (FAO, 2020; Free *et al.*, 2019; Pauly and Zeller, 2016). This is exacerbated by ocean warming, acidification, and deoxygenation. Overexploitation is supported by government subsidies, allowing commercial fishing fleets to expand geographically and into deeper waters which were previously not financially viable to exploit (Sumaila *et al.*, 2019; IPBES, 2019). Biodiversity is in dramatic decline globally (IPBES, 2019), and loss from overexploitation is an important compounding factor, as heavily exploited species will have a diminished gene pool and reduced resilience to environmental change. Importantly, the United Nations Food and Agriculture Organization (FAO) has reported that MLRs exploited on the high seas were overfished at twice the rate of the same species within national jurisdictions (FAO, 2020). Consequently, these MLRs may be more negatively affected by climate change than less exploited or unexploited species (Free *et al.*, 2019; Dulvy *et al.*, 2008).

A final key impact on pelagic ecosystems is pollution. In pelagic environments, pollution from plastics is pervasive (see Chapter 22). The United Nations estimates that between 4.8 and 12.7 tonnes of plastic enter the ocean per year (UNEP/EA.3/L.19, 2017: 1; Jambeck *et al.*, 2015: 770). The vast majority (around 80%) comes from land-based sources, while the rest comes from maritime activities such as fishing (Isensee and Valdes, 2015: 2).

The above impacts all have severe consequences on pelagic ecosystems. They cause changes in ecosystem structure and function and may weaken resilience and ability to withstand or respond to environmental change (IPCC, 2019a). Importantly, the ranges of many marine species are predicted to shift across management jurisdictions, such as moving between the regulatory areas of an RFB on the high seas due to climate change (IPCC, 2018, 2019a). This can cause a number of governance issues for transboundary fish stocks, such as lower predictability of scientific research on stocks, and allocation of quota. Adaptive action by the relevant institutions is necessary to avoid these issues (Rayfuse, 2019). Indeed, there are several multi-dimensional challenges across scales for ocean governance brought about by environmental change, and these will continue to become increasingly complex over time, further emphasising the need for adaptive action (IPCC, 2019b).

The high seas governance framework

The governance framework applicable to the high seas has been developed through an *ad hoc*, piecemeal approach, and has governance gaps and difficulties with enforcement. This section briefly outlines this framework, focusing on the key governance instruments applicable for adapting to global environmental change. Historically, the high seas were not open to appropriation or acquisition by any State through individual or collective means. The policies of freedom of navigation and fishing were argued from the seventeenth century onwards (Grotius, 1916). The freedom of the high seas is now a general principle of international law, 'a policy or concept from which particular rules can be inferred' (Crawford and Brownlie, 2019: 282). This freedom and other rules of international law applicable to the high seas were initially codified in an international treaty: the 1958 Convention on the High Seas.

Governance framework

The framework of the modern high seas governance regime is based in the 1982 United Nations Convention on the Law of the Sea (UNCLOS) (see Chapter 2). However, UNCLOS does not cover the whole of high seas governance; there are other treaties that govern particular sectors of high seas activities. Atmospheric pollution and dumping from ships are governed by, for example, the 1972 London Convention and its 1996 Protocol on dumping in the ocean, as adopted by the International Maritime Organization, as well as the 1973 International Convention for the Prevention of Pollution from Ships (MARPOL), as modified by the 1978 MARPOL Protocol. There are also species- and sector-specific treaties with regards to the conservation and management of marine species that apply on the high seas, such as the 1948 International Convention for the Regulation of Whaling, and the 1992 Convention on Biological Diversity.

UNCLOS is widely considered as the constitution for the oceans (see Chapter 2). It can be understood as a 'living treaty' (Barrett and Barnes, 2016), and its provisions are broadly drafted to adapt to changes, and to be supplemented by other legally binding and non-legally binding instruments (Boyle and Chinkin, 2007) UNCLOS prescribes the rights and duties of States and their nationals regarding activities on the high seas. Part VII of UNCLOS concerns the high seas and sets out the key applicable governance principles. Article 86 of which defines the high seas as all parts of the sea beyond a coastal States' 200 nautical mile exclusive economic zone (EEZ), or, if an EEZ is not declared by a coastal State, then beyond the 12nm territorial sea, or beyond the archipelagic waters of an archipelagic State. In certain regions (e.g., the Mediterranean), the high seas can be close to shore as coastal states may not have declared their EEZ. Article 87 reiterates the principle of the freedom of the high seas and explains that they are open to all States. There are six 'freedoms' including those of navigation, fishing (UNCLOS, Arts. 86 and 116), and scientific research. Exercising these freedoms is subject to some conditions under UNCLOS and other rules of international law. This includes the duty to protect and preserve the marine environment under UNCLOS Part XII, and the prohibition of claims to sovereignty over the high seas (UNCLOS, Art. 89). An important principle in high seas governance is that ships have a nationality, and every State has 'whether coastal or land-locked, has the right to sail ships flying its flag on the high seas' (UNCLOS, Art. 90). In doing so, the State where the ship is registered and whose flag it flies becomes what is known as the 'flag state'. All ships are under the *exclusive* jurisdiction of the flag State while on the high seas (UNCLOS, Art. 92(1)). Flag States are essentially responsible for ensuring vessels flying their flag refrain from undermining international law on the high seas. Including, for example, conservation and management measures (CMMs) of RFBs (*Fisheries Advisory Opinion*, 2015: 36). However, some flag States are unable or unwilling to comply with their duties and adopt a very lax approach in enforcing international maritime and environmental obligations. The FAO has focused on improving flag State performance by adopting non-legally binding guidelines which elaborate on what is required of States in exercising their duties as a flag State.

High seas fisheries

The governance regime for fisheries on the high seas is fragmented and complex (Barnes and Massarella, 2016). Obligations under UNCLOS for fisheries on the high seas are bolstered by the 1995 UN Fish Stocks Agreement (UNFSA), an array of legally binding and non-legally binding instruments adopted by the FAO, and portions of UN General Assembly Resolutions. While UNCLOS and other legally binding instrument do not specifically mention climate change, this framework does not prevent States from taking action to adapt to the challenges of global environmental change described above (Molenaar, 2021; Caddell and Molenaar, 2019; Barnes and Massarella, 2016). On the high seas, the freedom to fish is enshrined by way of Articles 87(1)(e) and 116. Obligations on high seas fishing under UNCLOS are limited and ambiguous (UNCLOS, Arts. 116–120). Since no State has exclusive jurisdiction over MLRs outside their EEZ, it is a necessity that States must cooperate with other States in high seas fisheries management. This cooperation is operationalised through participation in RFBs. However, UNCLOS failed to provide provisions to coordinate these organisations, and the provisions on transboundary stocks are very general. These have been further clarified in the UNFSA.

The UNFSA supplements, clarifies, and modernises the UNCLOS provisions on transboundary fish stocks. For example, UNFSA furthers the UNCLOS obligation to use the best scientific evidence in fisheries management (UNFSA, Art. 5(b); UNCLOS, Arts. 61(2) and 119(1)(a)). It established obligations for States to ensure sustainable fishing activities, safeguard and mitigate environmental impacts of fishing on the marine environment and biodiversity through the ecosystem approach (UNFSA, Arts. 5(d–e)), as well as to protect marine biodiversity through CMMs (UNFSA, Art. 5(g)). Parties must also apply the precautionary approach whenever there is scarce or inadequate scientific information on the status of high seas fish stocks (UNFSA, Art. 6). UNFSA obliges States fishing on the high seas to cooperate in the management of transboundary fish stocks either directly or through RFBs to 'to ensure effective conservation and management of such stocks' (UNFSA, Art. 8). This aims to ensure a collaborative and regional approach to fisheries management and conservation on the high seas. Provisions on the function of RFBs are provided in Arts. 8–12 including the establishment of CMMs, and how the RFB will obtain scientific advice. Non-members of RFMOs are prohibited from fishing stocks managed by that organisation, unless they agree to abide by that organisation's CMMs, further limiting the freedom of high seas fishing (UNCLOS, Art. 84; Harrison, 2011). Duties of the flag state and compliance and enforcement obligations are also outlined in UNFSA Arts. 19–23. Indeed, while these provisions do not directly account for global environmental change, they allow for CMMs to be informed by scientific evidence which takes into account, for example, shifting stocks due to global environmental change

Regional fisheries bodies

Article 8 of the UNFSA has confirmed the international community's preference for regional fisheries bodies (RFBs), as the appropriate international vehicles for high seas fisheries governance. There are currently over 60 RFBs. They have different geographical and substantive mandates; those that have a management mandate that allows them to adopt legally binding management measures, such as to determine a total allowable catch (TAC) or allocate the TAC between the participating States in their agreed maritime area, are called regional fisheries management organisations (RFMOs). RFMOs are 'at the forefront of international efforts to achieve the conservation and sustainable utilisation of fish stocks' (Harrison, 2011: 226). An issue is that RFMOs often use single-species management, meaning that 'an influx of additional species lies beyond their individual remits' (Pinsky *et al.*, 2018: 1190). Few organisations have developed a clear posture on regulating new fisheries. Many RFBs take account of associated and dependent species as required by UNCLOS and UNFSA, but only a handful, such as the Commission on the Conservation of Antarctic Marine Living Resources (CCAMLR) take a more holistic, ecosystem-oriented approach. Those which do the latter may be regarded as in better stead to adapt to the negative effects of global environmental change. The next section will outline adaptive practices by some RFBs on the high seas.

Adapting to environmental change on the high seas

As has been highlighted in this chapter, there are a number of pervasive threats from global environmental change to pelagic ecosystems. Since the high seas are a commons, addressing these threats requires cooperation between States through the existing governance framework, as has been illustrated through the issue of high seas fisheries and warming seas. The high seas governance framework has been developed in an *ad hoc* manner and has led to a fragmented and piecemeal web of instruments which are not designed in the first instance to account for environmental changes and pressures, but they do have the capacity to adapt to them. Existing policies and governance arrangements – including international agreements on transboundary stocks – will need to be modified to facilitate permitting, monitoring, and regulation of fisheries as distributions of stocks shift across static management boundaries. Another option suggested is transformative ocean governance on the high seas, such as closing the high seas to fishing, which could increase fish catches in coastal States' exclusive economic zones by 10% by 2050 (Cheung *et al.*, 2017: 257). This could build climate resilience for coastal States, including ensuring food security. Although one can view such an action as radical and politically unworkable, there are other potential and more realistic ideas for transformative high seas governance. Bearing in mind the warnings from the IPCC that adaptation to climate change is the only viable option in many circumstances (IPCC, 2019b; IPCC, 2018; IPCC, 2007), it is imperative that much of the adaptive action needed to address the impacts of environmental change on the high seas will need to be taken through the existing governance framework described above and operationalised through RFMOs and other regional organisations.

Operationalizing adaptation on the high seas: institutional change

It is evident from the discussions above that RFMOs must lead the way in adapting management of MLRs in response to global environmental change. Some RFMOs are in the process or have recently updated their constitutive instruments or replaced them to provide appropriate frameworks to address current critical fisheries management issues in response to climate-induced effects, such as stocks moving between management jurisdictions due to warming seas (Rayfuse, 2019). Failure to adapt to this results in 'newly fished stocks to be heavily exploited before meaningful standards are developed' (Pinsky *et al.*, 2018: 1190; see also Caddell, 2018). Other concerns that stymie adaptive action include weak management, lack of inter-organisation cooperation, and failure to elaborate on regulations for new fisheries, and poor enforcement by flag States, thus facilitating loopholes that allow for overfishing (Pinsky *et al.*, 2018). Moreover, there are criticisms over the limited application of the ecosystem approach to fishing by RFMOs 'including limited consideration of impacts on species not directly managed by the RFMOs' (Pinsky *et al.*, 2018: 1190). RFMOs may develop regulations on 'new or exploratory fisheries' by way of Article 6(6) of UNFSA, these could include newly exploitable species in the deep sea and those species that may have arrived in RFMO management areas as a consequence of climate change, with the purpose of eventually transitioning into a commercial fishery (Caddell, 2018). The existing new or exploratory regulations by RFMOs, particularly CCAMLR, have recently been analysed by Caddell (2018), who concluded that the vast majority of organisations were unprepared for the pervasive response of marine species and other issues brought about by climate change. Further, RFMOs have been criticised for their inability to successfully achieve sustainable management of transboundary MLRs (Gjerde, 2006).

Suggested institutional improvements for RFMOs to respond to the governance challenges include, among others, (i) actively anticipating stressors from environmental change in their scientific research; (ii) integrating those changes into their decision making; and (iii) adapting their management measures to address environmental changes (Rayfuse, 2019; A/RES/73/125). Several RFMOs have responded well to the anticipatory component; however, only, CCAMLR and the North Atlantic Fisheries Organization have taken reasonable steps to implement the integration and adaptation component (Molenaar, 2021; Rayfuse, 2019). Greater pressure from member States within RFMOs and RFBs is needed to achieve institutional improvements and achieve adaptive responses to environmental change on the high seas through the adoption of CMMs which account for climate change.

To adapt to global environmental changes on the high seas in the short amount of time available, it is necessary to use the existing governance framework, and there have been international developments to develop new instruments applicable to the high seas beyond the BBNJ Agreement. In a world first, the 2019 Agreement to Prevent Unregulated High Seas Fisheries in the Central Arctic Ocean (CAOF Agreement) is the first international fisheries instrument to be developed in response to global environmental change (Molenaar, 2021). This international legal instrument was negotiated in response to the fact that sea ice in the central Arctic Ocean has receded in recent years due to climate change, opening up new high seas fishing opportunities for States in the Arctic. The instrument features a temporary prohibition on high seas fishing opportunities for States in the central Arctic (Molenaar, 2019). This is an example of the key governance principles to adapt to global environmental change on the high seas: cooperation, use of the best available science, and exercising precaution.

Operationalizing adaptation on the high seas: marine protected areas

The evidence is compelling that the implementation of an ecologically connected network of MPAs on the high seas would aid in the conservation of biodiversity and help build resilience to climate change (Hammond and Jones, 2021; Roberts, 2012; Worm *et al.*, 2006; Kimball, 2005). For example, pre-emptive protected areas could alleviate some of the pressures on MLRs should they move to new areas due to climate change. However, the high seas governance framework provides limited guidance on the establishment of MPAs and other area-based management tools (ABMT). In practice, some RFMOs have established MPAs and other ABMT. For example, CCAMLR implemented the Ross Sea Regional MPA in 2017 after scientific and diplomatic discussion. The MPA protects large-scale ecosystem processes, protects ecologically important areas, and conserves biodiversity. This MPA is 'time-bound', meaning that it will remain in force until the end of its review period in 2052. While this time-bound condition is a result of political negotiation, and was not implemented due to climate change, it is still an interesting development for two reasons. First, it means that MPAs could have 'flexible' or 'temporary' status, allowing them to be moved or enlarged to protect particular species if their distribution shifts due to climate change. Second, it allows for the balancing of interest of members of RFMOs who may not be willing to conserve an area or species permanently, but it could be persuaded to do so for a limited period of time. This effectively 'buys time' for marine species which might be at risk of overexploitation due to climate change.

Indeed, the operationalisation of RFMOs to help deliver adaptive governance on the high seas has the advantage that they can adopt CMMs that are targeted at the ecological and political makeup of a particular region. Importantly, RFMOs can adopt measures binding on their parties, which is not the case for ABMT adopted through RFBs or the Regional Seas Programme under the UN Environment Programme (see Chapter 5). Further, many RFMOs have the ecosystem and precautionary approach included in their mandate, which can be further operationalised to include the impacts of climate change when making management decisions. While this does not necessarily allow them to regulate non-fishing activities on the high seas, they can undertake cumulative impact assessments which can evaluate the combined effects of human activities in their regulatory area (Erinosho *et al.*, 2022; Diz and Ntona, 2018). The output of these assessments can be made available to the international community and help inform practices by other RFMOs. Scientific guidance for the development of networks of representative MPAs was adopted by Parties to the CBD in 2008 (UNEP/CBD/COP/DEC/IX/20/Annex, 2008). Additionally, States have designated 'ecologically or biologically significant marine areas' (EBSAs). These range from small areas to large oceanographic features. They can then benefit from the implementation of enhanced CMMs, impact assessments and protected status (Erinosho *et al.*, 2022). RFMOs should make use of the CBD MPA Guidelines, and the available criteria for declaring EBSAs, to help inform the declaration of ecologically representative and climate adaptive and responsive MPAs or ABMTs.

The FAO has developed several scientific and technical documents on climate change impacts on fisheries. As noted above, non-legally binding instruments by FAO help inform the high seas governance structure and improve the practice of RFMOs and flag States fishing on the high seas. Considering this, the development of FAO Guidelines, Plan of Action, or advice on climate change and fisheries on the high seas would surely be welcomed by the international community and aid adaptive management of fisheries in response to global environmental change (Molenaar, 2021). Ultimately, it is down to States – particularly flag States – to exercise proper enforcement and control of their vessels on the high seas. RFMOs and other institutional arrangements on the high seas do not have the enforcement capacity to 'police' vast swathes of the ocean. As such, exercising flag State duties and responsibilities by States (and, by doing so, exercising the obligation to cooperate) is crucial to ensure adaptation to changes in pelagic ecosystems is not undermined.

Conclusion

Global environmental change, particularly from climate change and its associated effects, like warming and ocean acidification, creates a number of threats to pelagic ecosystems. Such impacts, in combination with overexploitation and pollution, undermine the ability of those ecosystems to function normally and to regulate the global climate. The high seas governance framework has its basis in UNCLOS and is supported by a number of other instruments related mainly to pollution and fisheries. The regime is fragmented and there are attempts underway at the international level to address this through the negotiation of a new legally binding instrument on biodiversity beyond national jurisdiction. The protection and restoration of the oceans is fundamental for ensuring ecological resilience and regulation of the global climate. The existing institutions, especially RFMOs, can help achieve this on the high seas. RFMOs can help to improve resilience on the high seas through the declaration of MPAs and other ABMT, especially in response to the impacts of ocean warming, such as shifting fish stocks. However, governance action on fisheries and climate change needs further development, and there is scope to increase effectiveness overall for ocean governance on the high seas. This requires an increase in global cooperation. Indeed, some areas of governance require working with the tools that are available, such as through cooperation between RFMOs, and RFMOs adopting ABMT where possible to increase the resilience of pelagic ecosystems to global environmental changes.

Acknowledgements

The author is very grateful to Daniela Diz and Kate McKenzie for comments and feedback on earlier versions of this chapter.

References

Angel, M. V. (1993) 'Biodiversity of the pelagic ocean' *Conservation Biology*, vol 7, 760–772.

Barlow, J., *et al.* (2018) 'The future of hyperdiverse tropical ecosystems' *Nature*, vol 559, 517–526.

Barnes, R., and Massarella, C. (2016) 'High seas fisheries' in Morgera, E., and Kulovesi, K. (Ed.), *Research Handbook on International Law and Natural Resources*. Edward Elgar, Cheltenham, 369–389.

Barrett, J., and Barnes, R. (2016) *Law of the Sea: UNCLOS as a Living Instrument*, British Institute of International and Comparative Law, London.

Beaugrand, G. (2014) 'Pelagic ecosystems and climate change' in Freedman, B. (Ed.), *Global Environmental Change*. Springer, New York, 141–150.

Boyle, A., and Chinkin, C. (2007) *The Making of International Law*. Oxford University Press, Oxford.
Caddell, R. (2018) 'Precautionary management and the development of future fishing opportunities: The international regulation of new and exploratory fisheries' *The International Journal of Marine and Coastal Law*, vol 33, 199–260.
Caddell, R., and Molenaar, E. (2019) *Strengthening International Fisheries Law in an Era of Changing Oceans*. Hart, London.
Cheung, W. W. L., *et al.* (2013) 'Shrinking of fishes exacerbates impacts of global ocean changes on marine ecosystems' *Nature Climate Change*, vol 3, 207–213.
Cheung, W. W. L. *et al.* (2017) 'Transform high seas management to build climate resilience in marine seafood supply' *Fish and Fisheries*, vol 18, 254–263.
Crawford, J., and Brownlie, I. (2019) *Brownlie's Principles of Public International Law*. 9th edition. Oxford University Press, Oxford.
Dickey-Collas, M. *et al.* (2017) 'Pelagic habitat: exploring the concept of good environmental status' *ICES Journal of Marine Science*, vol 74, 2333–2341.
Diz, D., Ntona, M. (2018) *Background report for the Second Meeting of Sustainable Ocean Initiative Global Dialogue with Regional Seas Organizations and Regional Fisheries Bodies on Accelerating Progress towards the Aichi Biodiversity Targets and Sustainable Development Goals*. On file with author.
Dulvy, N. K., *et al.* (2008) 'Climate change and a deepening of the North Sea fish assemblage: a biotic indicator or regional warming' *Journal of Applied Ecology*, vol 45, 1029–1039.
Erinosho, B., *et al.* (2022) 'Transformative governance for ocean biodiversity' in Visseren-Hamakers, I. J., and Kok, M., (Eds.) *Transforming Biodiversity Governance*. Cambridge University Press, Cambridge.
FAO (2020). *State of World Fisheries and Aquaculture 2020. Sustainability in Action*. FAO, Rome, Italy.
Free, C. M., *et al.* (2019) 'Impacts of historical warming on marine fisheries production' *Science*, vol 363, 979–983.
Game, E. T., *et al.* (2009) 'Pelagic protected areas: the missing dimension in ocean conservation' *Trends in Ecology and Evolution*, vol 24, 360–369.
Gjerde, K. (2006) 'High seas fisheries management under the Convention on the law of the sea' in Freestone, D., Barnes, R., and Ong, D. (Eds.), *The Law of the Sea: Progress and Prospects*. Oxford University Press, Oxford, 281–307.
Grotius, H. (1916) *The Freedom of the Seas of the Right Which Belongs to the Dutch to Take Part in the East Indian Trade*. Oxford University Press, Oxford.
Hammond, A., and Jones, P. S., (2021) 'Protecting the 'blue heart of the planet': Strengthening the governance framework for marine protected areas beyond national jurisdiction' *Marine Policy*, vol 127, 104260.
Harrison, J., (2011) *Making of the Law of the Sea*. Cambridge University Press, Cambridge.
IPBES, (2019) *Summary for Policymakers of the Global Assessment Report on Biodiversity and Ecosystem Services of the Intergovernmental Science-Policy Platform on Biodiversity and Ecosystem Services*. Díaz, S., *et al.*, (Eds.). IPBES Secretariat, Bonn, Germany.
IPCC, (2007) *Climate Change 2007: Impacts, Adaptation and Vulnerability: Contribution of Working Group II to the Fourth Assessment Report of the Intergovernmental Panel on Climate Change*. Parry, M. L., *et al.*, (Eds.). Cambridge University Press, Cambridge.
IPCC, (2018) *Global Warming of 1.5C - Summary for Policymakers*. Masson-Delmontte, V., *et al.*, (Eds.). World Meteorological Organization, Geneva, Switzerland.
IPCC, (2019a) *Special Report on the Ocean and Cryosphere in a Changing Climate, Summary for Policy Makers*. Pörtner, H-O *et al.*, (Eds.). In press.
IPCC, (2019b) *IPCC Special Report on the Ocean and Cryosphere in a Changing Climate*. Pörtner, H-O *et al.*, (Eds.). In press.
Isensee, K., and Valdes, L., (2015) *GSDR 2015 Brief: Marine Litter: Microplastics*. IOC/UNESCO. Available online at <https://sustainabledevelopment.un.org/content/documents/5854Marine%20 Litter%20-%20Microplastics.pdf>
Jambeck, J. R., *et al.*, (2015) 'Plastic waste inputs from land into the ocean' *Science* vol 347, 768–771.
Kimball, L., (2005) *The International Legal Regime of the High Seas and the Seabed Beyond the Limits of National Jurisdiction and Options for Cooperation for the Establishment of Marine Protected Areas (MPAs) in Marine Areas Beyond the Limits of National Jurisdiction*. CBD, Montreal.
Molenaar, E. J., (2019) 'Participation in the central arctic ocean fisheries. Agreement' in Shibtata, A., *et al.*, (Eds.), *Emerging Legal Orders in the Arctic: The Role of Non-arctic Actors*. Routledge, London, 132–161.

Molenaar, E. J., (2021) 'Integrating climate change in international fisheries law' in Johansen, E., Busch S. V., and Jakobsen, I. U., (Eds.) *The Law of the Sea and Climate Change - Solutions and Constraints*. Cambridge University Press, Cambridge, 263–288.

Norse, E., (2006) 'Pelagic protected areas: the greatest park challenge of the 21st century' *Parks*, vol 15, 33–40.

Pauly, D. (2002) 'Towards sustainability in world fisheries' *Nature*, vol 418, 689–695.

Pauly, D., and Zeller, D., (2016) 'Catch reconstructions reveal that global marine fisheries catches are higher than reported and declining' *Nature Communications*, vol 7, 10244.

Pinsky, M. L., *et al.* (2018) 'Preparing ocean governance for species on the move' *Science*, vol 360, 1189–1191.

Pörtner, H. O., and Peck, M. A. (2010) 'Climate change effects on fishes and fisheries: towards a cause-and-effect understanding' *Journal of Fish Biology*, vol 77, 1745–1779.

Rayfuse, R. 2019 'Addressing climate change impacts in regional fisheries management organizations' in Caddell, R. and Molenaar, E. (Eds.) *Strengthening International Fisheries Law in an Era of Changing Oceans* Hart, Oxford, United Kingdom.

Roberts, C. (2012) 'Marine ecology: Reserves do have a key role in fisheries', *Current Biology*, vol 22, no 11, R444–R446.

Sabine, C. L., *et al.* (2004) 'The ocean sink for anthropocentric CO_2' *Science*, 305, 367–371.

Worm, B., *et al.* (2016) 'Impacts of biodiversity loss on ocean ecosystem services', *Science*, vol 314, 787–790.

PART 5

Emerging issues in environmentally sustainable marine governance

22

PLASTIC POLLUTION

The challenges of uncertainty and multiplicity in global marine governance

Judith van Leeuwen, Tony R. Walker, and Joanna Vince

Marine plastic pollution has been reported in every marine environment on the planet, from densely populated coastal cities to remote Arctic (Mallory et al., 2021) and Antarctic shorelines (Walker et al., 1997). Due to unsustainable production, use, and mismanagement, coupled with low degradation rates, plastic pollution has become a severe transboundary threat to marine ecosystems and human health (Geyer et al., 2017; Worm et al., 2017). Plastic pollution is now so ubiquitous in the marine environment that plastics have been reported floating in the open ocean, within the water column, on the seafloor, in marine biota and even in seafood for human consumption (Karbalaei et al., 2019; Goodman et al., 2020; Sequeira et al., 2020). Marine plastic pollution has become a tragedy of the global environmental commons for modern society and therefore requires a multiplicity of governance solutions from the global community (Vince and Hardesty, 2018).

The rapid proliferation of marine plastic pollution has given rise to a plethora of scientific studies related to the problem along with increased media and public attention (Walker et al., 2021). There is a growing body of evidence of plastic's impacts on local and global economies (Beaumont et al., 2019), and on its effects upon individual organisms, as shown in laboratory toxicity testing (Karbalaei et al., 2021) and as demonstrated by entanglements (Jepsen and de Bruyn, 2019). However, knowledge gaps remain over the long-term effects of plastic pollution on ecological and human health (Bucci et al., 2020; Prata et al., 2020b). Moreover, marine plastic pollution does not only come solely from marine/sea-based sources and activities (e.g., fishing, cruises, beach tourism, and shipping), but also leaks into rivers through plastic production, use and waste on land. It is estimated that 80% of marine litter comes from land-based sources, while the other 20% is from sea-based sources (Goodman et al., 2019; Richardson et al., 2021). Plastic waste generation was further exacerbated during the global COVID-19 pandemic due to unprecedented use and disposal of personal protective equipment and single-use plastics (SUPs) (Prata et al., 2020a; Silva et al., 2021).

Global awareness of marine plastic pollution has given rise to changing consumer behaviour (bottom-up) and government policy (top-down) approaches to reduce land- and sea-based sources (see, e.g., Dauvergne, 2018; Schnurr et al., 2018; Vince and Hardesty, 2018; Clayton et al., 2020). Two trends are being debated by policymakers as well as scholars. First, a plethora of initiatives at the global level have emerged, including from multiple intergovernmental organisations, and the call for an international agreement to reduce marine

DOI: 10.4324/9781315149745-27

litter is growing. Second, there is a global movement around the transition to a circular economy, by extending product stewardship, in which both governments and the private sector are engaged. However, critical voices are emerging that these plastic reduction strategies, so far, remain ineffective in curbing marine plastic pollution, as well as production and consumption rates (Dauvergne, 2018; Borrelle et al., 2020).

This chapter will review the multiplicity of global governance efforts and challenges associated with preventing and mitigating marine plastic pollution as well as the growing critique of the ineffectiveness of current governance approaches. The current uncertainty related to the sources, and short- and long-term ecological and health effects of marine plastic pollution, inhibit setting priorities (as elaborated in section 2). Concerns about plastic pollution have risen rapidly over the last ten years, giving rise to a plethora of state and non-state efforts at the global level to address and protect the marine environment from plastic (sections 3 and 4). The scale and visibility of plastic pollution and its integration with daily life justify this multiplicity of initiatives. At the same time, concerns and scholarly debates over whether these efforts are incoherent and ineffective are increasing too. In the conclusion of this chapter, we therefore stress that it is important to ensure that the multiplicity of efforts should contribute to transformative change not only in waste and pollution of plastic but also in its production and use.

Plastic pollution and uncertainty

Plastic pollution has no international boundaries and has been detected in the marine environment for decades (Barnes et al., 2009). It is now widely recognised that most consumer plastics are designed to be only for a single use, with limited recyclability (Schnurr et al., 2018). The last decades have seen an increased demand for single-use consumer products along with increased global production, unprecedented waste generation and widespread marine pollution of plastics (Borrelle et al., 2020). In 2019, global plastic production reached 368 million metric tonnes (Mt), and it is estimated to double within 20 years (PlasticsEurope, 2020).

Researchers have estimated that millions of tonnes of plastic waste and trillions of plastic particles have been released into the world's oceans, but these are likely underestimates (Borrelle et al., 2020). A broadly cited study by Jambeck et al. (2015) estimated that 4.8 to 12.7 million metric tonnes (Mt) entered the ocean in 2010. In an updated study, it was estimated that 19 to 23 Mt of plastic waste entered aquatic and marine ecosystems globally in 2016, and that this might reach up to 53 Mt annually by 2030 (Borrelle et al., 2020). The global COVID-19 pandemic resulted in a further increase of SUP and personal protective equipment use and waste (Prata et al., 2020a; Silva et al., 2021). Prata et al. (2020a) estimated that personal protective equipment mismanagement resulted in the use of 129 billion face masks and 65 billion gloves each month during the early months of the COVID-19 pandemic. Thus, plastic production, use, and widespread plastic mismanagement are unsustainable and threaten the ability of the global community to effectively govern and manage plastic pollution (Walker, 2021).

As of April 2021, there were 3,699 aquatic species documented to have been impacted by plastic pollution (Litterbase, 2021). Plastic pollution has led to mortality or injury to thousands of species, and there is also a growing body of evidence that recognises plastic as a contaminant that threatens individual species (Bucci et al., 2020; Karbalaei et al., 2021). However, many of these studies are conducted in the laboratory, and thus our knowledge about the effects of plastics or microplastics (<5 mm) and associated chemicals on populations, communities, and ecosystems at environmentally relevant concentrations remains understudied (Bucci et al., 2020). Plastics and microplastics in the marine environment also lead to inevitable human exposure (see Danopoulos et al., 2020; Prata et al., 2020b). Exposure may occur by ingestion in seafood (Karbalaei et al., 2019; Danopoulos et al., 2020; Sequeira et al., 2020), inhalation and skin contact with microplastics in products, foodstuffs, and the air (Prata et al., 2020b). Microplastics have been found in the human body, including the human placentas (Ragusa et al., 2021). Microplastics entering the human body through ingestion or inhalation can be taken up in various organs and impact human health by damaging cells or inducing inflammatory and immune reactions (Vethaak and Legler, 2021). Although there is evidence that plastic leaches endocrine disrupters (Flaws et al., 2020), large knowledge gaps remain in our understanding of the potential impacts that microplastics have on human health. Thus, knowledge on microplastic toxicity is still limited and largely influenced by exposure concentration, particle properties, adsorbed contaminants, tissues involved and individual susceptibility, requiring further studies (Prata et al., 2020b).

Although marine plastic pollution is receiving attention globally, including through the UN Sustainable Development Goals (SDGs) under Goal 14 (see Chapter 25), an internationally accepted index of marine plastic pollution does not yet exist (Walker, 2021). This also goes for the mapping of riverine plastic pollution. Over 1,000 rivers release between 0.8 million and 2.7 million tonnes of plastic annually, contributing 80% to global annual releases (Meijer et al., 2021), yet a common framework to study riverine plastic pollution is lacking (Vriend et al., 2020). International guidelines for monitoring marine and riverine plastic debris to help inform government plastic reduction policies have been released only recently, specifically in 2019 and 2020, respectively (Kershaw et al., 2019; UNEP, 2020). In the meantime, however, many jurisdictions have already developed national marine plastics databases (Kershaw et al., 2019). Understanding the extent and magnitude of plastic pollution, including characterising types and categories of problem plastics (e.g., SUPs) found in the marine and riverine environment can help inform government policies and governance strategies and in turn help reduce sources of plastic pollution (Ambrose et al., 2019; Vriend et al., 2020, 2021).

Multiplicity in (inter)governmental efforts to reduce marine plastic pollution

Despite many unknowns about the impacts of plastic pollution, plastic pollution has been recognised by many international organisations as a global problem needing urgent attention. Legally binding agreements have addressed marine pollution since the 1960s. For example, the Convention on the Prevention of Marine Pollution by Dumping of Wastes and other Matter (The London Convention of 1972) and the International Convention for the Prevention of Pollution from Ships 1973 (MARPOL) are designed to limit pollution dumped from boats into marine ecosystems, though there is no standing equivalent that limits land-based sources. In 2019, during the fourteenth meeting of the Conference of the Parties to the Basel Convention (1989), amendments to Annexes II, VIII and IX were adopted that added mixed, unrecyclable, and contaminated plastic waste exports to the list of substances that require import country approval. These amendments enhance the control of transboundary movements of plastic waste and clarify the scope of the Convention regarding this type of waste. While the Basel Convention is the only global legally binding instrument that addresses plastic waste, its technical guidelines are not legally binding, and compliance is difficult to enforce (Raubenheimer and McIlgorm, 2018). Also in 2019, a Plastic Waste Partnership was established under the Basel Convention, and its first meeting was held in Seychelles in March 2020. Its goal is to promote the environmentally sound management of plastic waste on all governance levels and 'to prevent and minimise its generation so as to, among other things, reduce significantly and in the long-term eliminate the discharge of plastic waste and microplastics into the environment, in particular the marine environment' (UNEP, 2021).

Many other instruments that address marine plastic pollution in the global domain require political commitment but are soft law, meaning that they are not legally binding and do not require compliance (Ferraro and Failler, 2020). While such instruments strongly encourage their members to act, it is up to nation-states to implement policies and laws that reflect arrangements on a global level. Such soft law instruments include the Honolulu Strategy, which resulted from a collaboration between the United Nations Environment Programme (UNEP) and the US National Oceanic and Atmospheric Administration in 2011; the Manilla Declaration on Furthering the Implementation of the Global Programme of Action for the Protection of the Marine Environment from Land-based Activities (2012); and the G7 Action Plan to Combat Marine Litter (2015). The UNEP Global Partnership on Marine Litter was launched at the United Nations Conference on Sustainable Development (also known as 'Rio+20') in June 2012 in response to the Manilla Declaration. It is led by a Steering Committee and UNEP, and the partners include governments, NGOs and other international organisations. The UN addresses plastic pollution through its SDGs, in particular goal 14.1, which aims for the prevention and reduction of all marine debris by 2025. However, what is not included is a percentage of how much marine debris will be reduced in that time. Despite many countries having national goals to reduce plastic waste, no international standard exists yet. The UN's Environment Assembly adopted non-binding resolutions on plastic pollution in 2017 (Res 71/312) and in 2019 (Res 4/6) in which the importance of prevention and reduction of marine litter was stressed. As a response to the 2019 resolution, members pledged to 'significantly reduce' the use of plastics by 2030. As a result of resolution (Res 71/312), an Ad Hoc Open-ended Expert Group on Marine Litter and Microplastics was established and the mandate for the expert group was extended in resolution (Res 4/6) in 2019. In the fourth meeting of the expert group in 2020, two-thirds of members declared that they were interested in considering a new global agreement on plastic pollution.

The call for a global treaty on plastic waste is not new, and numerous global NGOs and environmental groups are campaigning for its establishment. Concerns around weak international institutions, uneven regulations, uncoordinated policies, and business-oriented solutions are fuelling this call (Dauvergne, 2018; Ferraro and Failler, 2020). Large corporations such as Coca-Cola, Nestle, and Colgate-Palmolive, among others, have contributed to the debate about a global treaty and how it could help transition the current linear economy to a circular economy to help reduce plastic leakage into the marine environment. In a manifesto to the UN, these organisations 'call on UN Member States to develop a global treaty on plastic pollution' (Plastic Pollution Treaty, 2021). The number of academic sources calling for a legally binding global plastics treaty has demonstrated a large gap in international law (see, e.g., Borrelle et al., 2017; Simon and Schulte, 2017; Worm et al., 2017; Raubenheimer and McIlgorm, 2018; Vince and Hardesty, 2018; Vince and Stoett, 2018; Tessnow-von Wysocki and Le Billon, 2019; Galaiduk et al., 2020; Simon et al., 2021; Raubenheimer and Urho, 2020). The scholarly debate is based on what needs to be addressed by such a treaty to effectively address marine plastic pollution and whether it has the regulatory support to be implemented at the national level.

The literature states that there are a number of factors that need to be included when drafting a global plastics treaty. Tessnow-vin Wysocki and Le Billon (2019) argue that these include common but differentiated responsibilities; whole-of-lifecycle focus and scope; issue linkage between plastics and other issues such as climate change; a financial mechanism to ensure developing countries can cover their costs; monitoring, reporting and review procedures; enforcement; and flexibility. Simon and Schulte (2017), on the other hand, argue the treaty should be built on five pillars: (1) a clear and binding goal to eliminate plastic contamination of water, food, air, land, and oceans; (2) strategic options for national implementation; (3) supporting mechanisms, with a strong multilateral financing component; (4) stringent monitoring, review, compliance, and enforcement processes; and (5) strong involvement of nongovernmental stakeholders. A global treaty will need a holistic, integrated approach that recognises regional, national, and local efforts while including industry, civil society and community groups (Vince and Hardesty, 2018). It will also require the coordination of international organisations with a specified lead agency (Ferraro and Failler, 2020). Given the difficulties associated with keeping plastics on the political agenda during COVID-19 (Prata et al., 2020a), a global agreement will take some time to be negotiated and, in the meanwhile, action is imminently needed.

Circular economy and product stewardship

As part of the quest to end plastic waste resulting in marine pollution, two interrelated policy concepts have gained global traction: product stewardship and circular economy. While circular economy presents a future ideal in which products and materials regain their value through cycling between production, (re)use and recycling, product stewardship refers to managing the environmental impacts of a product throughout its life cycle. Through both concepts, governments and private actors worldwide have started to engage with plastic products and packaging, and with innovative ways to reduce plastic use, waste, and pollution (Mah, 2021).

Both governments and private actors have published long-term ambitions for transforming to a circular economy. For example, the EU published its first Circular Economy Action Plan in 2015 and an updated one in March 2020. In 2018, it published its Strategy for Plastics in the Circular Economy. Legislation is in development that will set mandatory requirements on recycled plastic content and plastic waste reduction measures for key products such as packaging, construction materials, and vehicles. Similarly, in Asia, members of the Association of Southeast Asian Nations (ASEAN) have expressed commitment to work towards a circular economy for plastics. In 2019 a first analysis was done to identify the challenges related to, as well as possible initiatives ASEAN could take to support, transforming to a circular plastics economy (Akenji et al., 2019). One of the ways in which governments seek to initiate circular change is through the product stewardship approach of Extended Producer Responsibility (EPR). As part of the first Circular Action Plan, the EU continued to promote and incentivise Members States to further strengthen the implementation of EPR for plastic packaging. With a 32.5% recycling rate for plastic post-consumer waste (and 42% for plastic packaging) in 2018 (PlasticsEurope, 2020), the EU has indeed the highest recycling rate in the world. Recently, countries in Asia, such as China, Vietnam, and Indonesia are adopting and implementing EPR schemes to improve plastic waste separation, collection, and recycling.

A key focus of national and local regulatory activities is on reducing the use of SUPs. As of 2018, over 127 countries had enacted regulatory measures relating to single-use plastics and the majority of these related to the distribution and/or banning of plastic bags (UNEP, 2018). Other measures implemented include plastic bag taxes, plastic bag fees for consumers, extended producer responsibility policies, regulations around the use of intentionally added microplastics, including microbeads, and product bans (Dauvergne, 2018; Schnurr et al, 2018; UNEP, 2018; Raubenheimer and Urho, 2020). Within the EU, a single-use plastic ban has been in place since July 2021 for 10 SUP items commonly found on European beaches. An EU initiative to substitute single-use packaging and cutlery is pending. However, since COVID-19 pandemic and within a short amount of time, many plastics laws and policies in numerous jurisdictions around the world had either been postponed or reversed (da Costa et al., 2020; Prata et al., 2020a).

In a response to global marine plastic pollution, the Ellen MacArthur Foundation (EMF) has played an important role in mobilising government and industry in adopting visions and innovations for a circular economy. EMF defines a circular economy as an economic system that is restorative and regenerative by design (Ellen MacArthur Foundation, 2013: 14). The EMF initiated the New Plastics Economy Initiative in 2018, uniting 1,000 organisations worldwide. Part of this initiative are the New Plastics Economy Global Commitment and the Plastic Pact. While the former unites organisations at the global level to set individual targets related to the reduction, reuse, and recycling of SUP in 2025, the latter unites organisations at the regional and national level to implement circular solutions. Many multi-national companies producing plastic or packaged products, such as Unilever and Coca-Cola, have indeed set circular economy targets, such as reduction of packaging used, increasing the use of recycled matter in plastic packaging, as well as improving the recyclability of their packaging. Biodegradable packaging is also put forward as a solution. According to EMF, the signatories to the New Plastics Economy Global Commitment have, between 2018 and 2019, increased the recycled content in packaging by 22% leading to a 0.1% reduction in total volume of virgin plastic used (Ellen MacArthur Foundation, 2020). Moreover, more and more signatories are testing and piloting reuse models and collectively plan more than 10 billion United States Dollars of investments to achieve 2025 targets set (Ellen MacArthur Foundation, 2020).

With the emergence of the circular economy concept, scholarly attention to define and operationalise circular economy and circular strategies across countries and value chains has proliferated (Geissdoerfer et al., 2017; Hobson, 2021; Kirchherr et al., 2017). For example, many scholars have operationalised circular economy further by referring to creating 'closed material loops' for keeping products, components and materials at their highest utility and value (Geissendorfer et al., 2017). What is common across the proposed definitions is that the concepts propose a future of a regenerative and restorative system of production and consumption that will solve problems of resource scarcity, biochemical disruption, and climate change, while offering economic opportunities at the same time (Calisto Friant et al., 2020). Although parallels exist, the concept circular economy deviates from the concept sustainable development through its emphasis on the economic system rather than economic, social, and environmental sustainability (Geissendorfer et al., 2017; Korhonen et al., 2018). However, it is questionable whether circular economy will indeed be able to achieve environmental sustainability in general (Korhonen et al., 2018; Boldoczki et al., 2021; Hobson, 2021).

When applied to plastics, this idea of a future circular economy implies that by reducing plastic use and closing plastic material loops, plastic pollution to the environment will be reduced as well. Two sets of critique are notable. First, circular economy offers a whole suite of possible measures, the so-called R-strategies of reduce, reuse, and recycling (Reike et al., 2018). Assessment of circular strategies are, however, more focused on identifying technical and economic feasibility and challenges than on how much pollution, material use or other environmental issues they are able to mitigate (Zink and Geyer, 2017; Boldoczki et al., 2021). Similar critique exists towards EPR, as EPR comes in multiple forms and does not necessarily divert plastic away from landfill or the environment (Amin et al., 2021). It is therefore uncertain to what extent and under what conditions circular economy measures as well as product stewardship contributes to fostering environmental sustainability and reduce the flow of plastic entering the marine environment in particular. Signs that this flow of plastic is indeed reducing do not exist yet.

In fact, despite efforts to reduce plastic waste and improve its management, global plastic pollution is likely to increase in the future (Borrelle et al., 2020). Even if all current government and industry commitments were implemented by 2040, the amount of plastic that leaks from land into the oceans would still increase from 11 Mt in 2016 to 27 Mt per year in 2040 (an increase of 147%) (The Pew Charitable Trusts and Systemiq, 2020). There is a need for a system change to curb marine plastic pollution from land and governance efforts, including a future global treaty on plastic, should recognise that. While the circular economy concept is brought forward as having the potential to transform plastic production, consumption and leakages, the implementation of circular economy is currently characterised by a business as usual and incremental approach (Hobson, 2021). The circular economy concept is operationalised in various ways, sometimes in a more segmented and technocratic way, while others see circular economy as being more holistic and transformational in nature (Calisto Friant et al., 2020). The petrochemical industry in the plastic value chain has jumped on the concept of circular economy, but the industry uses it to future-proof capitalism by legitimising the continuation of unsustainable plastic production practices (Mah, 2021) and by focusing in large part on recycling as a key circular economy strategy. In the end, the central issue is whether secondary production will prevent primary production and the use of virgin materials (Zink and Geyer, 2017) and will stem the flow of plastic pollution. The outlook is, however, not promising for plastics as the global governance of governing marine plastic pollution is currently failing (Dauvergne, 2018).

Conclusion

Continued uncertainties about the sources and impacts of plastic pollution for the world's seas and oceans have not prevented a multiplicity of global efforts to curb marine plastic pollution. Transnational governmental and private actors' efforts focus not only on marine pollution from sea-based activities but also on that from land-based leakage, making the scope of these efforts unprecedented. Given the scale and urgency of plastic production, as well as the long-term transformation that is required to reduce it, this multiplicity is sorely needed. It allows for experimentation and matches the scope of sectors, scales, and actors involved. However, critique over the effectiveness to not only reduce marine pollution but also to reduce production and use of plastic is growing. There is a risk that the multiplicity of governance efforts will lead to fragmentation and policy incoherence. To reduce these risks, there is a need to develop and continuously reflect upon several issues: a shared vision and a set of high-level objectives to provide long-term direction for change that is focused not only on reducing marine pollution but also on reducing production and use of plastic; soft and hard law coordination mechanisms between governing institutions to avoid contradictions and build synergy in efforts and focus within and between regions; building a strong scientific basis to identify sources and effects of plastic pollution; and developing platforms for the sharing and monitoring of information, best practices, and progress (see Chapter 8). All in all, the fragmented set of governance efforts should, through experimentation, critical monitoring, and learning, lead to transformative change to reduce production, use, and waste of plastics, and thereby limit plastic in the marine environment.

References

Akenji, L., Bengtsson, M., Kato, M., Hengesbaugh, M., Hotta, Y., Aoki-Suzuki, C., Gamaralalage, P.J.D. and Liu, C. (2019) *Circular Economy and Plastics: A Gap-Analysis in ASEAN Member States*. European Commission Directorate General for Environment and Directorate General for International Cooperation and Development, Brussels and Association of Southeast Asian Nations (ASEAN), Jakarta.

Ambrose, K.K. Box, C. Boxall, J. Brooks, A. Eriksen, M. Fabres, J. Fylakis, G. and Walker, T.R. (2019) 'Spatial trends and drivers of marine debris accumulation on shorelines in South Eleuthera, The Bahamas using citizen science' *Marine Pollution Bulletin*, vol 142, pp145–154.

Amin, S. Strik, D. and van Leeuwen, J. (2021) 'A reflexive policy design methodology to aid circularity: Coupling extended producer responsibility scenarios with material flow mapping of PET plastic in Jakarta, Indonesia', unpublished research paper.

Barnes, D.K. Galgani, F. Thompson, R.C. and Barlaz, M. (2009) 'Accumulation and fragmentation of plastic debris in global environments' *Philosophical Transactions of the Royal Society B: Biological Sciences*, vol 364, no 1526, pp1985–1998.

Beaumont, N.J. Aanesen, M. Austen, M.C. Börger, T. Clark, J.R. Cole, M. Hooper, T. Lindeque, P.K. Pascoe, C. and Wyles, K.J. (2019) 'Global ecological, social and economic impacts of marine plastic' *Marine Pollution Bulletin*, vol 142, pp189–195.

Boldoczki, S., A. Thorenz, and A. Tuma. (2021) 'Does increased circularity lead to environmental sustainability?' The case of washing machine reuse in Germany' *Journal of Industrial Ecology*, vol 25, pp864–876.

Borrelle, S.B., Rochman, C.M., Liboiron, M., Bond, A.L., Lusher, A., Bradshaw, H., & Provencher, J.F. (2017) 'Opinion: Why we need an international agreement on marine plastic pollution' *Proceedings of the National Academy of Sciences*, vol 114, no 38, pp9994–9997.

Borrelle, S.B. Ringma, J. Law, K.L. Monnahan, C.C. Lebreton, L. McGivern, A. Murphy, E. Jambeck, J. Leonard, G.H., Hilleary, M.A. and Eriksen, M. (2020) 'Predicted growth in plastic waste exceeds efforts to mitigate plastic pollution' *Science*, vol 369, no 6510, pp1515–1518.

Bucci, K. Tulio, M. and Rochman, C.M. (2020) 'What is known and unknown about the effects of plastic pollution: A meta-analysis and systematic review' *Ecological Applications*, vol 30, no 2, p02044.

Calisto Friant, M., Vermeulen, W.J. and Salomone, R. (2020) 'A typology of circular economy discourses: Navigating the diverse visions of a contested paradigm' *Resources, Conservation and Recycling*, vol 161, pp1–19.

Clayton, C.A. Walker, T.R. Bezerra, J.C. and Adam, I. (2020) 'Policy responses to reduce single-use plastic marine pollution in the Caribbean' *Marine Pollution Bulletin*, vol 162, p111833.

da Costa, J.P. (2021) 'The 2019 global pandemic and plastic pollution prevention measures: Playing catch-up' *Science of The Total Environment*, vol 774, p145806.

Danopoulos, E., Jenner, L.C., Twiddy, M. and Rotchell, J.M., (2020) 'Microplastic contamination of seafood intended for human consumption: A systematic review and meta-analysis' *Environmental Health Perspectives*, vol 128, no 12, p126002.

Dauvergne, P. (2018) 'Why is the global governance of plastic failing the oceans?' *Global Environmental Change*, vol 51, pp22–31.

Ellen MacArthur Foundation. (2013) *Towards the Circular Economy. Economic and Business Rationale for an Accelerated Transition*. Ellen MacArthur Foundation, Cowes, UK.

Ellen MacArthur Foundation. (2020) *The Global Commitment 2020 Progress Report*. Ellen MacArthur Foundation, Cowes, UK.

Ferraro, G. and Failler, P. (2020) 'Governing plastic pollution in the oceans: Institutional challenges and areas for action' *Environmental Science & Policy*, vol 112, pp453–460.

Flaws, J., P. Damdimopoulou, H.B. Patisaul, A. Gore, L. Raetzman and L.N. Vandenberg. (2020) *Plastics, EDCs & Health: A Guide for Public Interest Organizations and Policy-Makers on Endocrine Disrupting Chemicals and Plastic*. Endocrine Society and International Pollutants Elimination Network.

Galaiduk, R., Lebreton, L., Techera, E. and Reisser, J. (2020) 'Transnational Plastics: An Australian Case for Global Action' *Frontiers in Environmental Science*, vol 8, p115.

Geyer, R. Jambeck, J.R. and Law, K.L. (2017) 'Production, use, and fate of all plastics ever made' *Science Advances*, vol 3, no 7, p1700782.

Goodman, A.J., Brillant, S., Walker, T.R., Bailey, M. and Callaghan, C. (2019) 'A ghostly issue: Managing abandoned, lost and discarded lobster fishing gear in the Bay of Fundy in Eastern Canada' *Ocean & Coastal Management*, vol 181, p104925.

Goodman, A.J. Walker, T.R. Brown, C.J. Wilson, B.R. Gazzola, V. and Sameoto, J.A. (2020) 'Benthic marine debris in the Bay of Fundy, eastern Canada: Spatial distribution and categorization using seafloor video footage' *Marine Pollution Bulletin*, vol 150, p110722.

Hobson, K. (2021) 'The limits of the loops: Critical environmental politics and the circular economy' *Environmental Politics*, vol 30, no 1–2, pp161–179.

Jambeck, J.R., Geyer, R., Wilcox, C., Siegler, T.R., Perryman, M., Andrady, A., Narayan, R. and Law, K.L. (2015) 'Plastic waste inputs from land into the ocean' *Science*, vol 347, no 6223, pp768–771.

Jepsen, E.M. and de Bruyn, P.N. (2019) 'Pinniped entanglement in oceanic plastic pollution: A global review' *Marine Pollution Bulletin*, vol 145, pp295–305.

Karbalaei, S., Golieskardi, A., Hamzah, H.B., Abdulwahid, S., Hanachi, P., Walker, T.R. and Karami, A. (2019) 'Abundance and characteristics of microplastics in commercial marine fish from Malaysia' *Marine Pollution Bulletin*, vol 148, pp5–15.

Karbalaei, S., Hanachi, P., Rafiee, G., Seifori, P. and Walker, T.R. (2021) 'Toxicity of polystyrene microplastics on juvenile Oncorhynchus mykiss (rainbow trout) after individual and combined exposure with chlorpyrifos' *Journal of Hazardous Materials*, vol 403, p123980.

Kershaw, P., Turra, A. and Galgani, F. (2019) *Guidelines for the Monitoring and Assessment of Plastic Litter in the Ocean*. GESAMP reports and studies no. 99. GESAMP, London.

Kirchherr, J., Reike, D. and Hekkert, M. (2017) 'Conceptualizing the circular economy: An analysis of 114 definitions' *Resources, Conservation and Recycling*, vol 127, pp221–232.

Korhonen, J., Nuur, C., Feldmann, A. and Birkie, S.E. (2018) 'Circular economy as an essentially contested concept' *Journal of Cleaner Production*, vol 175, pp544–552.

Litterbase. (2021) 'Species interaction graph' https://litterbase.awi.de/interaction_graph

Mah, A. (2021) Future-proofing capitalism: The paradox of the circular economy for plastics' *Global Environmental Politics*, vol 21, no 2, pp121–142.

Mallory, M.L. Baak, J. Gjerdrum, C. Mallory, O.E. Manley, B. Swan, C. and Provencher, J.F. (2021) 'Anthropogenic litter in marine waters and coastlines of Arctic Canada and West Greenland' *Science of the Total Environment*, vol 783, p146971.

Meijer, L.J. van Emmerik, T. van der Ent, R. Schmidt, C. and Lebreton, L. (2021) 'More than 1000 rivers account for 80% of global riverine plastic emissions into the ocean' *Science Advances*, vol 7, no 18, p5803.

Plastic Pollution Treaty. (2021) 'Manifesto: The business call for a UN treaty on plastic pollution' https://www.plasticpollutiontreaty.org/

PlasticsEurope. (2020) *Plastics – The Facts 2020 an Analysis of European Plastics Production, Demand and Waste Data*. PlasticsEurope, Brussels.

Prata, J.C. da Costa, J.P. Lopes, I. Duarte, A.C. and Rocha-Santos, T. (2020b) 'Environmental exposure to microplastics: An overview on possible human health effects' *Science of the Total Environment*, vol 702, p134455.

Prata, J.C. Silva, A.L. Walker, T.R. Duarte, A.C. and Rocha-Santos, T. (2020a) 'COVID-19 pandemic repercussions on the use and management of plastics' *Environmental Science & Technology*, vol 54, no 13, pp7760–7765.

Ragusa, A., Svelato, A., Santacroce, C., Catalano, P., Notarstefano, V., Carnevali, O., Papa, F., Rongioletti, M.C.A., Baiocco, F., Draghi, S. and D'Amore, E. (2021) 'Plasticenta: First evidence of microplastics in human placenta' *Environment International*, vol 146, p106274.

Raubenheimer, K. and McIlgorm, A. (2018) 'Can the Basel and Stockholm Conventions provide a global framework to reduce the impact of marine plastic litter?' *Marine Policy*, vol 96, pp285–290.

Raubenheimer, K. and Urho, N. (2020) 'Rethinking global governance of plastics–The role of industry' *Marine Policy*, vol 113, p103802.

Reike, D., Vermeulen, W.J. and Witjes, S. (2018) 'The circular economy: New or Refurbished as CE 3.0? — Exploring Controversies in the Conceptualization of the Circular Economy through a Focus on History and Resource Value Retention Options' *Resources, Conservation and Recycling*, vol 135, pp246–264.

Richardson, K., Wilcox, C., Vince, J. and Hardesty, B.D. (2021) 'Challenges and misperceptions around global fishing gear loss estimates' *Marine Policy*, vol 129, p104522.

Schnurr, R.E., Alboiu, V., Chaudhary, M., Corbett, R.A., Quanz, M.E., Sankar, K., Srain, H.S., Thavarajah, V., Xanthos, D. and Walker, T.R. (2018) 'Reducing marine pollution from single-use plastics (SUPs): A review' *Marine Pollution Bulletin*, vol 137, pp157–171.

Sequeira, I.F., Prata, J.C., da Costa, J.P., Duarte, A.C. and Rocha-Santos, T. (2020) 'Worldwide contamination of fish with microplastics: A brief global overview' *Marine Pollution Bulletin*, vol 160, p111681.

Silva, A.L.P., Prata, J.C., Walker, T.R., Duarte, A.C., Ouyang, W., Barcelò, D. and Rocha-Santos, T. (2021) 'Increased plastic pollution due to COVID-19 pandemic: Challenges and recommendations' *Chemical Engineering Journal*, vol 405, p126683.

Simon, N. and Schulte, M.L. (2017) *Stopping Global Plastic Pollution: The Case for an International Convention*. Volume 43 of the Publication Series Ecology. Heinrich Böll Foundation, Berlin.

Simon, N., Raubenheimer, K., Urho, N., Unger, S., Azoulay, D., Farrelly, T., Sousa, J., van Asselt, H., Carlini, G., Sekomo, C. and Schulte, M.L. (2021) 'A binding global agreement to address the life cycle of plastics.' *Science*, vol 373, no 6550, pp43–47.

Tessnow-von Wysocki, I. and Le Billon, P. (2019) 'Plastics at sea: Treaty design for a global solution to marine plastic pollution' *Environmental Science & Policy*, vol 100, pp94–104.

The Pew Charitable Trusts and Systemiq. (2020) *Breaking the Plastic Wave: A Comprehensive Assessment of Pathways towards Stopping Ocean Plastic Pollution*. The Pew Charitable Trusts.

UNEP. (2018) 'Single-use plastics: A roadmap for sustainability' https://www.unenvironment.org/resources/report/single-use-plastics-roadmap-sustainability

UNEP. (2020) *Monitoring Plastics in Rivers and Lakes: Guidelines for the Harmonization of Methodologies*. UNEP, Nairobi.

UNEP. (2021) 'Basel convention overview' http://www.basel.int/Implementation/Plasticwastes/PlasticWastePartnership/tabid/8096/Default.aspx

Vethaak, A.D. and Legler, J. (2021) 'Microplastics and human health' *Science*, vol 371, no 6530, pp672–674.

Vince, J. and Hardesty, B.D. (2018) 'Governance solutions to the tragedy of the commons that marine plastics have become' *Frontiers in Marine Science*, vol 5, p214.

Vince, J. and Stoett, P. (2018) 'From problem to crisis to interdisciplinary solutions: plastic marine debris' *Marine Policy*, vol 96, pp200–203.

Vriend, P., Hidayat, H., van Leeuwen, J., Cordova, M. R., Purba, N. P., Löhr, A. J.,... & van Emmerik, T. (2021) 'Plastic pollution research in Indonesia: State of science and future research directions to reduce impacts' *Frontiers in Environmental Science*, vol 9, p692907.

Vriend, P., Roebroek, C.T. and van Emmerik, T. (2020) 'Same but different: A framework to design and compare riverbank plastic monitoring strategies' *Frontiers in Water*, vol 2, p563791.

Walker, T.R. (2021) '(Micro)plastics and the UN sustainable development goals' *Current Opinion in Green and Sustainable Chemistry*, vol 30, p100497.

Walker, T.R., McGuinty, E., Charlebois, S. and Music, J. (2021) 'Single-use plastic packaging in the Canadian food industry: consumer behavior and perceptions' *Humanities and Social Sciences Communications*, vol 8, no 1, pp1–11.

Walker, T.R., Reid, K., Arnould, J.P. and Croxall, J.P. (1997) 'Marine debris surveys at Bird Island, South Georgia 1990–1995' *Marine Pollution Bulletin*, vol 34, no 1, pp61–65.

Worm, B., Lotze, H.K., Jubinville, I., Wilcox, C. and Jambeck, J. (2017) 'Plastic as a persistent marine pollutant' *Annual Review of Environment and Resources*, vol 42, pp1–26.

Zink, T. and Geyer, R. (2017) 'Circular Economy Rebound' *Journal of Industrial Ecology*, vol 21, no 3, pp593–602.

23

MARITIME COMMERCE AND TRANSPORT

The imperfect match between climate change and the International Maritime Organization

Judith van Leeuwen and Jason Monios

Maritime commerce and transport are the lifeline of our global economy, with 90% of global goods traded by sea (UNCTAD, 2020). Shipping is also a source of a range of water-polluting discharges and air-polluting emissions. A growing concern in response to climate change is the reduction of greenhouse gas (GHG) emissions from shipping and the industry's eventual decarbonisation. Shipping contributes roughly 2.8% of global anthropogenic GHG emissions, and this is projected to increase in the decades to come (IMO, 2020). Because of the global nature of shipping, and the subsequent difficulty of assigning GHG emissions from particular ships to individual countries, emissions from shipping were excluded from the United Nations Framework Convention on Climate Change (UNFCCC) and the 2015 Paris Agreement on climate change (see Chapter 3). The main regulator of the shipping industry, the UN International Maritime Organization (IMO), was asked to develop climate change regulation instead. The IMO has been regulating the navigation, safety, and pollution of shipping since 1958. Regulating shipping at the global level allows for a harmonised approach and a level playing field, reducing regulatory differences for ships travelling the world's oceans (Van Leeuwen, 2015). The road to developing a global approach to climate change within the IMO has, however, been challenging and – as many would argue – slow. While regulatory measures have been adopted, these mostly focus on increasing the energy efficiency of ships, not the reduction of GHG emissions (Psaraftis, 2019a; Monios and Ng, 2021). In 2018, the IMO adopted an initial GHG reduction strategy, which put forward the 'ambition' of at least 50% GHG emissions reduction by 2050 (IMO, 2018).

Pressure on the IMO to act proactively to decarbonise shipping is increasing. However, this pressure comes largely from external actors and has led to growing fragmentation of regulatory initiatives to incentivise shipping to reduce GHG emissions. The IMO itself is an intergovernmental body in which political dynamics are informed by a very powerful industry lobby of commercial interests. Technological uncertainty over future possibilities to reduce GHG emissions and fully decarbonise shipping is preventing the industry and many maritime states from supporting ambitious climate change policies. The IMO and climate change are an imperfect match, and we question whether the IMO will be able to rise to the challenge to successfully guide the transformation to zero-carbon shipping. To further understand the challenging position in which the IMO finds itself, this chapter will first

DOI: 10.4324/9781315149745-28

introduce the global nature of the shipping industry and the environmental challenges associated with shipping, including climate change. It will then outline both the regulatory responses taken by the IMO and other organisations as well as the main technologies available for reducing GHG emissions from shipping. We then discuss why the IMO is imperfectly matched to the challenge of regulating climate change before suggesting future pathways for the decarbonisation of shipping.

Climate change challenges for the maritime transport sector

In 2019, 11.08 billion tonnes of goods were transported by sea, a growth of 85% since 2000 (see Figure 23.1). Growth was faster in the first decade of the century while global GDP boomed, but slower since the global financial crisis and its aftermath in 2008–2012. Key areas of growth have been in dry bulk commodities such as coal and iron ore, and containers carrying consumer goods, both linked to the rise of rapidly developing Asian countries as global exporters of manufactured goods. In 2019, 29% of total transported material was liquid bulk (primarily oil and gas products), 29% was major dry bulk (such as coal, iron ore and grain) and 17% was consumer goods transported by container ships. As a result of COVID-19, 2020 saw both world GDP and seaborne trade contract by 4%; however, UNCTAD predicted that trade would bounce back in 2021 with growth of 4.8% in maritime transport (UNCTAD, 2020).

Asia continues to dominate the geography of global trade, with 41% of total loaded and 62% of unloaded goods occurring in Asian ports, representing not just trade between Asia and the west, but intra-Asian trade as well (UNCTAD, 2020). To illustrate, in 2019, nine of the top ten container ports were in Asia, where 65% of all port container handlings took place, 50% in China alone. Europe was next with 15%, followed by North America with 8%. The major container trade lanes continue to be East-West. Three main routes, representing 39% of global containerised trade, exist: Asia to Europe via the Suez Canal, trans-Pacific Asia to West Coast US, and Transatlantic Europe to East Coast US. However, East-West trade is slowing and fell by 1.8% in 2019. Stronger growth is now occurring in intra-regional and South-South trade lanes (see Figure 23.2).

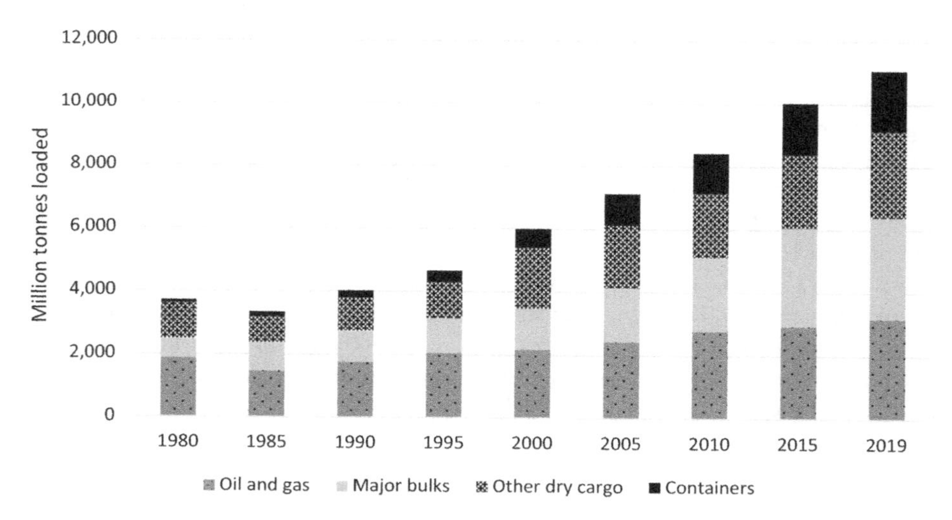

Figure 23.1 Global maritime trade 1980–2019, millions of tonnes loaded.
Source: Authors, based on data from UNCTAD (2020).

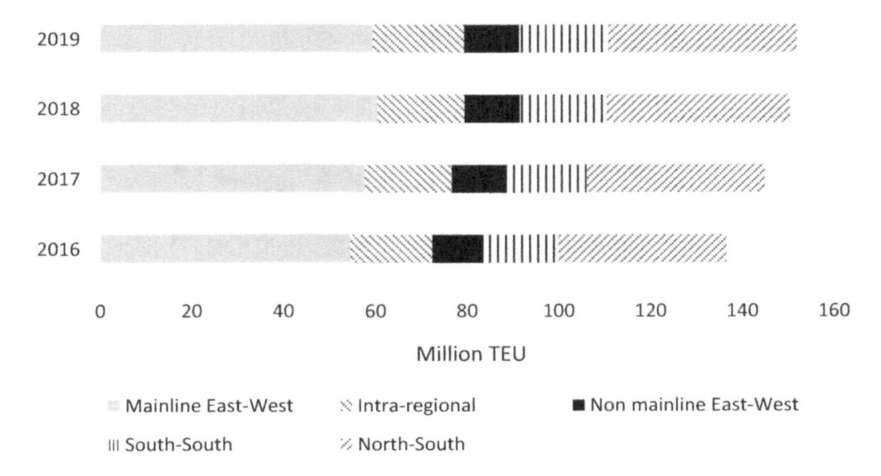

Figure 23.2 Annual containerised cargo per trade lane, 2016–2019.
Source: Authors, based on data from UNCTAD (2020).

Shipping produces a range of negative impacts on the environment, including ballast water discharge (often containing invasive organisms); pollution from toxic hull coating; accidents that lead to oil spills, cargo and containers falling overboard; water pollution from chemicals, sewage, and waste; noise pollution that harms marine life; and additional pollution from the ship scrapping process. Air emissions remain an ongoing concern too, given that almost the entirety of maritime transport is powered by fossil fuels. Heavy fuel oil (HFO), essentially refinery waste with high levels of sulphur, provided 79% of total fuel in the sector in 2018, the remainder being primarily Marine Diesel Oil, Marine Gas Oil and Liquefied Natural Gas (LNG, i.e., cooled methane) (IMO, 2020). Emissions from the combustion of these fuels can be broadly divided into those that cause climate change (e.g., Carbon Dioxide [CO_2] and methane) and air pollution that harms human health and the environment (e.g., Sulphur Oxides [SO_x], Nitrogen Oxides [NO_x], Particulate Matter [PM]), every one of which continues to increase (IMO, 2020). In 2018, total GHG emissions from shipping (in CO_2e) reached 1,076 m tonnes, 88% more than 1990 (562 m tonnes). The share of shipping emissions in global anthropogenic GHG emissions increased from 2.76% in 2012 to 2.89% in 2018, with all types of emissions experiencing increases (IMO, 2020).

Despite this growth, 2018 carbon emissions from shipping were still 10% below the peak of 2008, the start of the global financial crisis. The reason why more cargo was shipped in 2018 than in 2008, but emissions were lower, was because the carbon intensity (the fleet's CO_2 emissions per tonne mile (a measure of transport work, calculated as tonnes lifted multiplied by miles travelled) improved from 2008 to 2018. Figure 23.3 shows that, since 2000, total goods loaded, tonne miles, and carbon emissions have increased, while carbon intensity has declined. Figure 23.3 also shows that more than half of this improvement occurred before 2012, largely due to the adoption of slow steaming to save fuel. As a result of the financial crisis of 2008–2012, carriers were fighting to stay in business due to the overcapacity of ship tonnage available. Only about 1–2% efficiency improvements have been achieved per year since 2015. GHG emissions are projected to increase by anything up to 50% by 2050 for a range of six long-term economic and energy scenarios, with even the IMO's best-case scenario showing no reduction in carbon emissions from 2018 to 2050 (IMO, 2020).

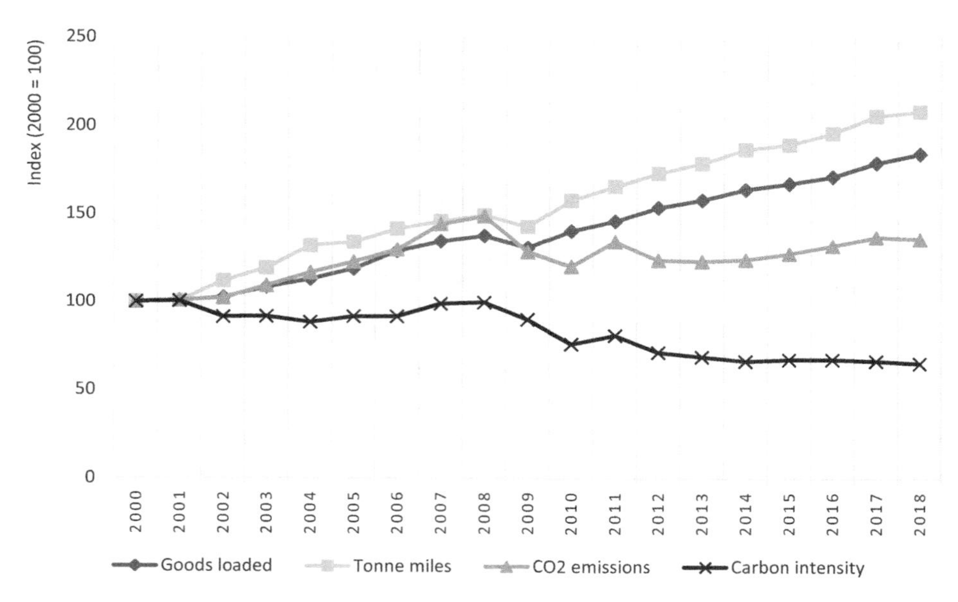

Figure 23.3 Carbon intensity vs goods loaded, tonne miles, and CO_2 emissions, 2000–2018 (2000 = 100).
Source: Authors, based on data from UNCTAD and IMO, several years.

While no readily available alternative fuel or technology exists to decarbonise the sector at present (Gilbert et al., 2018), several options are being explored, from changing the primary fuel source to various means that can provide some percentages of propulsion (for a review see Bouman et al., 2017; OECD, 2018). Wind power can be harnessed in different ways, from kites that assist ship propulsion to flettner rotors (vertical cylinders that are rotated to act like sails and provide additional propulsion), to new designs for entirely wind-powered vessels. For example, the Wallenius Ocean Bird, a 200m long car carrier using 'wing sails' (with a similar construction to aeroplane wings), is due for delivery in 2024. Several small companies (e.g., TOWT in France) have emerged that ship goods entirely by traditional sailboats. Full electric propulsion is advancing in land-based modes such as trains (using overhead catenary systems) and trucks (using batteries). For shipping, small vessels can be partly or entirely battery-powered (charged from land-based sources), although the vessel size and range are severely limited, constraining them at present to short-range ferries and supply vessels. A recent development along these lines is the *Yara Birkeland*, which is a fully autonomous electric ship delivered in November 2020 that is 80 metres in length, 15 metres wide, and with a capacity of 120 20-foot equivalent units (TEU; 1 TEU is a container of about 6 metres in length).

With regard to alternative fuel sources, there is a divide between fuels – such as methanol and biofuels – which can be blended with existing fuels and entirely different fuels that require system-level change. As it is questioned whether the supply of methanol and biofuels can ever be sufficient to scale up to the entire global fleet of ships (OECD, 2018), new fuels will be needed in large volumes. The challenge is that switching to alternative fuels is not only a matter of modifying vessel engines but also requires an overhaul of production, transportation, and refuelling networks, all of which are currently based around fuel oil. LNG has received significant attention in recent decades. While it is very attractive for reducing air pollution, producing relatively little NO_x, SO_x, and PM, it is not an option for decarbonisation. When burned, LNG produces around 25% lower GHG emissions than HFO, but, as some methane is released before it is burned as fuel (known as 'methane slip'), this potential difference is limited to around 20%. Furthermore, when the entire life cycle of emissions of LNG production are considered, the total GHG reduction of LNG compared to traditional fuels is negligible, and in certain cases may be negative (Lindstad and Rialland, 2020). While the growth of LNG remains slow, some shipping lines are investing in it. For example, the first of nine 22,000-TEU vessels equipped to run on LNG was delivered to French carrier CMA CGM in September 2020, and aims for 26 LNG vessels by 2022.

The leading options for decarbonised shipping are hydrogen and ammonia, both zero-carbon fuels that are attracting a significant amount of industry and government attention (Englert et al., 2021). Hydrogen and ammonia are both energy carriers; therefore, like electricity, they are only as green as the power source used to create them. Ammonia (NH_3) is in fact a second-order energy carrier as it is produced from hydrogen in order to make it safer and cheaper to handle, not requiring the same low temperature and high pressure. Two key challenges exist. First, although trials are ongoing, carriers will not move away from fossil fuels while they remain abundant and cheap. Second, it is not clear whether and when sufficient hydrogen or ammonia will be available for shipping. Countries already face difficulties switching their current energy systems to renewables by 2050. When we add huge additional demand for other areas of society, such as electrified domestic transport (i.e., cars, trucks, and trains), it is questionable whether this demand can be met. Moreover, incentives to create the capacity to produce hydrogen and ammonia for shipping are lacking because maritime emissions are not counted as part of national GHG figures. There is a serious danger that shipping switches to hydrogen produced from fossil fuels rather than 'green' hydrogen produced from renewables.

The International Maritime Organization's policy responses to climate change

The International Convention for the Prevention of Pollution from Ships (MARPOL) was adopted by the IMO in 1973 and entered into force in 1983. It contains six annexes, with the first five covering pollution and waste in the sea, while the sixth covers air pollution. MARPOL Annex VI, added to MARPOL in 1997 and in force since 2005, initially addressed SO_x, NO_x, and PM, but not GHG emissions. The UNFCCC excluded GHG emissions from seaborne trade and asked the IMO to develop global GHG emissions-reduction strategies for international shipping. The IMO commenced its discussions on this in 2003. Since then, several measures have been discussed and adopted through amendments of MARPOL Annex VI.

The first measures that the IMO adopted (in 2011) were the Energy Efficiency Design Index (EEDI) and the Ship Energy Efficiency Management Plan (SEEMP). Since 2013, the EEDI and SEEMP have been mandatory for new ships above 400 gross tonnes (gross tonnage refers to the total enclosed volume of a ship). The EEDI imposes a minimum energy efficiency level per capacity mile, which differs across ship types and size segments. In addition, the minimum energy efficiency level per capacity mile for new ships is tightened every five years to ensure continual improvements in ship design to help reduce GHG emissions. The EEDI is therefore a technical measure that allows designers and shipbuilders to choose the most cost-efficient technologies to meet the EEDI requirements (Shi and Gullett, 2018). Compliance with EEDI is ensured through the International Energy Efficiency Certificate authorised by a mandated Classification Society, which can be inspected during Port State Control. This certificate, which shows the energy efficiency of the ship's design (rather than its operations), is valid for the whole life of a ship. Psaraftis (2019a) suggests that EEDI is deficient because it only applies to new ships and its targets can easily be met by reducing the design speed. Energy efficiency-enhancing technologies still need to be developed and breakthroughs are currently lacking (Shi and Gullett, 2018). Instead, energy efficiency gains have been achieved through the adoption of slow steaming, partly in response to overcapacity and high oil prices (Cariou et al., 2019), although it seems that the limits of this strategy have been reached (see Figure 23.3).

A SEEMP is based on the concept of continual improvement of a ship's energy use through the management cycle of planning, implementing, monitoring, self-evaluation, and improvement. A range of measures are available, including optimising voyage planning, ship handling, hull maintenance, waste-heat recovery, and cargo management. An additional tool is the Energy Efficiency Operational Indicator (EEOI), which can be used to provide insight into the amount of CO_2 emitted per unit of transport work within a given time period. A challenge in reaping benefits from the SEEMP is that fuel-consumption management is not well developed within the shipping industry (Taudal Poulsen and Johnson, 2016). The industry generally sees itself as the most environmentally friendly mode of transport because it has lower air emissions per transported goods or per mile (Van Leeuwen and Van Koppen, 2016; Lister et al., 2015), even though this efficiency does not translate into a reduction in total GHG emissions (see Figure 23.3). More importantly, while the SEEMP is developed by the ship owner and/or operator, energy-efficiency gains do not necessarily flow back to these actors. In fact, it is often the hirer (the charterer and/or cargo owner) who benefits from fuel efficiencies and associated fuel-cost reductions (including through slow steaming) and not the ship owner itself (Scott et al., 2017; Taudal Poulsen et al., 2021).

Other types of measures that the IMO has discussed are market-based mechanisms (MBMs). MBMs should provide an economic incentive to reduce GHG emissions and are therefore seen as a cost-effective and business-friendly measure. While various proposals were put forward, ranging from a fee on bunker fuel to emissions-trading schemes (see, for an overview, Shi, 2016; Van Leeuwen and Van Koppen, 2016), no consensus has been reached and MBMs were dropped off the IMO agenda in 2013. A general fear of the industry is that increased (fuel) costs flowing from an MBM cannot be passed on to the hirer or customer, reducing the margins of what is already a very cost-competitive industry. Moreover, it was not only the form of the MBMs that was contested; a debate has emerged about whether the IMO has the mandate and competence to regulate MBMs and collect associated taxes (Shi and Gullett, 2018). Alternative proposals include the adoption of a mandatory speed limit (Psaraftis, 2019b). More recently, the debate around the adoption of MBMs has been reinvigorated because of an industry proposal to impose a small bunker levy of 2USD per tonne of fuel to create a research fund (Psaraftis and Kontovas, 2021).

In response to the EU's Monitoring, Reporting and Verification (MRV) regulation for ships' CO_2 emissions (see the next section), the IMO adopted a mandatory Data Collection System for Fuel Consumption in 2016. Similar to the EU's MRV regulations, ships above 5,000 gross tonnes are required to collect and submit data on fuel consumption to their flag state, which then aggregates this data for its entire fleet and submits it to the IMO. While being a response to the EU's MRV, important differences remain, of which the most prominent is that the EU gathers data on individual ships, while for the IMO it is on flag states' fleets. The MRV is seen as more comprehensive because it requires higher data granularity and public disclosure (Taudal Poulsen et al., 2021). The data collection system in itself will not lead to GHG reductions, but it might enhance energy management on ships, for example through the SEEMP (Taudal Poulsen and Johnson, 2016; Shi and Gullet, 2018).

Finally, it was not until the adoption of the Paris Agreement in 2015 that the IMO was put under pressure to adopt an actual emissions-reduction target. This target, a 50% reduction of annual GHG emissions by 2050 compared to 2008, was adopted in 2018 through the initial IMO strategy on the reduction of GHG emissions from ships (IMO, 2018). This initial strategy represents a framework for further action; further short-, mid- and long-term measures are still to be adopted. Short-term measures (up to 2023) concern the continued development of existing technological and operational measures, mid-term measures (up to 2030) include the possibility for alternative fuels and market-based mechanisms, while long-term measures remain undefined. As a result of IMO postponing mid-and long-term measures to guide the transition to alternative fuels to 2030 or beyond, technological uncertainty continues to prevail. This is problematic because ships that are ordered today and delivered in a few years' time will be in operation until at least 2050 and some until even 2060 (because of their expected life span of 25–35 years). This means that alternative propulsion and fuel systems should already be part of decisions about orders for new ships.

Other policy responses

In 2013, after a decade of IMO decision-making, the European Commission (EC) published a strategy to integrate CO_2 emissions from shipping into its legislation. Despite the entry into force of EEDI and SEEMP, the EC found progress within the IMO to develop global measures for reducing GHG emissions from shipping to be too slow (Shi, 2016; Monios, 2020). The EU developed a stepwise approach, which started in 2015 with the adoption of the Regulation on Monitoring, Reporting and Verification (MRV) of CO_2 emissions from maritime transport. Next on the horizon would be establishing reduction targets and measures, including MBMs. The MRV regulation requires ships over 5,000 gross tonnes calling at EU ports to annually collect and publish verified data on CO_2 emissions and associated parameters, including distances travelled, time at sea and cargo carried. The EU publishes this data once a year in June. At the time of writing, amendments were being negotiated after the European Parliament proposed, in 2019, to amend the EU MRV Regulation to take into account the global data collection system of the IMO and to reduce the administrative burden of dealing with two different systems. The MRV increases transparency over GHG emissions associated with seaborne trade while also helping ship owners to identify emissions-reduction potential and cargo owners to charter energy-efficient ships. However, while the MRV targets ship owners and managers, in practice, the commercial decisions of charterers and cargo owners, as well as weather circumstances, have more influence over actual fuel consumption (Taudal Poulsen et al., 2021). In addition to amending the MRV regulation, the European Parliament voted, in 2020, to include maritime emissions from ships over 5,000 gross tonnes in its existing emissions-trading scheme (ETS) by 2022. It also proposed to establish a maritime transport decarbonisation fund (the Ocean Fund), to be financed by revenues from the auctioning of ETS allowances.

Alongside these measures, various industry groups have made voluntary pledges and developed schemes to stimulate reduction of GHG emissions from shipping (see Table 23.1). While initially coming from either cargo owners or ports, recent initiatives show a broader range of engagement by maritime supply chain actors, such as financiers. These initiatives signal that pressure to reduce GHG emissions no longer comes only from formal regulation. However, the use of these tools is limited to front runners and to companies that are

Table 23.1 Voluntary initiatives focused on shipping and climate change

Initiative	*Coordination and partnership*	*Year of launching*	*Instruments developed/used*
Clean cargo working group	Major brand owners, ship owners, and charterers; supported by Business for Social Responsibility	2003	• Provides a methodology for CO_2 emissions calculations and benchmarking • Is developing a Sustainable Freight Procurement Framework
Environmental ship index	Developed by the International Association of Ports and Harbours, Ports and ships can register to participate	2011	• Identifies seagoing ships that perform better in reducing air emissions than current emissions standards of the IMO by evaluating the amount of NO_X, SO_X, and CO_2 that is released by a ship
World ports climate action programme	Twelve leading ports, supported by multiple port associations	2018	• Has five groups working on energy efficiency, policy, renewable 'power2ship', low-carbon fuels and decarbonising port facilities
Poseidon principles	Signatories are financial institutions involved in financing the shipping industry, supported by Global Maritime Forum and experts	2019	• Provides a framework for integrating climate considerations into lending decisions to promote international shipping's decarbonisation • Assesses and discloses the climate alignment of ship finance portfolios
Sea cargo charter (for bulk ships)	Developed by cargo owners and experts. Signatories include cargo owners, charterers, and ship owners	2020	• Provides a framework for aligning chartering activities with responsible environmental behaviour to promote international shipping's decarbonisation by the annual reporting of the GHG emissions intensity and total GHG emissions of charter's activities and an assessment of a charterer's climate alignment relative to established decarbonisation trajectories

susceptible to external pressure and demands to build a greener profile (Lister et al., 2015; Van Leeuwen, 2019). Most of the shipping industry is very reactive in its environmental attitude and sees no added value in participating (Van Leeuwen and Van Koppen, 2016).

Finally, action is also being taken at other levels, such as port authorities requiring the use of shore-side power while at berth (enabling berthed vessels to connect to shore-based electric power to avoid running their engines to provide on-board power needs, also known as 'cold ironing') or providing discounts on port dues for more environmentally efficient ships (Styhre et al., 2017), and regulations coming from the sub-state level (e.g., California cold ironing regulations). However, these actions apply only to ships' emissions within port areas. Such emissions are minimal compared to emissions while at sea. Some ports located in cities are also applying various incentives to stimulate reduced emissions on the landside of operations, such as encouraging booked pick-up times and engine standards for trucks, and increasing the use of rail transport (Gonzalez Aregall et al., 2018).

The International Maritime Organization and climate change: an imperfect match

The challenge to reduce GHG emissions from shipping in the coming decades is daunting as progress is slow and many technical and political obstacles exist. The IMO's slow progress has spurred other actors to unilaterally initiate regulatory measures, leading to a fragmented and polycentric approach to decarbonisation. As a result, the IMO is subject to two competing pressures: the pressure to act and the pressure to keep business as usual for the industry (Monios and Ng, 2021). The IMO is a platform for 174 member states that has developed environmental protection measures for ships for almost 50 years. As with any intergovernmental organization, it is highly dependent on the political will to adopt and amend international agreements (see Chapter 23). The IMO aims for consensus rather than voting (although the EEDI and SEEMP were adopted by votes). Political challenges derive not only from differing positions on environmental regulations in the waters of particular countries but also from flag states. A flag state is a country in which a ship is registered and, as such, is responsible for regulating that ship. In recent decades, many ship owners changed the country of registration for their vessels, a process known as 'flagging out' or using 'flags of convenience', in order to save on taxes and benefit from lower labour regulations as well as the fact that some flag states are generally less inclined to inspect vessels and enforce regulatory compliance for IMO regulations (Roe, 2013; DeSombre, 2006). Countries with large vessel registries tend to resist strong environmental measures.

With regard to climate change, the power of flag states is further strengthened because of the UNFCCC principle of 'common but differentiated responsibilities' (CBDR). Developing countries, including some large flag states, such as China, India, and Brazil, resist regulations that they believe will damage their economies. Developed nations insist that IMO regulations, including those related to climate change, should follow the IMO's policy of 'No More Favourable Treatment' and apply to all countries (Van Leeuwen and Van Koppen, 2016; Scott et al., 2017). Selin et al. (2021) suggest that this deadlock could be broken by transferring the IMO's mandate for decarbonisation of shipping to the UNFCCC, which would bring shipping emissions into national accounts, and they propose allocating these emissions to each country according to the location of the shipowners rather than the flag states. They argue that the UNFCCC has a larger membership than the IMO and could allow for differentiation of responses according to CBDR.

While front runners within the industry are aware and are taking efforts to become more energy-efficient and to experiment with different ways to decarbonise, most of the industry resists ambitious climate change measures (see Chapter 5). Shipping industry organisations such as the International Chamber of Shipping (ICS), the World Shipping Council (WSC) and the Baltic and International Maritime Council (BIMCO) argue that they prefer a unified global approach to maritime policymaking, yet they have actively obstructed the development of climate change policies by the IMO (Lister et al., 2015; InfluenceMap, 2017). It is also common practice that countries are represented at the IMO's Marine Environment Protection Committee (MEPC) meetings by industry representatives, which allows them to directly influence negotiations within the MEPC.

The resistance of the industry to environmental regulation exhibits two further underlying economic dynamics. First, the separate incentive systems between ship owners and hirers (charterers or cargo owners) inhibit ship owners from investing in energy efficiency and other decarbonisation options (Scott et al., 2017; Rayner, 2021). Fuel costs are a major part of daily operating costs but are often borne by the hirer. Not only the ship owner, but multiple other actors are involved in, and co-determine, ship management and operations around the use of fuel, including ship operators or managers, charterers, and cargo owners (see for an overview Taudel Poulsen et al., 2021). This means that ship owners, even if they are able to lower the fuel use of their ships, are not incentivised to do so. Second, frontrunning ship owners aside, environmental considerations do not play a role in creating a competitive position in the market (Lister et al., 2015; Van Leeuwen and Van Koppen, 2016). Shipping is a service industry and is considered a cost by cargo owners who hire ships to transport their cargo. In addition, except for big brands, cargo owners do not consider the environmental performance of this transport as part of their company and product's environmental footprint (Lister et al., 2015).

These political and economic dynamics explain why decision-making within the IMO is usually characterised by a reactive and technocratic approach. The IMO responds to confirmed ecological risks with command-and-control policies that set emissions/discharge standards based on available technology. The MEPC traditionally focuses on design of ships and equipment and procedures on board, including the use of logbooks, because that is where it is possible to achieve consensus. They are not used to dealing with long-term threats like climate change, which require proactive approaches to guide technological innovation and transformation at the sector level. Monios and Ng (2021) argue that, while the IMO is facing political pressure to take decisive action, the maritime sector is to some degree isolated from direct social influence as maritime emissions remain outside of national responsibility and are invisible to the wider public. They demonstrate that this isolation allows a continued business-as-usual technocratic approach whereby climate policy is diverted into endless technical debates in the MEPC that foster increased inertia and ambiguity and vague long-term targets with no action in the short term. Rayner (2021) indeed confirms how the IMO scores low to medium in terms of the five general functions that international institutions can perform to contribute to effective decarbonisation: providing guidance and signals; setting rules to facilitate collective action; transparency and accountability to enhance monitoring and enforcement; support implementation (through capacity building, technology and finance); and by knowledge and learning. Instead, it is due to strong external pressure (i.e., the Paris Agreement and the EU) that the IMO has moved at all.

It remains to be seen whether the emergence and further development of the multi-level or polycentric approach to decarbonisation will have positive or negative effects (Roe, 2009; Van Leeuwen, 2015; Monios, 2019). Several influential actors at a variety of governance levels, from cities and ports to voluntary initiatives and trials to individual countries and the EU, can encourage best practice in industry and incentivise more proactive policymaking when the IMO is slow to act, as it has been so far. On the other hand, the IMO's legitimacy is diluted and weakened, with the risk that neither the IMO nor other actors are able to enact sufficiently stringent targets or measures, leading to an ongoing impasse (Monios and Ng, 2021).

Conclusion

Slow progress by the IMO in curbing climate change has spurred policymakers at other levels and various industry actors to take unilateral actions to go beyond IMO regulations. However, this will have only a small effect at the global level because the IMO remains the primary venue for maritime policymaking. There are three key weaknesses in the IMO approach to decarbonisation, with the risk that inertia will continue to characterise the decarbonisation of shipping. First, the IMO's goal is a 50% reduction in GHG emissions by

2050, rather than the IPCC's clear call for full decarbonisation by that date. The anticipated reduction is, in any case, relative to 2008 rather than 1990 (the base year for most other GHG reduction objectives). Second, the IMO's policies are focused on efficiency, aiming to decrease the carbon *intensity* of shipping. While decreasing the carbon output per unit of work is laudable and necessary, total shipping emissions continue to rise and are forecast to continue doing so. Third, political deadlock and regulatory capture mean that the IMO is not capable of rising to the challenge of enacting decarbonisation through transforming ship propulsion. There is a risk that, even while research and trials for alternative fuels such as hydrogen and ammonia are being pursued by industry, they will remain small-scale and proceed alongside continued long-term use of fossil fuels. This inhibits the adoption of maritime policies in the *short* term that are needed to effectively phase out fossil fuels in the *long* term.

Acknowledgements

We thank the Social Sciences and Humanities Research Council of Canada, Green Shipping Partnership Project (Grant No. 13R75917), for support and particularly for acting as an interdisciplinary and transdisciplinary platform to discuss and share ongoing insights into the challenges of governing the decarbonisation of shipping.

References

Bouman, E. A., Lindstad, E., Rialland, A. I. and Strømman, A. H. (2017) 'State-of-the-art technologies, measures, and potential for reducing GHG emissions from shipping–A review', *Transportation Research Part D*, vol 52, pp408–421.

Cariou, P., Parola, F. and Notteboom, T. (2019) 'Towards low carbon global supply chains: A multi-trade analysis of CO_2 emission reductions in container shipping', *International Journal of Production Economics*, vol 208, pp17–28.

DeSombre, E. R. (2006) *Flagging Standard; Globalization and Environmental, Safety and Labor Regulations at Sea*. MIT Press, Cambridge (MA)/London.

Englert, D., Losos, A., Raucci, C. and Smith, T. (2021) 'The Potential of Zero-Carbon Bunker Fuels in Developing Countries'. World Bank, Washington DC. Available at https://openknowledge.worldbank.org/handle/10986/35435 License: CC BY 3.0 IGO.

Gilbert, P., Walsh, C., Traut, M., Kesieme, U., Pazouki, K. and Murphy, A. (2018) 'Assessment of full life-cycle air emissions of alternative shipping fuels', *Journal of Cleaner Production*, vol 172, pp855–866.

Gonzalez Aregall, M., Bergqvist, R. and Monios, J. (2018) 'A global review of the hinterland dimension of green port strategies', *Transportation Research Part D: Transport and Environment*, vol 59, pp23–34.

IMO. (2018) *Adoption of the Initial IMO Strategy on Reduction of GHG Emissions from Ships and Existing IMO Activity Related to Reducing GHG Emissions in the Shipping Sector*. IMO: London, UK.

IMO. (2020) *Fourth IMO GHG Study 2020*. IMO: London, UK.

InfluenceMap. (2017) 'Corporate capture of the International Maritime Organization: How the shipping sector lobbies to stay out of the Paris Agreement'. Available at https://influencemap.org/report/Corporate-capture-of-the-IMO-902bf81c05a0591c551f965020623fda.

Lindstad, E. and Rialland, A. (2020) 'LNG and cruise ships, an easy way to fulfil regulations—Versus the need for reducing GHG emissions', *Sustainability*, vol 12, no 5, pp2080.

Lister, J., Poulsen, R. T. and Ponte, S. (2015) 'Orchestrating transnational environmental governance in maritime shipping', *Global Environmental Change*, vol 34, pp185–195.

Monios, J. (2019) 'Polycentric port governance', *Transport Policy*, vol 83, pp26–36.

Monios, J. (2020) 'Environmental governance in shipping and ports: Sustainability and scale challenges', in A. K. Y. Ng, J. Monios, C. Jiang (eds) *Maritime Transport and Regional Sustainability*. Elsevier, Cambridge, MA, pp13–29.

Monios, J. and Ng, A. K. Y. (2021) 'Competing institutional logics and institutional erosion in environmental governance of maritime transport', *Journal of Transport Geography,* vol 94, p. 103114.
OECD. (2018) *Decarbonising Maritime Transport: Pathways to Zero-Carbon Shipping by 2035.* OECD, Paris.
Poulsen, R. T. and Johnson, H. (2016) 'The logic of business vs. the logic of energy management practice: Understanding the choices and effects of energy consumption monitoring systems in shipping companies', *Journal of Cleaner Production*, vol 112, part 5, pp3785–3797.
Poulsen, R. T., Ponte, S., van Leeuwen, J. and Rehmatulla, N. (2021) 'The potential and limits of environmental disclosure regulation: A global value chain perspective applied to tanker shipping', *Global environmental politics*, vol 21, no 2, pp99–120.
Psaraftis, H. N. (2019a) 'Decarbonization of maritime transport: To be or not to be?', *Maritime Economics & Logistics*, vol 21, no 3, pp353–371.
Psaraftis, H. N. (2019b) 'Speed optimization vs speed reduction: The choice between speed limits and a bunker levy', *Sustainability*, vol 11, no 8, p2249.
Psaraftis, H. N. and Kontovas, C. A. (2021) 'Decarbonization of maritime transport: Is there light at the end of the tunnel?', *Sustainability*, vol 13, no 1, p237.
Rayner, T. (2021) 'Taking the slow route to decarbonisation? Developing climate governance for international transport', *Earth System Governance*, vol 8, p100100.
Roe, M. (2009) 'Multi-level and polycentric governance: Effective policymaking for shipping', *Maritime Policy & Management*, vol 36, no 1, pp39–56.
Roe, M. (2013) 'Maritime governance and policy-making: The need for process rather than form', *The Asian Journal of Shipping and Logistics,* vol 29, no 2, pp167–186.
Scott, J., Smith, T., Rehmatulla, N. and Milligan, B. (2017) 'The promise and limits of private standards in reducing greenhouse gas emissions from shipping', *Journal of Environmental Law*, vol 29, no 2, pp231–262.
Selin, H., Zhang, Y., Dunn, R., Selin, N. E. and Lau, A. K. (2021) 'Mitigation of CO_2 emissions from international shipping through national allocation', *Environmental Research Letters*, vol 16, no 4, p045009.
Shi, Y. (2016) 'Reducing greenhouse gas emissions from international shipping: Is it time to consider market-based measures?', *Marine Policy*, vol 64, pp123–134.
Shi, Y. and Gullett, W. (2018) 'International regulation on low-carbon shipping for climate change mitigation: Development, challenges, and prospects', *Ocean Development & International Law*, vol 49, no 2, pp134–156.
Styhre, L., Winnes, H., Black, J., Lee, J. and Le-Griffin, H. (2017) 'Greenhouse gas emissions from ships in ports – Case studies in four continents', *Transportation Research Part D: Transport and Environment,* vol 5, pp212–224.
UNCTAD. (2020) 'Review of Maritime Transport 2020'. United Nations Publication. Available at https://unctad.org/system/files/official-document/rmt2020_en.pdf
Van Leeuwen, J. (2015) 'The regionalization of maritime governance: Towards a polycentric governance system for sustainable shipping in the European Union', *Ocean & Coastal Management*, vol 117, pp23–31.
Van Leeuwen, J. (2019) 'Capturing a moving target', in Harris, P. G. (ed) *Climate Change and Ocean Governance: Politics and Policy for Threatened Seas*, Cambridge University Press, Cambridge, pp376–392.
Van Leeuwen, J. and van Koppen, C. (2016) 'Moving sustainable shipping forward: The potential of market-based mechanisms to reduce CO_2 emissions from shipping', *The Journal of Sustainable Mobility*, vol 3, no 2, pp42–66.

24

GLOBAL CHANGE AND THE DEVELOPMENT OF SUSTAINABLE FLOATING CITIES

Regulatory and legal implications

Otto Spijkers and Chuxiao Yu

The sea level is rising rapidly as a direct consequence of climate change. Millions of people, some of whom are living in small island developing states (SIDS), already live in coastal areas that are less than 10 meters above sea level, and this number will increase dramatically in the next decades (Lanovoy & O'Donnell, 2021). Drastic measures might thus be needed as the sea continues to rise. Relocating people from submerged coastal zones to floating cities could be part of the solution. Indeed, already in 2019, a prominent international law of the sea expert predicted that 'the concept of floating cities may very well become the new reality as low-lying coastal States and SIDS seek ways to adapt to the rising sea level and maintain their legal, political, and economic existence' (Oral, 2019: 430; see Chapters 12–14). Floating cities are thus already seen as a realistic solution to the adverse effects of global environmental change on coastlines. Consequently, we need to situate these floating cities in the context of existing international law to provide tools for effective marine governance. This chapter aims to do exactly that.

Floating cities consist of large platforms resting on the sea surface that can adapt easily and smoothly to sea-level rise. Floating cities are made up of two elements: a substructure and various superstructures. The superstructures are the parts situated above the sea surface, such as residential houses, office buildings, shopping malls, restaurants, and so on. The substructure is the city's foundation, situated at the sea level and below. Both elements can be made elsewhere, towed to the installation site, and assembled there. The substructure is normally connected and moored to the seabed, which ensures the floating city's stability and safety. The superstructures of a floating city resemble city-buildings on land, and their legal qualification is not the concern of this chapter.

We shall focus on the question of how to properly qualify the substructure in the language of existing international law. Different legal labels have been proposed for this substructure. Which of these labels is the most fitting for a floating city? And what legal consequences do these different labels have? The first part of this chapter answers these and related questions. The second part examines whether, given the need to find answers to the challenges posed by global environmental change, the development of floating cities can be classified as marine scientific research, and what legal consequences this classification could have. We focus on relevant provisions of the United Nations Law of the Sea Convention (hereafter referred to as the Convention or LOSC) (see Chapter 2).

DOI: 10.4324/9781315149745-29

Qualifying floating cities in the language of the Law of the Sea

The term 'floating city' is not found anywhere in the LOSC. Which of the currently existing labels to attach to a floating city depends on its precise characteristics and purposes. We might call it an 'island', 'artificial island', 'installation', 'structure', 'permanent harbor work', 'ship' or 'vessel'. Floating cities could also be characterized as 'platforms', but this is generally considered not to be a separate category in the LOSC. Platforms may be qualified under one of the other categories just mentioned, depending on whether they are fixed to the seabed or floating (more on this below). From the perspective of international law, this choice of label is not without consequences: what a floating city is, determines the related legal rights and obligations of the associated coastal State (Lin, Spijkers, and Plank, 2021; Zohourian, 2018). Here we elaborate on the various potential labels for floating cities.

Island

Is a floating city an 'island'? Article 121 of the LOSC defines an island as a 'naturally formed area of land, surrounded by water, which is above water at high tide' (Nordquist, 2012). Since islands must be *naturally* formed – and thus cannot be *artificially* constructed by human beings – we can quickly conclude that this is not the appropriate label for our floating cities.

Artificial island

Is a floating city then perhaps an *artificial* island? An 'artificial island' is constructed by human beings, so this puts us on the right track (Oude Elferink, 2013). It is important to note that the distinction between natural and artificial islands does not relate to the materials of which the feature consists, but to the process of its becoming. In other words, islands consisting of materials provided by nature, such as sand, gravel, and stone, but manufactured out of those materials by human beings, are not naturally formed, and thus belong to the category of 'artificial islands'. Artificial islands are areas of land, surrounded by water, above water at both high and low tide (which means that they are always above water, except perhaps in extreme sea conditions), and made by human beings. So, does this make our floating cities 'artificial islands'? Is that the proper label to attach to them? One reason to doubt whether a city constructed on a floating platform can be said to constitute an artificial island, is that it is not made of land-like materials dumped on the seafloor, and thus it might not be appropriate to call it an 'area of land'.

Installation or structure

Can a floating city then be termed an 'installation' or a 'structure'? The term 'installation' – but not the term 'structure' – was used in Article 5 of the Convention on the Continental Shelf, a treaty concluded in Geneva in 1958. This treaty can be regarded as the predecessor to the LOSC. When the 1958 Convention was negotiated and drafted, the delegation of the Netherlands felt that the term 'installation' should be used only to refer to *fixed* structures, and that *floating* structures should thus not be regarded as such.

The LOSC does not define the terms 'installation' or 'structure'. Some definitions were proposed during the drafting process of the LOSC. For example, the United States wanted to define installations as 'all offshore facilities, installations, or devices *other than* those which are *mobile* in their normal mode of operation at sea' (United States of America, 1973: 1236, emphasis added). The United States, like the Netherlands had done in the 1950s, wished to exclude *floating* platforms from the category of installations, reserving this category only to *fixed* platforms. Belgium agreed and proposed to regard a floating platform as a 'ship' instead (Belgium, 1973).

These documents, no matter how interesting and informative they might be, cannot finally determine today's meaning of the terms used in the LOSC. Indeed, Articles 31 and 32 of the Vienna Convention on the Law of Treaties tell us that a treaty must be interpreted in good faith, in accordance with the ordinary meaning to be given to the terms of the treaty. This means that the remarks made by the Netherlands, Belgium, and US delegations at the time the treaty was being drafted cannot be decisive in determining today's ordinary meaning of the terms 'installation' and 'structure' as used in the LOSC.

Already in 1974, Fred Soons distinguished 'artificial islands' from 'installations' and 'structures'. He believed that artificial islands should be regarded as 'constructions which have been created by the dumping of natural substances like sand, rocks and gravel' on the seabed (Soons, 1974: 3). On the other hand, installations should be seen as 'constructions resting upon the seafloor by means of piles or tubes driven into the bottom' or 'concrete structures' (Soons, 1974: 3). This description should not be read too narrowly. Most importantly, Soons believed that constructions could be attached to the seafloor by methods other than with piles or tubes. The point he wished to make, rather, was that installations and structures should not be regarded as artificial 'land areas' because they did not consist of natural substances dumped on the seafloor. Soons' approach thus suggests that we should focus on the *materials* of which our floating cities are made, and the way in which they are *connected to* the seafloor. If we follow Soons' distinction, the label of 'installation' seems more suitable for our floating city than that of 'artificial island'.

The LOSC does not regard 'platforms' as a special category, with its own legal regime. This does not mean, however, that the term is absent in the Convention. Most importantly, in Article 1 of the LOSC mention is made repeatedly of 'platforms or other man-made structures at sea', which suggests that platforms are best seen as a sub-category of 'structures'. The LOSC specifically refers to several subcategories of installations, such as 'lighthouses' (Articles 7 and 47), 'port installations' (Article 129), and 'scientific research installations' (Articles 249 and 258–262; more on these below). It further distinguishes 'installations' from 'equipment' (Articles 249(1)(g) and 258–262). All these examples of specific kinds of installations provide us with further guidance and hints on how to interpret and understand the term 'installation' itself. From the above, it can be concluded that the category of 'installation' and 'structure' encompasses *fixed* platforms. That much is uncontroversial. It might also include *floating* platforms that have been (temporarily) anchored into the seafloor, although there is some room for different opinions on this. The main difference of opinion is on the question whether 'ship' or 'vessel' might not be the better or more appropriate category for such floating platforms, of which floating cities are a sub-category.

Before we continue to look at other categories, it might be worthwhile to say a little bit more about an interesting legal puzzle, which applies primarily to a floating city as an 'installation' or 'structure'. Indeed, an installation or structure situated in a State's exclusive economic zone (hereafter referred to as EEZ) can have its own harbor, and this might very well be the case for our floating cities. This raises the question of whether such a harbor is legally part of the floating city. Article 60(5) of the LOSC, which regulates the use of installations and structures in the EEZ, makes clear that the safety zone of installations shall be 'measured from each point of their outer edge' (Pesch, 2015; Kaye, 2007; Esmaeili, 2000, 1999; Meer

Mohr, 1988). It could be argued that this outer edge refers to the low-water line along the installation, as according to Article 5 of the LOSC. This would mean that the waters of the port of a floating city, situated on a fixed platform in the exclusive economic zone, are part of its safety zone. This means that the coastal State does not have full jurisdiction over these waters, but that it can only exercise the limited rights that the coastal State has in safety zones around the city, provided such safety zones have been established. These limited rights are enumerated in Article 60 of the LOSC. However, this appears to be a rather unsatisfactory situation.

A better view is to see the port of a floating city as part of that city, and the waters of the port as part of the city itself and not part of the city's safety zone. There are three reasons to support this view. First, it can be assumed that the port of a floating city is normally constructed together with the city itself. Second, ordinary ports – ports situated on a State's land territory or on a naturally formed island – are considered to constitute an integral part of the coastal State's territory. Thus, it does not seem farfetched to conclude from this, applying analogous reasoning, that ports of a floating city are also an integral part of that city. And there is a third argument to support the conclusion that the harbor of a floating city should be considered an integral part of that city. Article 11 of the LOSC states that permanent harbor works 'are regarded as forming part of the coast' (see also Articles 25(2), 50 and 218–220). There is no reason to approach the harbor works of a floating city any differently than permanent harbor works on land.

In the future, some floating cities may have their own airports, which raises lots of interesting legal issues as well (Wassenbergh, 1999). However, we shall leave an in-depth discussion of the status and use of an airport on a floating city for further research.

Permanent harbor works

Some floating cities might be situated in a State's EEZ, but most floating cities will probably be placed much closer to the State's land territory. In fact, floating cities drifting in a State's territorial sea might very well be located very close to an existing land port. This raises the question whether, and under what circumstances, they might constitute an integral part of the port, instead of being regarded as stand-alone installation or structure. Article 11 of the LOSC suggests that 'the outermost permanent harbor works' still 'form an integral part of the harbor system'. However, the same article states that 'off-shore installations and artificial islands shall not be considered as permanent harbor works'. There is no definition of 'permanent harbor works' in the LOSC. The United Nations Office for Ocean Affairs and the Law of the Sea published a most helpful glossary of technical terms. In this glossary, 'permanent harbor works' are defined as 'permanent man-made structures built along the coast which form an integral part of the harbor system such as jetties, moles, quays or other port facilities, coastal terminals, wharves, breakwaters, sea walls, etc.' (UN Office for Ocean Affairs, 1989: Appendix I, 56). From this, we can conclude that offshore loading and unloading areas meant to service ships that are too large to enter the port are not to be considered harbor works, and therefore they must fall within the category of either 'artificial islands' or 'installations/structures'. It is worth noting that there is no definition of 'off-shore', and thus this term can be assumed to simply mean 'away from or at a distance from the coast'. There is no reason to suppose that the distance must be considerable, or that 'off-shore' is an implicit reference to a particular maritime zone, such as the EEZ or high seas.

But what does this tell us about our floating cities? What consequence does the above analysis have for them? Since floating cities have an entirely different purpose than ports, it seems unlikely that they can be qualified as 'permanent harbor works', no matter how closely they are situated to an existing port on land.

The difference between artificial islands and installations/structures, on the one hand, and permanent harbor works, on the other, is important from a legal point of view, primarily because the latter, which form an integral part of the harbor system and are regarded as forming part of the coast, can cause a seaward shift of the coastal State's baseline. The same effect can be achieved by artificially enlarging a coastal State's land territory, as was done, for example, with the construction of the Maasvlakte in the Netherlands. This is relevant for present purposes, because some floating cities situated in a State's territorial sea may be regarded as such, namely as land reclamation, if their connection with the coast is sufficiently dense. The core question, in both cases, is how to distinguish between a floating city as installation and a floating city being an artificial extension of a port or natural coast. Only the latter leads to a shift in the coastal State's low-water line. Off-shore artificial islands, installations, and structures are not considered to be part of the coast or a port system and do not have such an effect (11 LOSC). When a floating city is artificially connected to the mainland, for example by means of a bridge, tunnel, or land road (e.g., dirt road), then the question arises whether it has thereby become an integral part of that mainland. The LOSC does not provide much guidance here. This lack of clarity is unfortunate, because this may turn out to be a crucial question for floating cities situated close to a State's land territory and in some way connected with it. It is thus a topic for much needed further research.

Ship or vessel

Can a floating city be qualified as a 'ship'? We have already alluded to this possibility, primarily when distinguishing *fixed* and *floating* platforms. The most important characteristic of a 'ship' or 'vessel' is not that it floats, but rather its purpose: ships are used to *navigate* from one place to another (Schmidtke, 2019: 14). In other words, ships are designed and intended for *transportation* on water, and they must be used for that purpose. This raises the question what the precise meaning might be of 'transportation' as the prime purpose of a ship. Transportation can best be described as the conveyance of things or persons from one place to another. Floating oil platforms and similar structures do not have this purpose and are therefore not considered to be ships or vessels. That much is not very controversial. Applying the same reasoning to floating cities, it can be safely concluded that 'ship' is not the appropriate label because a floating city is not meant to move its residents from one place to another. Floating cities are not gigantic cruise ships, slowly sailing around the world. Instead, they are meant to provide people with a stable home.

This conclusion has far-reaching legal consequences. Indeed, the difference between installations/structures, on the one hand, and ships and vessels, on the other, is so important, because ships do not fall under the regime of Article 60 of the LOSC, which regulates the use of installations and structures situated in the EEZ. Instead, objects made by humankind, which move (independently) in the marine environment, and which serve to navigate from one place to another, fall under the freedom of shipping or navigation. This is an entirely different legal regime. When a 'ship' is moored, submerged, or anchored for purposes other than what falls within 'the normal activities of ships' as provided for in Article 58 of the LOSC, it no longer falls under Article 58, but under Article 60, and it can then be regarded as an installation or structure. This distinction between vessels/ships and installations/structures is crucial for floating cities and might provide us with a solution to our *problématique* (more on this below).

A new category

When determining whether 'installation' or 'structure' is the best label to attach to a floating city, we concluded that the former term primarily refers to constructions resting upon or firmly attached to the seafloor. In other words, they are *fixed,* and this is often contrasted with *floating* constructions. It thus appears doubtful whether a floating city, anchored (temporarily) into the seafloor, can properly be qualified as an 'installation'. As noted, the label of 'ship' also seems doubtful, because floating cities are not intended to navigate from one place to the other. Most other qualifications discussed above also appear inappropriate, for one reason or another. And thus, we are running out of terms to apply. It could very well be that none of the labels listed above can be properly attached to floating cities. Instead of fitting within one of the existing categories, they might constitute an entirely new category, one that is not yet regulated under the LOSC. They could, for example, be qualified as 'barge', which is a term not used in the LOSC (Schmidtke, 2019: 14–15), or simply as 'floating city'. That is, we could introduce an entirely new category *sui generis*.

Because the LOSC was meant to regulate all uses of the sea in a comprehensive way, we should proceed on the assumption that even floating cities must fall within the existing legal regime, no matter how difficult this might seem, and no matter how much creativity of interpretation this requires. And thus, the most elegant solution is probably to regard floating cities as 'ships' or 'vessels' when they are being moved (towed) to their locations, and as 'installations' or 'structures' when moored. This is because they float when moving, and they become fixed – at least in some way or another – when moored. Indeed, as soon as floating cities are moored, they become non-movable constructions, which makes them fixed, and thus from then on they are probably best labelled as 'installations'. The term 'non-movable' does not mean that the floating city remains completely static for the rest of its life. If the floating city is allowed a very limited degree of motion, for example to rise along with the rising sea levels, this is not considered movement. Referring to floating cities as 'non-movable' also does not mean that they lose all capacity to move. It simply means the floating city is attached to the seafloor in such a way that it is, until it has been detached again, non-movable (Ranganathan, 2019; Binder, 2016: 790; Fateh, 2013: 909; Morris and Kindt, 1979).

From the above analysis, it can be concluded that a floating city does not fit neatly, smoothly, and comfortably in any of the existing categories. It is not an 'island' because it is not made by nature itself. It is not an 'artificial island' because it is not made of land-like materials dumped on the seafloor. It appears not to be an 'installation', 'structure' or 'platform' because it is not permanently fixed to the seafloor. It is clearly not a 'permanent harbor work' because it is established for a different purpose than that of a port. It appears not to be a 'vessel' or 'ship' because it does not serve as mode of transportation. However, because the LOSC is meant to regulate all uses of the ocean, an attempt was made to fit the floating city into the existing legal framework by arguing that it can probably best be seen as a 'ship' when moved to its location and as 'installation' when moored.

The development of floating cities as marine scientific research

Another question that arises is how the *development* of floating cities fits into the rules of the law of the sea. This will become especially important as coastal areas are affected by sea-level rise and demand for floating cities grows. The LOSC does not contain norms that are specifically designed to regulate the development of floating cities. This, however, does not mean the LOSC is not applicable to such activities, especially considering that the Convention is drafted with a view to settling 'all issues relating to the law of the sea' (Preamble of the LOSC).

It is noted that when large scale marine geoengineering activities were first introduced – marine geoengineering defined as the 'deliberate intervention in the marine environment to manipulate natural processes, including to counteract anthropogenic climate change and/or its impacts' (Scott, 2015: 452; Contracting Parties, 2013) – it was submitted that, at the stage of testing and modelling, such activities meet the requirements of marine scientific research in the context of the LOSC since they aim at increasing the knowledge about the marine environment and the ocean's reaction to marine geoengineering activities (Haomiao, 2017: 108; Proelss, 2012: 5). In the same vein, it is pertinent to assess whether and to what extent the establishment of floating cities, seen as novel and innovative method to adapt to global environmental change, in particular climate change and rising sea levels which are a consequence of climate change, qualifies as marine scientific research under the LOSC (Yu, 2021).

Definition of marine scientific research

The LOSC does not provide a definition of the term 'marine scientific research'. The ordinary meaning of the term 'scientific research' is 'the application of the scientific method to investigate any relationships amongst natural phenomena or to solve a technical or medical problem' (Black's Law Dictionary, 2021). 'Research' as a noun is defined as 'the systematic investigation into and study of materials and sources to establish facts and reach new conclusions' (Oxford Dictionary of English, 2015). 'Science' means 'the intellectual and practical activity encompassing the systematic study of the structure and behavior of the physical and natural world through observation and experiment' (Oxford Dictionary of English, 2015). According to its ordinary meaning, activities that constitute marine scientific research should have some scientific attributes. Besides, the ordinary meaning of the term encompasses the idea of systematically studying the research object using scientific methods.

During the Third United Nations Conference on the Law of the Sea, definitions of the term 'marine scientific research' were included in some official proposals that were put forward, and in several informal draft texts of the Convention (Nordquist, Rosenne, and Yankov, 1991: 438–450). Although many differences are noticeable among these definitions, one commonality can be identified, which is that activities covered by the term 'marine scientific research' should aim at the *furtherance of knowledge* of the marine environment. At the moment, the development of floating cities – perceived as a possible solution to rising sea levels and other changes in the marine environment – is still at its experimental stage. Much further research needs to be done to evaluate the feasibility and effect of this activity, and the possible impact of establishing floating cities on the marine environment. It is thus arguable that, at this stage, the development of floating cities qualifies as marine scientific research under the LOSC since it contributes to increasing knowledge of the marine environment.

The implications of the LOSC Marine Scientific Research Regime for the establishment of floating cities

Although the LOSC does not define the term 'marine scientific research', it contains a comprehensive set of rules for the conduct and regulation of such activities. These rules provide the legal basis for the development of sustainable floating cities as a form of marine scientific research. First and foremost, as stipulated by the Convention, all marine scientific research shall be conducted following the general principles outlined in Article 240. These principles require that all marine scientific research must be conducted exclusively for peaceful purposes, with appropriate scientific methods and means, in a way that does not unjustifiably interfere with other legitimate uses of the sea compatible with the Convention, and in compliance with all relevant regulations, including those for the protection and preservation of the marine environment. Of particular relevance for the current discussion is the provision that subjects the establishment of floating cities to several environmental protection obligations. The interrelation between marine scientific research and marine environmental protection has long been discussed (Hubert, 2011; Warner, 2008; Verlaan, 2007: 210–216). With the development of technology and the expansion of the scale of research activities, the tension is likely to become more intense. Promoting responsible research, for instance through the adoption of a code of conduct for marine scientific research, is of relevance for relieving the tension. As to the establishment of floating cities, the importance of marine environmental protection is clear, since such activities aim exactly at the provision of a solution to global environmental change, in particular rising sea levels caused by climate change. It thus runs contrary to the very objective of such activities if the development of floating cities itself poses a threat to that very same marine environment.

Second, according to Article 244(1) of the LOSC, States and competent international organizations carrying out marine scientific research are under an obligation to publish and disseminate 'through appropriate channels information on proposed major programmes and their objectives'. In the view of Soons, the term 'major programmes' refers to all marine scientific research programmes that involve the conduct of research activities in areas that are not under the jurisdiction of the researching State itself, or which 'are of more than purely local interest' (Soons, 1982: 243). Since the development of floating cities may contribute to the welfare of humanity, this activity can certainly be regarded as 'major programme' under Article 244(1) of the LOSC. According to this provision, States or competent international organizations that establish floating cities are, at the experimental stage, obliged to publish and disseminate information on the proposed activity and its objectives 'through appropriate channels'. These 'appropriate channels' clearly include international channels but are not necessarily limited to official international channels (Nordquist, Rosenne, and Yankov, 1991: 486).

Third, if and to the extent that floating cities are to be considered as 'installations', provisions set out under Section 4 of Part XIII of the LOSC, which is entitled 'Scientific research installations or equipment in the marine environment', are also to be applied. According to these provisions, several requirements need to be met. For instance, the development and use of floating cities, not being islands (Article 259), 'shall not constitute an obstacle to established international shipping routes' (Article 261), and the floating cities 'shall bear identification markings indicating the State of registry or the international organization to which they belong and shall have adequate internationally agreed warning signals to ensure safety at sea and the safety of air navigation' (Article 262). States and competent international organizations must create around the floating cities safety zones 'of a reasonable breadth not exceeding 500 metres' (Article 260; see Article 60).

Last, as the LOSC prescribes different sets of rules for the conduct and regulation of marine scientific research in different maritime zones, depending on the location of floating cities, the respective rules are to be applied. In areas within national jurisdiction, Part XIII of the LOSC provides that in its internal waters and territorial sea, the coastal State has the exclusive right to regulate, authorize and conduct marine scientific research, and that foreign research may be conducted 'only with the express consent of and under the conditions set forth by the coastal State' (Article 245). In the EEZ and on the continental shelf, the Convention similarly provides the coastal State with the right to regulate, authorize, and conduct marine scientific research (Article 246(1)). Consent from the coastal State shall be obtained by other States before the latter can engage in marine scientific research in the coastal State's EEZ or on its continental shelf (Article 246(2)). In areas beyond any State's national jurisdiction, at the experimental stage, the establishment of floating cities can be carried out freely without the obligations of notification or consent because no such requirements are contained in the relevant provisions of the LOSC.

Although no rules under the LOSC were drafted specifically with the establishment of floating cities in mind, we may conclude that, at its experimental stage, the development of sustainable floating cities can be considered as marine scientific research in the context of the LOSC. As a result, provisions concerning the conduct and regulation of marine scientific research, which are mainly set out in Part XIII of the LOSC, apply. The application of these provisions entails several legal implications in relation to, among others, the protection of marine environment, the publication and dissemination of research information and its objectives, and the development and use of floating cities that qualify as installations and equipment.

Conclusion

Floating cities are no longer science-fiction. More and more States consider the development of sustainable floating cities as a realistic means to adapt to global environmental change, in particular rising sea levels which are a consequence of climate change. At the same time, the international legal regime applicable to activities at sea has not been adapted to this global environmental change. And thus, the development of floating cities does not (yet) fit comfortably within the existing legal regime of marine governance. In this chapter, we have looked at two of the most important legal puzzles. First, we examined which of the currently existing labels can best be attached to floating cities, and we concluded that they are best referred to as a 'ship' when towed to their destination and 'installation' when moored. We then looked at whether the development of sustainable floating cities, especially considering that such development is still in an experimental phase, can be regarded as marine scientific research. We concluded that the development of floating cities does indeed fall under the notion of marine scientific research as used in the LOSC. These conclusions may help to guide the development and management of floating cities as the need for them becomes more important in the future because of global environmental change.

References

Belgium. (1973) 'Selected Documents from the Meetings Held from July 20 to August 24, 1973 (Artificial Islands, Land-Locked States, Settlement of Disputes, Territorial Sea, Continental Shelf, Straits, Fisheries, Economic Zones, Archipelagos)', *International Legal Materials*, vol 12, pp 1210–1213.

Binder, M. (2016) 'Taking to the Sea: The Modern Seasteading Movement in the Context of Other Historical Intentional Communities', *Indiana Journal of Global Legal Studies*, vol 23, no 2, pp 765–794.

Black's Law Dictionary. (2021), Online Legal Dictionary, 2nd edition, available at https://thelawdictionary.org/scientific-research/.

Contracting Parties to the 1996 Protocol to The Convention on The Prevention of Marine Pollution by Dumping of Wastes and Other Matter. (2013), *Resolution LP.4(8) on the Amendment to the London Protocol to Regulate the Placement of Matter for Ocean Fertilization and Other Marine Geoengineering Engineering Activities*, adopted on 18 October 2013.

Esmaeili, H. (1999) 'The Protection of Offshore Oil Rigs in International Law (Part I)', *Australian Mining and Petroleum Law Journal*, vol 18, pp 241–252.

Esmaeili, H. (2000) 'The Protection of Offshore Oil Rigs in International Law (Part II)', *Australian Mining and Petroleum Law Journal*, vol 19, pp 35–43.

Fateh, R. H. (2013) 'Is Seasteading the High Seas a Legal Possibility: Filling the Gaps in International Sovereignty Law and the Law of Seas', *Vanderbilt Journal of Transnational Law*, vol 46, no 3, pp 899–932.

Haomiao, D. (2017) *An International Legal Framework for Geoengineering: Managing the Risks of an Emerging Technology*. Routledge, London, UK.

Hubert, A. M. (2011) 'The New Paradox in Marine Scientific Research: Regulating the Potential Environmental Impacts of Conducting Ocean Science', *Ocean Development & International Law*, vol 42, no 4, pp 329–355.

Kaye, S. (2007) 'International Measures to Protect Oil Platforms, Pipe-lines, and Submarine Cables from Attack', *Tulane Maritime Law Journal*, vol 31, pp 377–423.

Lanovoy, V., and O'Donnell, S. (2021) 'Climate Change and Sea-Level Rise: Is the United Nations Convention on the Law of the Sea up to the Task?', *International Community Law Review*, vol 23, pp 133–157.

Lin, F. Y., Spijkers, O., and Plank, P (2021) 'Legal Framework for Sustainable Floating City Development: A Case Study of The Netherlands', *Conference Proceedings of Paving the Waves – World Conference on Floating Solutions*.

Meer Mohr, P. (1988) 'Measures to Prevent Collisions with Off-shore Installations on the Dutch Continental Shelf', *Leiden Journal of International Law*, vol 1, pp 222–230.

Morris, M. K., and Kindt, J. W. (1979) 'The Law of the Sea: Domestic and International Considerations Arising from the Classification of Floating Nuclear Power Plants and Their Breakwaters as Artificial Islands', *Virginia Journal of International Law*, vol 19, no 2, pp 299–320.

Nordquist, M. H. (2012) 'Textual Interpretation of Article 121 in the UN Convention on the Law of the Sea', in Hestermeyer, H., and Wolfrum, R. (eds) *Coexistence, Cooperation and Solidarity: Liber Amicorum Rudiger Wolfrum*. Brill Nijhoff, Leiden, Netherlands, pp 991–1035.

Nordquist, M. H., Rosenne, S, Yankov, A., and Grandy Neal, R. (eds) (1991) *United Nations Convention on the Law of the Sea, 1982: A Commentary (Vol. IV)*. Martinus Nijhoff Publishers, Dordrecht, Netherlands.

Oral, N. (2019) 'International Law as an Adaptation Measure to Sea-level Rise and Its Impacts on Islands and Offshore Features', *International Journal of Marine and Coastal Law*, vol 34, pp 415–439.

Oude Elferink, A. (2013) 'Artificial Islands, Installations and Structures', *Max Planck Encyclopaedia of Public International Law*.

Oxford Dictionary of English (2015), 3rd edition, Oxford University Press, Oxford, UK.

Pesch, S. (2015) 'Coastal State Jurisdiction around Installations: Safety Zones in the Law of the Sea', *International Journal of Marine and Coastal Law*, vol. 30, pp 512–532.

Proelss, A. (2012) 'International Legal Challenges Concerning Marine Scientific Research in the Era of Climate Change' in *Securing the Ocean for the Next Generation*. Papers from the Law of the Sea Institute, UC Berkeley–Korea Institute of Ocean Science and Technology Conference, Seoul, South-Korea.

Ranganathan, S. (2019) 'Seasteads, Landgrabs and International Law', *Leiden Journal of International Law*, vol 32, pp 205–214.

Schmidtke, R. C. (2019) 'Artificial Islands of the Future: The Seasteading Movement and the International Legal Regimes Governing Seasteads in EEZs and on the High Seas', *Asian-Pacific Law & Policy Journal*, vol 21, no 1, pp 1–28.

Scott, K. N. (2015) 'Geoengineering and The Marine Environment,' in Rayfuse, R. (ed.) *Research Handbook on International Marine Environmental Law*. Edward Elgar, Cheltenham, UK, pp 451–472.

Soons, A. H. A. (1974) *Artificial Islands and Installations in International Law*. Law of the Sea Institute, University of Rhode Island.

Soons, A. H. A. (1982) *Marine Scientific Research and the Law of the Sea*. Kluwer, Deventer, Netherlands.

United Nations Office for Ocean Affairs and the Law of the Sea. (1989) *Baselines: An Examination of the Relevant Provisions of the United Nations Convention on the Law of the Sea*. United Nations, New York.

United States of America. (1973) 'Selected Documents from the Meetings Held from July 20 to August 24, 1973 (Artificial Islands, Land-Locked States, Settlement of Disputes, Territorial Sea, Continental Shelf, Straits, Fisheries, Economic Zones, Archipelagos)', *International Legal Materials*, vol 12, pp 1200–1273.

Verlaan, P. A. (2007) 'Experimental Activities That Intentionally Perturb the Marine Environment: Implications for the Marine Environmental Protection and Marine Scientific Research Provisions of the 1982 United Nations Convention on the Law of the Sea', *Marine Policy*, vol 31, no 2, pp 210–16.

Warner, R. M. (2008) 'Protecting the Diversity of the Depths: Environmental Regulation of Bioprospecting and Marine Scientific Research Beyond National Jurisdiction', *Ocean Yearbook*, no 22, pp 411–43.

Wassenbergh, H. (1999) 'The Status and Use of an Airport on an Artificial Island', *Air & Space Law*, vol XXIV, no 4/5, pp 177–180.

Yu, C. (2021) 'The Legal Regulation of 'Modern' Ocean Data Collection Activities under UNCLOS - With A Focus on the Marine Scientific Research Regime', PhD thesis, Utrecht University at Utrecht, Netherlands.

Zohourian, M. A. (2018) 'The Real Nature of Artificial Islands, Installation and Structures from Perspective of Law of the Sea', *Asia-Pacific Journal of Law, Politics and Administration*, vol 2, no 1, pp 13–26.

25

OCEANS AND SEAS FOR SUSTAINABLE DEVELOPMENT

Challenges of global environmental change for SDG14

Shailly Kedia and Kapil Narula

Oceans cover more than 70% of the surface of the planet and form 95% of the biosphere (WOA, 2021a). According to the World Ocean Assessment, oceans are Earth's largest ecosystem, regulating global climate and driving weather-related systems (WOA, 2021a). They are also extremely crucial for 'blue carbon', which refers to carbon captured by the world's ocean and coastal ecosystems, including mangroves, sea grasses, and salt marshes (NOAA, 2021a). Scientists estimate that 50–80% of the earth's oxygen is produced by phytoplankton in the ocean; one species, *Prochlorococcus* (which is the smallest photosynthetic organism), can produce up to 20% of the oxygen in the biosphere (NOAA, 2021b). The ocean absorbs around 23% of the annual CO_2 emissions generated by anthropogenic activity and helps mitigate the impacts of climate change (UN, 2020). When marine systems are damaged, an enormous amount of carbon can be emitted into the atmosphere, where it augments the factors that lead to climate change (NOAA, 2021c). Undoubtedly, life on the biosphere is dependent on the oceans. Conservation and sustainable use of the seas are crucial for attaining social, environmental, and economic goals and hence, for sustainable development.

In September 2015, by way of Resolution 70/1, the United Nations (UN) General Assembly adopted 'Transforming our world: the 2030 Agenda for Sustainable Development' (UN, 2015). This resolution of the UN Sustainable Development Summit contained 17 Sustainable Development Goals (SDGs), along with 169 targets. Goal 14 of the SDGs (SDG14) – 'Conserve and sustainably use the oceans, seas and marine resources for sustainable development' – is dedicated to oceans, seas, and marine resources. The SDGs framework provides the basis for governments and development agencies around the world to improve their policies and policy implementation regarding marine related activities. It is being used as a basis to frame or adapt policies and monitoring frameworks for SDGs at global, regional, national, and sub-national levels. Apart from having spatial and temporal implications, SDGs have policy and political significance for marine governance.

Understanding the monitoring and operationalization of Goal 14 has significant implications for marine environmental governance amidst global environmental change. There are several maritime activities that contribute to environmental issues; in turn, global environmental changes have implications, which pose a challenge to SDG14 and its targets. Realizing SDG14 would imply proactively addressing global environmental issues, including

DOI: 10.4324/9781315149745-30

climate change, ozone depletion, biodiversity loss, pollution, and waste management. This chapter uses SDG14 as a lens to understand the implications for ocean governance.

SDG14 and the implications of global environmental changes

After the adoption of SDGs and targets, a global indicator framework (GIF) was developed by the Inter-Agency and Expert Group on SDG Indicators. The GIF was agreed upon at the 48th session of the United Nations Statistical Commission held in March 2017. The GIF was then adopted by the General Assembly on 6 July 2017 (UN, 2017). Goal 14 of the SDGs has ten targets and ten indicators (UN, 2021a). These are listed in Table 25.1. For the purpose of monitoring, the UN has established a tier classification for indicators (UN, 2021a). Tier I Indicator is conceptually clear. It has an internationally established methodology; standards are available, and data are regularly produced by countries for at least 50% of countries and for the population in every region where the indicator is relevant. Tier II Indicator is also conceptually clear. It has an internationally established methodology and standards are available. However, data are not regularly produced by countries. Tier III Indicator has no internationally established methodology or standards for the indicator; however, these methodologies or standards are being (or will be) developed or tested.

About 40% of the world's population lives within 100 km of a coast, and many of them depend on marine and coastal biodiversity for their livelihood. About 80% of all tourism takes place in coastal areas with ocean-related tourism industry supporting over a third of

Table 25.1 Tier classification of indicators under SDG14

Targets	*Indicators*	*Tier classifications*
14.1 By 2025, prevent and significantly reduce marine pollution of all kinds, in particular from land-based activities, including marine debris and nutrient pollution	14.1.1 (a) Index of coastal eutrophication and (b) plastic debris density	Tier II
14.2 By 2020, sustainably manage and protect marine and coastal ecosystems to avoid significant adverse impacts, including by strengthening their resilience, and take action for their restoration in order to achieve healthy and productive oceans	14.2.1 Number of countries using ecosystem-based approaches to managing marine areas	Tier II
14.3 Minimize and address the impacts of ocean acidification, including through enhanced scientific cooperation at all levels	14.3.1 Average marine acidity (pH) measured at agreed suite of representative sampling stations	Tier II
14.4 By 2020, effectively regulate harvesting and end overfishing, illegal, unreported, and unregulated fishing and destructive fishing practices and implement science-based management plans, in order to restore fish stocks in the shortest time feasible, at least to levels that can produce maximum sustainable yield as determined by their biological characteristics	14.4.1 Proportion of fish stocks within biologically sustainable levels	Tier I

(*Continued*)

Targets	*Indicators*	*Tier classifications*
14.5 By 2020, conserve at least 10% of coastal and marine areas, consistent with national and international law and based on the best available scientific information	14.5.1 Coverage of protected areas in relation to marine areas	Tier I
14.6 By 2020, prohibit certain forms of fisheries subsidies which contribute to overcapacity and overfishing, eliminate subsidies that contribute to illegal, unreported, and unregulated fishing and refrain from introducing new such subsidies, recognizing that appropriate and effective special and differential treatment for developing and least developed countries should be an integral part of the World Trade Organization fisheries subsidies negotiation	14.6.1 Degree of implementation of international instruments aiming to combat illegal, unreported, and unregulated fishing	Tier I
14.7 By 2030, increase the economic benefits to small island developing States and least developed countries from the sustainable use of marine resources, including through sustainable management of fisheries, aquaculture, and tourism	14.7.1 Sustainable fisheries as a proportion of GDP in small island developing States, least developed countries, and all countries	Tier I
14.a Increase scientific knowledge, develop research capacity and transfer marine technology, taking into account the Intergovernmental Oceanographic Commission Criteria and Guidelines on the Transfer of Marine Technology, in order to improve ocean health and to enhance the contribution of marine biodiversity to the development of developing countries, in particular small island developing States and least developed countries	14.a.1 Proportion of total research budget allocated to research in the field of marine technology	Tier II
14.b Provide access for small-scale artisanal fishers to marine resources and markets	14.b.1 Degree of application of a legal/regulatory/policy/ institutional framework which recognizes and protects access rights for small-scale fisheries	Tier I
14.c Enhance the conservation and sustainable use of oceans and their resources by implementing international law as reflected in the United Nations Convention on the Law of the Sea, which provides the legal framework for the conservation and sustainable use of oceans and their resources, as recalled in paragraph 158 of 'The future we want'	14.c.1 Number of countries making progress in ratifying, accepting, and implementing through legal, policy and institutional frameworks, ocean-related instruments that implement international law, as reflected in the United Nations Convention on the Law of the Sea, for the conservation and sustainable use of the oceans and their resources	Tier II

Source: UN (2021a).

the labour force. Apart from livelihoods, oceans also provide the primary source of protein to over 50% of the population in least developed countries (UN, 2020). Hence, oceans are crucial for supporting the economic, social, and environmental needs of the global population (WOA, 2021b). A number of maritime activities contribute to global environmental changes. In turn, global environmental changes have implications for oceanic and coastal ecosystems, posing challenges to realizing SDG14 and its targets. A number of these global environmental issues are shown in Table 25.2. Realizing SDG14 would imply proactively addressing these issues, including climate change, ozone depletion, biodiversity loss, pollution, and waste management.

Table 25.2 Global environmental issues and implications for SDG14

Global environmental issues	*Contributing maritime activities*	*Implications for oceans and coastal ecosystems*	*Direct impacts on SDG14 targets*
Climate change	Fossil fuels used for shipping. Energy used for various maritime related activities, including shipbuilding, tourism, seafood processing, etc.	Increase in ocean heat and sea surface temperature leading to coral bleaching and changes in reef habitat leading to loss of breeding grounds for marine life. Increased frequency and intensity of harmful algal blooms, affecting marine life and biodiversity present in a specific location and threatening sensitive ecosystems. Changes in ocean current circulation patterns and increase in the amount of atmospheric water vapour over the oceans, influencing global climate patterns and increasing storm intensity. Reduction in the circulation patterns bringing nutrients from the deep sea to surface waters leading to declines in fish populations. Thermal expansion leading to sea-level rise and coastal flooding. Increase in ocean acidity leading to impact on marine organisms that rely on carbonate-based shells and skeletons and to the marine food chain. Deoxygenation that can affect fisheries and marine life by impacting growth, survival, and reproduction of marine life, affecting fishable biomass, movement of fish, and their quality, leading to poor yields.	14.3, 14.7
Air pollution	Use of high sulphur fuel in marine transport	Increased concentration of NOx, SOx , PM10, PM2.5, PM1 especially in port cities leading to decreased air quality having several health implications on human health. Increased emissions of black carbon in atmosphere, accelerating melting of ice. Increased probability of acid rain, causing acidification of soil and water. Increased probability of formation of ground level ozone from interaction with volatile organic compounds (VOCs). Loss of tourism.	14.3, 14.7

(*Continued*)

Global environmental issues	*Contributing maritime activities*	*Implications for oceans and coastal ecosystems*	*Direct impacts on SDG14 targets*
Stratospheric ozone depletion	Use of ozone depleting material in refrigerators and fire extinguishers on board ships	Possible changes to ocean circulation. Possible changes to phytoplankton production in oceans.	14.a
Loss of marine biodiversity and extinction of species	Overfishing and unsustainable fishing practices	Loss of marine ecosystems, fish stocks and marine life leading to stress on food production. Loss of ecosystem services such as flood control and waste detoxification with declining diversity. Reduced capacity to recover from perturbations of ocean biogeochemistry.	14.1, 14.4, 14.6, 14.7
Changes in global supply of freshwater entering oceans	Activities leading to loss of polar ice sheets in Greenland and Antarctic	Changes in ocean currents. Increase in oxygen-depleted 'dead zones'. Possible changes in rainfall patterns and global climate. Sea-level rise (SLR) leading to inundation, flood, and storm damage.	14.2, 14.7
Ocean and coastal degradation	Reclamation of coastal land, bottom trawling, dredging, deep-sea mining	Loss of marine habitat, including mangroves, seagrass, and coastal habitats. Threat of invasive species	14.2, 14.4, 14.5
Nutrient pollution, untreated waste disposal and use of hazardous substances and plastics	Discharge of effluents from marine transportation. Use of hazardous substances on board ships	Increased nutrient pollution (from land and agricultural runoffs) leading to deoxygenation, eutrophication, and hypoxic zones in oceans. Increased marine debris, plastic, and marine litter. Decreased marine biodiversity and marine life.	14.1, 14.4, 14.5

As noted in the first chapter of this volume, the oceans are facing a serious threat from climate change leading to sea-level rise and thermal expansion of the ocean, and from rapid increase in ocean pollution from human activities. Acidification of oceans has increased in response to higher amounts of carbon dioxide in the atmosphere, leading to decrease in pH levels, which has a direct impact on marine life (EPA, 2021). The oceans also absorb more than 90% of the excess heat in the climate system (UN, 2020). Climate change has caused widespread marine heatwaves, threatening the rich ecosystems of the oceans and killing coral reefs around the world. Coastal ecosystems are some of the most threatened in the world (see Chapter 13). Rising temperatures and ocean acidification due to climate change are concurrently impacting them. Along with rising temperature and acidification, increasing levels of debris, including 5 to 12 million metric tonnes of plastic (see Chapter 22) enter the ocean every year (UN, 2020). The Second World Ocean Assessment identifies drivers of global environmental change involving oceans, seas, and marine ecosystems, including social, demographic, and economic developments in societies, and changes in lifestyles and associated consumption and production patterns that pressure the oceans (WOA, 2021a). Marine and coastal habitats are linked to ecological integrity as well as socio-economic issues. Hence, realizing the goals of SDG14 is critical.

Marine governance challenges for SDG14

Ocean governance includes the processes, agreements, rules, and institutions developed to organize the way in which humans use the ocean and its resources (IUCN, n.d.). The European Union (EU) mentions that international ocean governance is about managing the world's oceans and their resources together, so that they are healthy and productive for the benefit of current and future generations (EU, undated). Over the years, marine governance has evolved at the international level and various organizations have played a key role, given the international and transboundary nature of marine related activities. As early as 1902, the International Council for the Exploration of the Sea (ICES) was created, which used the principle of scientific inquiry as the basis for a rational exploitation of marine resources. In the years that followed, goals of state sovereignty and protection of biodiversity emerged as important goals. The foundation of marine governance rests on the United Nations Convention on the Law of the Sea (UNCLOS), also frequently referred to as the 'constitution for the oceans' (see Chapter 2). UNCLOS links different legal and non-legal institutions spanning international and national governance, and in which national and intergovernmental organizations are the primary actors (Soares, 2017).

The challenges of multiple governance mechanisms

UNCLOS and International Maritime Organization (IMO) have taken the approach of regulation and hard law involving states and intergovernmental organizations to resolve maritime issues including marine pollution. In 1956, the International Convention for the Regulation of Whaling was signed in Washington for regulating whaling based on maximum sustainable yield and allocating quotas. In 1993, following the United Nations Conference on Environment and Development, the Sub-committee on Oceans and Coastal Areas of the Administrative Committee on Coordination was established. In 2003, following the World Sustainable Development Summit, UN-Oceans was established for greater inter-agency engagement. There has been an increased focus on the oceans in recent years, which is evident from the adoption of SDG14 as a stand-alone goal in the Agenda 2030 for Sustainable Development, the commencement of the United Nations Ocean Conference, adoption of the UN Decade of Ocean Science for sustainable development, appointment of the Special Envoy for the Ocean by the United Nations Secretary-General, release of an IPCC Special Report on the Ocean and Cryosphere (IPCC, 2019), formation of a High Level Panel for a Sustainable Ocean Economy, and release of the World Ocean Assessment II (HLPSOE, 2021). Apart from these, recent initiatives such as the Sustainable Ocean Business Action Platform of the UN Global Compact, and the World Bank's PROBLUE and Sustainable Blue Economy Finance Initiative indicate efforts to enhance engagement with non-state stakeholders, especially business and industry actors.

The fact that there is a specific target for marine governance in SDG14 shows the importance that the international community has accorded to this issue. Target 14.c on enhancing the conservation and sustainable use of oceans and their resources by implementing international law, as reflected in the UNCLOS, provides the legal framework for the conservation and sustainable use of oceans and their resources. It is evident that ocean governance is essential to protect oceanic ecosystems and to manage human activities in the oceans. It is also important to provide equitable access to ocean resources, both, for different countries and for people from different income categories within countries. Strengthening ocean governance is required to promote synergies among sectors and to avoid conflicts among different stakeholders. Moreover, ocean governance needs to consider political economy approaches (see Kedia and Gautam, 2020). A strong ocean governance framework also helps to de-risk investments that attract finance for maritime sectors.

Ocean governance is multi-layered. While there is some hierarchy, often there is an intermingling among global and regional multilateral agreements, political and voluntary commitments, and among various treaties. Therefore, it is important to coordinate different actors, including governments and non-state actors, through diplomatic arrangements such as bilateral or multilateral mechanisms and 'soft law' if ocean governance is to work effectively. Attainment of SDG14 is hinged on ocean governance, but there are many challenges. Coastal areas fall under national jurisdiction but there are areas beyond national jurisdiction (ABNJ) that require multi-level, integrated and coherent approaches to address the three pillars of sustainability: social, economic, and environmental development. There are several important gaps in ocean governance. The current ocean governance frameworks do not currently integrate the challenges of climate change, and the integration of land–sea interactions is weak. UNCLOS has large gaps, especially regarding biodiversity in ABNJ, and the lack of a legal framework for marine spatial planning (MSP) in many countries is evident. The current frameworks also do not cover several emerging maritime sectors, such as use of marine genetic resources, deep seabed mining, and ocean geo-engineering. There are knowledge gaps regarding scientific understanding about marine ecosystems, impact of maritime activities (especially in benthic regions and the deep sea) and impacts from installation of permanent infrastructure at sea and their decommissioning (see Chapter 24). Extension of terrestrial legislation to offshore activities, or lack of translation of international laws to national laws, is another common challenge that is seen across countries.

Failure to consider marine ecosystems as a whole, lack of adoption of ecosystem-based approaches, insufficient MSP, weak strategic environmental assessments, and failure to integrate traditional knowledge about marine ecosystems from local communities are some of the other reasons for lack of protection of marine spaces and degradation of oceans. A sectoral approach to governing marine spaces also implies failure to consider other sectoral activities and priority to certain activities, which leads to sectoral conflicts. Implementation and enforcement of laws remains weak due to lack of monitoring capacity and resources.

Apart from gaps, there are jurisdictional and regulatory overlaps in ocean governance for marine spaces where different actors and stakeholders compete due to adoption of sector-based approaches. The jurisdictional overlap between treaties and regulations; national, regional, and international legislation; and between sectoral use of oceanic resources is challenging. While the fragmented nature of ocean governance and overlapping measures may not be a bad thing, as they sometimes provide multiple layers of environmental protection, it is observed that these mechanisms are often dysfunctional, incoherent and fail to adopt an ecosystem approach to the maritime domain.

Multilateral actions, international processes, and forums for strengthening ocean governance that are taking shape include developing an international legally binding instrument under the UNCLOS on the conservation and sustainable use of marine biological diversity of areas beyond national jurisdiction. The negotiations on this new BBNJ-instrument or High Seas Treaty are now in the final stages after more than a decade of discussions. This new treaty has four main elements: marine genetic resources, including questions on the sharing of benefits; measures such as area-based management tools, including marine protected areas; establishing uniform requirements for conducting environmental impact assessments; and capacity building and the transfer of marine technology for management and conservation of maritime resources. Adoption of regional ocean governance through Regional Seas Conventions (RSCs) and Regional Fisheries Management Organisations (RFMOs) is also being strengthened (see Chapter 5).

SDG14 advocates for ecosystem-based approaches for which an important consideration is the interconnectedness of the oceans, wherein interconnectedness involves movement of water masses, migration of species and the dispersal of larvae. While countries can exercise territorial sovereign rights over exclusive economic zones (EEZs), these zones are ecologically and biologically connected. Applying governance in the context of an ecosystem approach will need to take into account ecosystems and species that cross political boundaries (Vierros, 2017). Thus, achieving SDG14 will need marine governance and institutions to go beyond political boundaries at the national and sub-national levels.

Challenges of monitoring SDG14 indicators

Five of the ten target indicators for SDG14 are Tier I, and the other five are Tier II. Half of the indicators under SDG14 are monitored at the global level. Data for Tier I indicators are reported in the Global SDG Indicators Database of the United Nations (UN, 2021b), which are shown in Figures 25.1–25.3. As shown in Figure 25.1, the proportion of fish stocks within biologically sustainable levels, that is, those that are not overexploited, has steadily declined over the years between 2000 and 2017. The coverage of protected areas in relation to marine areas (EEZs) at the world level has increased from 2% in 2000 to 17% in 2019, as shown in Figure 25.2. Figure 25.3 shows that sustainable fisheries as a proportion of GDP at the world level remained less than 0.1% between 2011 and 2017.

The progress by countries in the degree of implementation of international instruments aiming to combat illegal, unreported, and unregulated (IUU) fishing at the world level (SDG14.6.1) has improved from a score of 3 in 2018 to 4 in 2020 (where the level of implementation is lowest at 1 and highest at 5). Similarly, the degree of application of a legal/regulatory/policy/institutional framework that recognizes and protects access rights for small-scale fisheries at the world level (SDG14.b.1) has also improved from a score of 3 in 2018 to 4 in 2020. Table 25.3 shows the custodian and partner agencies involved with monitoring various targets under SDG14.

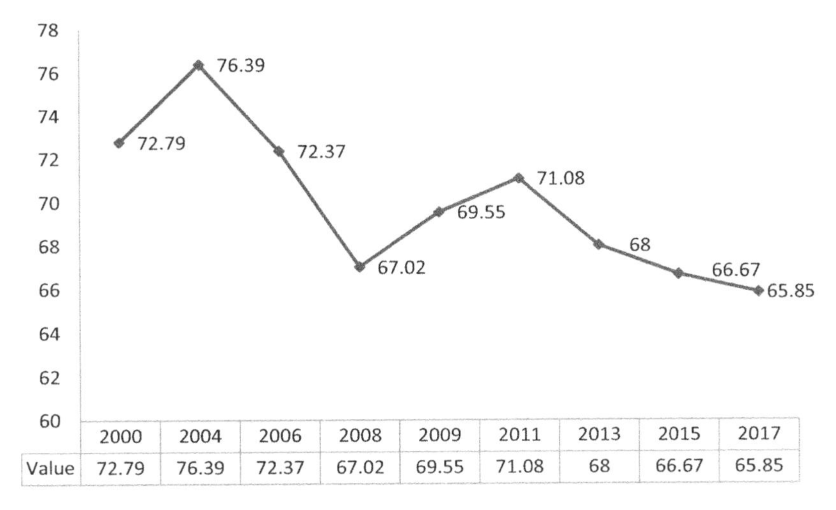

	2000	2004	2006	2008	2009	2011	2013	2015	2017
Value	72.79	76.39	72.37	67.02	69.55	71.08	68	66.67	65.85

Figure 25.1 Proportion of fish stocks within biologically sustainable levels (not overexploited) at the world level (%).

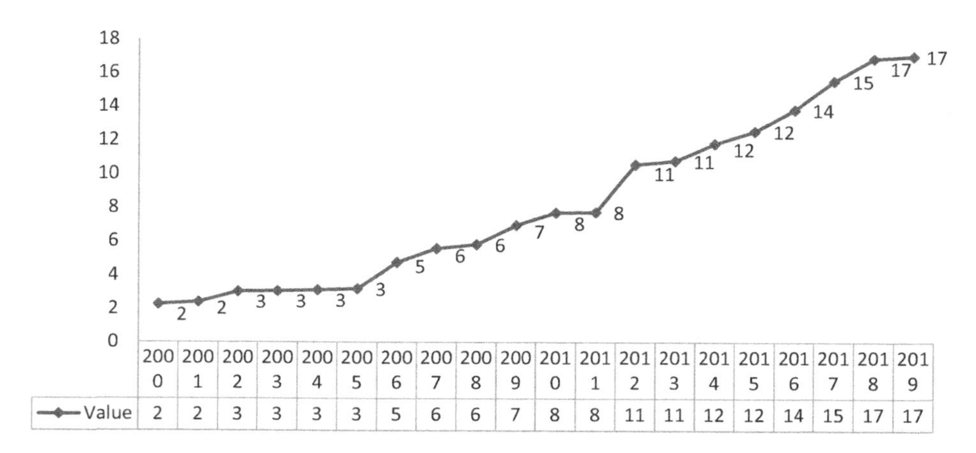

	2000	2001	2002	2003	2004	2005	2006	2007	2008	2009	2010	2011	2012	2013	2014	2015	2016	2017	2018	2019
Value	2	2	3	3	3	3	5	6	6	7	8	8	11	11	12	12	14	15	17	17

Figure 25.2 Percentage coverage of protected areas in relation to marine areas (exclusive economic zones).

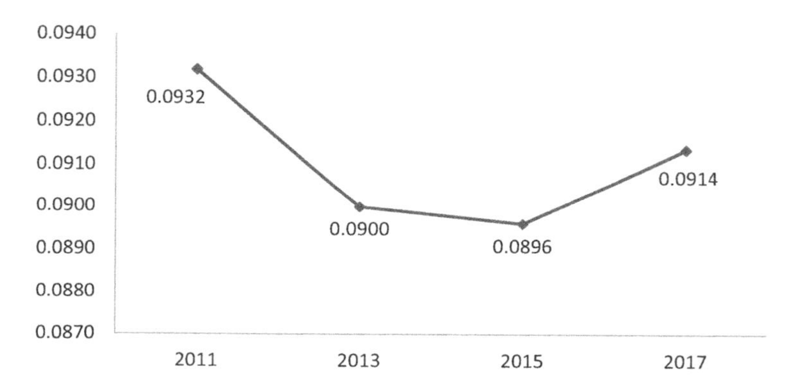

Figure 25.3 Sustainable fisheries as a percentage of gross domestic product at the world level.

Cormier and Elliott (2017) called for a review of SDG14 targets to extract and outline quantifiable 'SMART' (specific, measurable, achievable, realistic, and time-bounded) objectives, especially when it comes to developing national regulatory frameworks. According to Cormier and Elliott, indicators for SDG14 can be mapped using the 'DAPSI(W)R(M)' framework, which represents cause-and-effect pathways of risk whereby 'Drivers' generate 'Activities' that result in 'Pressures' as the mechanisms of adverse change. The latter can then change the 'State' of the natural environment causing 'Impacts' (to human welfare, including well-being) (Elliott et al., 2017). Following this, 'Responses' are required, representing 'Measures' that are needed to manage Drivers and Activities to reduce the likelihood that the Pressures will change the State that can cause Impacts (to human Welfare) (Elliott et al., 2017: 29–30). Cormier and Elliott (2017: 30) argue that the prevailing indicator framework focuses on 'state' and 'impact' and does not adequately cover 'drivers', 'activities' or 'pressures', and that this is one of the gaps in monitoring SDG14 indicators. In short, the SDG14 GIF is inadequate as it does not cover the root causes of oceans and global environmental change.

Furthermore, SDG14 and targets in their present form are not based on principles of fair sharing of benefits and burdens flowing from the ocean, nor do they address underlying

Table 25.3 Custodian and partner agencies for global monitoring of SDG14

Issues and SDG targets	*International agencies*
Marine pollution (Target 14.1)	UNEP IOC-UNESCO IMO FAO
Ecosystem-based approaches (Target 14.2)	UNEP IOC-UNESCO FAO
Ocean acidification and scientific cooperation (Target 14.3)	IOC-UNESCO UNEP
Unsustainable fishing (Target 14.4)	FAO
Conservation in marine areas (Target 14.5)	UNEP-WCMC UNEP IUCN Ramsar
Subsidies and overfishing (Target 14.6)	FAO
Economic activities in small island developing states and least developed countries (Target 14.7)	FAO UNEP-WCMC
Scientific knowledge and research (Target 14.a)	IOC-UNESCO UNEP
Small-scale artisanal fishers (Target 14.b)	FAO
Legal framework (Target 14.c)	UN-DOALOS FAO UNEP ILO UN-Ocean

Acronyms
IUCN: International Union for Conservation of Nature
IMO: International Maritime Organization
DOALOS: Division for Ocean Affairs and the Law of the Sea
FAO: Food and Agriculture Organization
UNEP: United Nations Environment Programme
ILO: International Labour Organization
IOC-UNESCO: Intergovernmental Oceanographic Commission of UNESCO
UNESCO: United Nations Educational, Scientific and Cultural Organization
UN-WCMC: United Nations Environment Programme World Conservation Monitoring Centre

Source: Based on UN (2021a).

causes of inequality in the ocean economy (Armstrong, 2020). The SDG14 monitoring framework also does not seriously engage with climate change issues, such as ocean acidification, and sea-level rise. For economic activities, the focus of SDG14 is on fishing, aquaculture, and tourism. The present monitoring framework excludes many important economic sectors, such as minerals, shipping, energy, and marine genetic resources. Access to capital and technology is also ignored; it is assumed that research alone will solve the problem of capital needed by countries. Fiscal instruments and legal or regulatory policies can be combined to achieve a specific balance of distributional outcomes (Steinbach et al., 2016). However, the focus of SDG14 is on subsidies rather than other policy instruments, such as spending on programmatic interventions. The framing of targets and indicators further reinforces the fragmented institutional context, which impedes holistic governance. Given the complexity involved in monitoring SDG14, especially at the global level, indicators need to be disaggregated for national jurisdictions as well as for ABNJ.

The lack of technical capabilities and resources

Data on SDG14 implementation indicate that 'Responses' are not translating to changes in the 'State' of the oceans. Thus, for example, the mere implementation of national policies, such as the increase in MPAs in EEZs, has not resulted in sustainable fisheries. This points to the lack of technical foresight as well as resources for decision-making for SDG14. Effective decision-making needs to be informed by evidence. There is a need for strengthened data-reporting systems to understand environmental as well as socio-economic impacts on the oceans, along with synergies and trade-offs between different marine policies (Mohammed, 2017). Ocean models are key to better overall climate change predictions. Scholars have identified the need for improved Arctic research, which is presently impaired by the limited ability to conduct observations (Sturesson et al., 2018). Technical capacity needs to be strengthened to address knowledge and data gaps concerning pollution, marine debris, plastics and micro-plastics, and hazardous substances (Recuero Virto, 2018). Companies also need to better incorporate 'downstream' and marine impacts into environmental assessments by considering the connectivity of ecosystems from watersheds to the ocean floor (SDSN, 2016).

Marine governance needs to holistically consider the land–sea nexus of interactions, and the various processes impacting on coastal zones, including the human dimensions (Neumann et al., 2017). Scholars have proposed strengthening MSP to operationalize ecosystem approaches by reconciling concepts of marine ecosystem services and facets of human well-being (Ntona and Morgera, 2018). Silo-based decision-making, often irrespective of ecosystem-dynamics and meaningful ecological boundaries, still prevails in many cases (Schmidt et al., 2017). The assessment of the conditions of fishermen and the capacity of RFMOs have to be strengthened in accordance with the ILO Convention No. 188 on labour standards for fishing vessels (Haward and Haas, 2021). Along with national organizations, the capacity of regional organizations needs to be strengthened so that they can better integrate social goals alongside environmental goals so that the interconnections between SDG14 and other SDGs can be better addressed.

Conclusion

Global environmental change and unsustainable marine activities present a challenge to realizing SDG14. Realizing SDG14 implies addressing root causes of global environmental change. Management approaches and tools for SDG14 require a holistic interdisciplinary approach that cuts across the natural and social sciences. The role of data collection using remote sensing and satellites is increasingly going to become important. To monitor fishing activities using the DAPSI(W)R(M) framework, automatic identification systems and vessel-monitoring system data can be used on a larger scale. There is also a need to strengthen implementation by building capacity at all levels, including the international, regional, national, and sub-national levels, and in coastal communities. A shared vision and common understanding for ocean governance, which is based on agreed scientific approaches and criteria, is essential to fill in ocean governance gaps. Further, states need to implement recommendations individually and collectively. Widespread adoption of science-based ecosystem approaches to ocean management, while upholding precautionary principles and advancement of knowledge about oceanic activities, will maximise healthy and safe oceans. Ocean literacy, involving the younger generation in activities, and widespread rollout of community-based programmes for ocean restoration are some of the measures that can be taken. Public availability of information, transparent and open decision-making processes, promotion of sustainable and equitable use of oceanic spaces, and the bearing of responsibility by states to protect the global marine environment, are some of the possible ways ahead. While realizing all of these objectives is unlikely, working toward them will make marine environmental governance in general, and implementation of SDG14 in particular, more likely even as global environmental changes become more significant.

Acknowledgement

The authors thank Jebah David for language support in terms of copy editing.

References

Armstrong, C. 2020. Ocean Justice: SDG14 and Beyond. *Journal of Global Ethics*, 16, 239–255.

Cormier, R., Elliott, M. 2017. SMART Marine Goals, Targets and Management – Is SDG14 Operational or Aspirational, is 'Life Below Water' Sinking or Swimming? *Marine Pollution Bulletin*, 123, 28–33.

Elliott, M., Burdon, D., Atkins, J. P., Borja, A., Cormier, R., De Jonge, V. N. and Turner, R. K. 2017. "And DPSIR begat DAPSI(W)R(M)!" - A Unifying Framework for Marine Environmental Management. *Marine Pollution Bulletin*, 118, 27–40.

EU. undated. *International Ocean Governance* [Online]. European Union. Available: https://ec.europa.eu/oceans-and-fisheries/ocean/international-ocean-governance_en [Accessed 10 July 2021].

Haward, M. and Haas, B. 2021. The Need for Social Considerations in SDG14. *Frontiers in Marine Science*, 8, 1–6.

HLPSOE. 2021. *High Level Panel for a Sustainable Ocean Economy* [Online]. World Resources Institute. Available: https://www.oceanpanel.org/ [Accessed 10 July 2021].

IPCC. 2019. *Special Report: Special Report on the Ocean and Cryosphere in a Changing Climate* [Online]. Intergovernmental Panel on Climate Change. Available: https://www.ipcc.ch/srocc/chapter/summary-for-policymakers/ [Accessed 10 July 2021].

IUCN. undated. *International Ocean Governance* [Online]. International Union for Conservation of Nature. Available: https://www.iucn.org/theme/marine-and-polar/our-work/international-ocean-governance [Accessed 10 July 2021].

Kedia, S. and Gautam, P. 2020. Blue Economy Meets International Political Economy: The Emerging Picture. *Maritime Affairs: Journal of the National Maritime Foundation of India*, 16, 46–70.

Mohammed, A. J. 2017. Mobilizing the Global Community to Achieve SDG14. *UN Chronicle*, 54, 1–1.

Neumann, B., Ott, K. and Kenchington, R. 2017. Strong Sustainability in Coastal Areas: A Conceptual Interpretation of SDG14. *Sustainability Science*, 12, 1019–1035.

NOAA. 2021a. What is the Carbon Cycle? [Online]. National Ocean Service, National Oceanic and Atmospheric Administration. Available: https://oceanservice.noaa.gov/facts/carbon-cycle.html [Accessed 10 July 2021].

NOAA. 2021b. How much Oxygen Comes from the Ocean? [Online]. National Ocean Service, National Oceanic and Atmospheric Administration. Available: https://oceanservice.noaa.gov/facts/ocean-oxygen.html [Accessed 10 July 2021].

NOAA. 2021c. What is Blue Carbon? [Online]. *National Ocean Service, National Oceanic and Atmospheric Administration*. Available: https://oceanservice.noaa.gov/facts/bluecarbon.html [Accessed 10 July 2021].

Ntona, M. and Morgera, E. 2018. Connecting SDG14 with the other Sustainable Development Goals through Marine Spatial Planning. *Marine Policy*, 93, 214–222.

Recuero Virto, L. 2018. A Preliminary Assessment of the Indicators for Sustainable Development Goal (SDG) 14 "Conserve and Sustainably Use the Oceans, Seas and Marine Resources for Sustainable Development". *Marine Policy*, 98, 47–57.

Schmidt, S., Neumann, B., Waweru, Y., Durussel, C., Unger, S. and Visbeck, M. 2017. SDG14-*Conserve and Sustainable Use the Oceans, Seas and Marine Resources for Sustainable Development. In A Guide to SDG Interactions: From Science to Implementation.* International Council for Science (ICSU). Available: https://council.science/wp-content/uploads/2017/03/SDGs-interactions-14-life-below-water.pdf [Accessed 21 February 2022].

SDSN. 2016. *SDG14- Life Below Water. Mapping Mining to the Sustainable Development Goals.* New York: Sustainable Development Solutions Network.

Soares, M. D. S. 2017. Achieving SDG14: The Role of the United Nations Convention on the Law of the Sea. *UN Chronicle*, 54, 1–1.

Steinbach, D., Mohammed, E. Y. and Steele, P. 2016. *A Sustainable Future for Fisheries: How Fiscal Policy can be Used to Achieve SDG14.* London: International Institute for Environment and Development.

Sturesson, A., Weitz, N. and Persson, Å. 2018. *SDG14: Life Below Water. A Review of Research Needs. Technical Annex to the Formas report Forskning för Agenda 2030: Översikt av forskningsbehov och vägar framåt.* Stockholm: Stockholm Environment Institute.

UN. 2015. 70/1. *Transforming our World: The 2030 Agenda for Sustainable Development, United Nations General Assembly Resolution.* New York: United Nations.

UN. 2017. 71/313. *Work of the Statistical Commission Pertaining to the 2030 Agenda for Sustainable Development, United Nations General Assembly Resolution.* New York: United Nations.

UN. 2020. *Life Below Water: Why It Matters?* [Online], New York, United Nations. Available: https://www.un.org/sustainabledevelopment/wp-content/uploads/2019/07/14_Why-It-Matters-2020.pdf [Accessed 10 July 2021].

UN. 2021a. *Tier Classification for Global SDG Indicators as of 29 March 2021.* New York: United Nations.

UN. 2021b. *Global SDG Indicators Database* [Online]. New York: United Nations. Available: https://unstats.un.org/sdgs/indicators/database/ [Accessed 10 July 2021].

Vierros, M. 2017. Global Marine Governance and Oceans Management for the Achievement of SDG14. *UN Chronicle*, 54, 1–1.

WOA. 2021a. *The Second World Ocean Assessment. World Ocean Assessment II, Volume I.* New York.

WOA. 2021b. *The Second World Ocean Assessment. World Ocean Assessment II, Volume II.* New York.

26

ETHICS, JUSTICE, AND HUMAN RIGHTS

Normative considerations in marine environmental change

Konrad Ott, Colin von Negenborn, and Nele Matz-Lück

Human interferences with ocean life and specific marine systems have sharply increased in recent decades, rightly being dubbed the 'great acceleration'. In *The Sea Around Us*, Rachel Carson wrote in 1951 about 'man': 'He cannot control or change the ocean as, in his brief tenancy of earth, he has subdued and plundered the continents' (1951: 20). Retrospectively, this quotation echoes the final illusion of planetary infinity. Yes, we can change the state of the ocean – for better or worse. Without doubt, humankind produces global environmental change within the marine realm as well. The Anthropocene has reached the ocean. Such a change implies paramount ethical challenges. In recent years, there have been many pleas for ocean literacy, ocean stewardship, ocean sustainability, ocean recovery, ocean ethics, and ocean health (Franke et al. 2020, Blaustein 2016, Visbeck et al. 2014, Mengerink et al. 2014, Auster et al. 2008), all of which implicitly rest on moral and ethical grounds. The ethical grounding itself, however, remains underdetermined.

In this chapter, we argue that normative considerations are unavoidable and implicitly present in any framing of the human-ocean relationship. Furthermore, we consider claims for justice in the context of global environmental change in and around the ocean. Crucially, we identify the moral responsibility of and towards collectivities of humans regarding ocean preservation and safeguarding against marine environmental change. We then show how such *pro toto* commitments can be implemented in legal practice. '*Pro toto*' means that a collective commitment refers to a larger good in its entirety (to 'all of it'). Such commitments cannot be derived from individual human liberty rights, as we shall argue. They hold with respect to collectives of humans (as 'future generations', 'Europeans') irrespectively of their composition (number, age, and individuality). Such commitments can be specified to moral and legal obligations. We argue that approaches to protecting the ocean based on human rights are insufficient for both conceptual and practical reasons. We discuss alternatives, such as the assignment of rights to humanity or to the ocean itself, and we show why those alternatives, too, fail to implement the obligations at hand. Instead, we call for a federal covenant of progressive states, pushing for reforms within the framework of the United Nations Convention on the Law of the Sea (UNCLOS) and following an agenda motivated by an environmental ethic and 'strong' sustainability.

DOI: 10.4324/9781315149745-31

Our ethical approach follows Habermas's framework of unconstrained practical discourse. It stands in contrast to approaches that adopt a constructivist approach to different framings of ocean governance (Hannigan 2016). We proceed by first describing the framework of marine environmental ethics and its axiological (value-based) and deontological (rights- and duties-based) dimensions, and then adding some principles of ocean justice. Subsequently, we bridge the gap between ethics and law, especially with regard to international law. We cast some doubts on the idea that 'ocean ethics' can be grounded in a theory of human rights. The final section illustrates how the law can help to substantiate these obligations. Finally, we argue for a political proposal of how humans may bring about both moral and legal progress in maritime affairs.

Groundings of ocean ethics

Ethics is reflective upon moral belief systems. Ethics distinguishes between axiology and deontology: axiology deals with values and protective goods, while deontology deals with rules, rights, and obligations. This distinction is of relevance here because any environmental ethic will be both axiological and deontological (Ott and Reinmuth 2021). Claims for justice in the light of global environmental change deal with the distribution of valuable goods according to some criteria and entitlements, thus presupposing axiological premises as well as normative ones. Any ocean ethics must address the different correlations between axiological and deontological significances.

Environmental ethics and values

To identify the 'right' course of action in the light of marine environmental change, one first needs to specify the criteria and theories according to which any action is to be evaluated. Here, three main lines of such ethical theories can be distinguished. *Consequentialists* focus on the net outcome of an action, aggregating all those affected by it. Doing good then amounts to choosing the action that furthers the particular objective the most, such as maximising overall happiness or minimising pain. In contrast, *deontologists* are concerned with the duties we have towards others, emphasising the preservation of – and infringement on – personal rights. They shift the focus from the outcome of an action to the rules governing the action. Finally, *virtue* ethicists ask about the character traits one exhibits by choosing an action, thus offering virtuous (or vicious) dispositions as a third criterion – besides consequences and rules – to assess our interaction with the marine environment. However, one does not necessarily need to commit to a specific ethical theory. Instead, one can maintain pragmatic pluralism within a liberal discourse-ethical paradigm. Such a paradigm insists on some basic implicit commitments of moral arguing but gives leeway for different kinds of substantial reasons by which complex matters might be properly addressed. Substantial reasons can be prudential, axiological, consequentialist, legal, moral, and even virtue-based. Since the discourse on ocean ethics is still in its infancy, we see such discourse as a 'laboratory of reasoning'. This chapter operates within this laboratory.

Anthropocentrism

An ethical consideration of the ocean requires us to specify an axiology, specifically to identify the relevant set of values. It is beyond doubt that the ocean provides many benefits to humans. Thus, the set of ocean values is richly textured and can be analysed axiologically or in terms of ecosystem services (supporting, providing, regulating, and cultural) (Ott and Reinmuth 2021). One should not underrate the discursive strength of enlightened environmental anthropocentrism (Ott 2016), which takes fully into account the many values and protective marine goods at different scales. Besides the question of nature's potential inherent moral value, nature also has *metabolic* and *reliant* value by providing goods and services essential to life. But we may also value the sea for non-instrumental reasons, such as due to its sheer beauty, the sense of belonging it provides, or the culture attached to it, all of which can be captured by its *eudaimonic* value (Ott 2016). Eudaimonic values can be categorised as recuperation and leisure, aesthetics (all senses); place-making and feeling at home, transformative (in terms of virtues), and resonance and inspiration. All types of values (reliance, metabolic, eudemonic) are present in coastal zones and at sea. They can be specified by empirical (economic) studies on 'cultural services' (contingent valuation, travel-cost analysis) or by cultural studies on ocean narratives, poems, paintings, and so forth. These values can also be, at least in principle, acknowledged by law. For example, the reference to the effects of pollution as a 'reduction of amenities' in the UNCLOS goes beyond measurable economic loss.

A deontological dimension is represented by *future ethics*, which ascribe value, such as to a fair legacy to later generations (bequest) and to an option of later availability (insurance). Legacies are complex bundles of goods to be bequeathed either to personal descendants or to future collectives, be they particular (as nations) or universal (as humankind). Although many marine policies operate on minor scales, the future ocean belongs to planetary goods as a 'common heritage of humankind' (for the conceptualisation of this heritage within environmental ethics, see Negenborn 2021; for its legal implications see van Doorn 2021). The crucial problem of determining fair collective legacies at different scales has an axiological and a deontological dimension. Both dimensions need to be addressed in concepts of sustainability. Commitments to future generations with respect to the ocean are twofold *pro toto*. On the one hand, the ocean is a single planetary entity, while, on the other hand, the chain of generations has no definite end, and one cannot know the individuality of future persons (Parfit 1984). The logic of twofold *pro toto* commitments being grounded in environmental values has rarely been properly conceived. Modern ethics, with its emphasis on individual rights, is at some pain to make planetary *pro toto* commitments fully explicit. We do not wish to conceive *pro toto* commitments in terms of a 'Gaian' ethics, referring to an obscure metaphysical entity and self-regulating system with its own teleology. *Pro toto* commitments do not refer to such entities but are relational and meaningful. Such meaningful relation must be actualised from the side of intelligible beings. The relation between humanity and the ocean is one in which intelligible terrestrial beings discover a large otherness full of life. The utilitarian side of the relation can be conceived in terms of ecosystem services. There are many studies that refer to different ocean services. The ocean provides food, regulates the global climate, and offers leisure, recreation, and beauty. Ocean relations, however, are meaningful beyond the side of ecosystem services. They are also existential and planetary (Chakrabarty 2015). To many people, attitudes such as awe and humility belong to this existential relation. *Pro toto* commitments are the best available ethical concept for such relations.

The term 'humanity' in this relation is not just an empirical term of reference for all members of our species in the past, present, and future; it is an ethical term. Being 'humane' means that one, being an embodied individual person, has access to a realm of axiological and deontological reasons and is, in principle, capable of shaping one's own life accordingly. The unity of humankind is twofold: physical and discursive. As discursive beings, humans are united under a realm of reasons which can be shared in common. Within this realm of reasons, there is the specific universe of discourse in environmental ethics by which values and commitments with respect to nature can be revealed (for an overview see Ott and Reinmuth 2021). As physically embodied and mortal beings, humans are united within their common habitat: planet Earth. Being terrestrial beings, humans must recognize that the ocean, in its otherness to terrestrial spaces, is part of the planetary habitat. Perhaps, this insight gains traction for the first time in history in the unprecedented era of global change.

In consequence, a collective entity called 'humankind' has a long-term responsibility against future instances of the same entity with respect to a huge global good called 'ocean' and its qualities and inhabitants. The principles governing such responsibility must become moral and legal traditions followed by each generation. The responsibilities must be shared in common and must be distributed according to different scales. Such responsibilities cannot be reduced to obligations against individual future members of humankind. Thus, it cannot be reduced to the familiar patterns of reasoning in moral and legal individualism. The question is: How can *pro toto* commitments be represented within international law?

The deontological dimension in environmental ethics also focuses on the demarcation problem: Should the category of inherent moral value be attributed to some (or all) natural beings, and, if so, to which ones? *Inherent moral value* assigns moral relevance to a being for a being's own sake, not because of its value for others. While anthropocentrism reserves this moral status for human beings, physiocentrism extends the moral community to include, for example, all beings capable of feeling pain or of experiencing themselves as subjects of a life. Biocentrism goes even further by assigning inherent moral value to all living organisms. The assignment of inherent value can take place both on the individual level and on the collective one. Approaches embracing the latter include ecocentrism, taking environmental collectivities such as entire ecosystems to be the relevant entity of moral concern (Callicott 1980), and holism, assigning inherent value both to all individuals and to all wholes (Gorke 2003). The discourse on inherent moral value in nature has resulted in the stipulation of intrinsic value of biological diversity in the preamble of the Convention on Biological Diversity (CBD). While not binding as such, the preamble guides the interpretation of the legal obligations and bridges moral value and law.

Extending moral significance beyond the anthropocentric realm does not necessarily imply equal moral status of all members of the community. One can also embrace a gradualist instead of an egalitarian position, allowing for the prioritisation of the needs or rights of certain members of the moral community over others. In marine ethics, moral grading allows recognition of the heterogeneities within marine life, ranging from plankton over jellyfish to marine mammals. Gradualism, for example, would not set individual sardines on a par with individual whales. The demarcation problem implicitly plays a role in the legal and political dispute, for example on whaling and killing seals for fur. It remains, however, questionable, whether the demarcation problem can become a focus in the law of the sea beyond the protection of 'higher' marine animals. Thus, a somewhat sceptical result emerges: How can intergenerational *pro toto* commitments and inherent moral values for nature be represented within the law? The combination of the many values (axiology) and *pro toto* commitments against future generations (deontology), augmented with animal welfare considerations for (higher) sentient beings, can serve as a sound and robust ethical framework for marine environmentalism. It can be applied to topics as diverse as fisheries, pollution, ocean warming, ocean acidification and so forth. If represented in law, such building blocks can also serve as a common moral ground for political and legal negotiations.

Dimensions of justice

The multitude of values in the marine environment and the threat to these values posed by environmental change give rise to many, often conflicting, claims for justice. First, there are claims for distributive justice. Such claims need to specify *scope* (who are the claimants and addressees of these claims?), *pattern* (what is the criterion of entitlement?) and *currency* (which goods or units are to be distributed?). Principles of environmental justice are (1) democratic procedures, (2) respect for human rights, (3) rule of law, (4) global sufficientarianism (i.e., the Rawlsian principle of difference), (5) non-victimization, and (6) not to live at the expense of others, especially at the expense of poor and disadvantaged persons (Ott 2020). The scope of the principles is about current and future humanity, the pattern is sufficientarianism, and currency might be either welfare or capabilities. Even if the principles are sound, it remains open how to apply them to maritime affairs. Often, problems of application are more sophisticated than problems of justification. An overview of unfair and uneven outcomes in the distribution of marine resources is given by Bennett et al. (2021). That study is, however, based on a very simple concept of injustice. We question the authors' conceptualisation of (social) injustice as defined in the inspiring paper. In particular, the authors define 'social injustices as unequal distribution of benefits and/or burdens as well as unfair decision-making and governance processes' (Bennett et al. 2021: 2). We take two main issues with this definition. First, the definition is simply circular in that it defines those processes as *unjust* which are *unfair*, leaving the dimension of *procedural* justice ill-specified. Second, concern for the *distributive* dimension of justice does by no means necessarily imply an *equal* distribution of benefits or burdens. Quite the contrary, ethical considerations (e.g., past actions, geography, and climate) may mandate uneven distributions to account for (marine) claims for justice. The equation of equality and justice is rejected in almost all theories of justice. Progress in marine justice could, however, be achieved if the case studies in Bennett et al. were reinterpreted by the six principles just mentioned.

A similar ethical shortcoming is to be found in Widener (2018, p. 3) who 'polarizes' marine justice according to the following question: 'Is the sea a natural, public space open to all and protected by all for all species, or is it a private pocket controlled or contained for the economic interests or pleasures of a few?' Widener proposes to understand conflicts over marine resources as located 'along a continuum of the two polarized ends' (2018, p. 3). One end is a (wishful, good) ecocentric-care-communist ideal (based on the three all-quantifiers 'to all', 'by all', and 'for all'), while the other end is (repugnant, bad) commercialisation of marine resources in favour of profits. Such polarisation is a highly generic and simplifying scheme which implicitly conveys morally loaded value judgements on either end, without a normative grounding of such judgements. No reason is given for the polarisation itself and for its generic adequacy to marine justice topics.

Some of the many problems for distributive marine justice in the ocean include the following: (1) Natural heterogeneities: There is no equal access to the ocean for all humans and no equal rights over the exploitation of marine resources because some nations are land-locked and exclusive maritime zones are dependent upon the length of coasts; (2) Inaccessibility of the ocean floor for most humans; (3) Technology: Only some nations have, for instance, options for ocean mining (Should they share the benefits of ocean mining with other nations less equipped?); (4) History, with some countries having been naval powers for centuries; (5) Encroaching long-term changes, such as ocean acidification (Böhm and Ott 2019), by which no particular person is directly victimized; (6) Empirical problems of how to identify economic and political structures which make one unintentionally live at the expense of the poor (as artisan fisheries) even if one wishes to follow the principle. Notwithstanding these pitfalls, the six principles may serve as guidelines for orientation and 'eye openers' for debates on ocean justice. States are political units being capable to institutionalise principles of justice. As ocean justice reaches beyond single states, the very idea of global environmental justice must, with necessity, ask for international legal authorities (more on this below).

Sustainability

In the context of marine environmental change, all dimensions of ethics and principles of justice have a temporal dimension, linking today's actions towards nature with our obligations towards future generations and asking how precisely we ought to preserve the environment for them. Different notions of sustainability provide different answers. A *weak* notion of sustainability accepts the substitution of natural capital by human-made capital. For example, declining stocks of certain marine species can be compensated for by new technological means of providing proteins and oxygen. A *strong* notion, on the other hand, mandates a 'constant natural capital rule', preserving natural capital and its constituents in order to maintain a constant flow of all ecosystem service indefinitely (Ott 2012). The most prominent approach to conceptualize sustainability is based on the UN's sustainable development goals (SDGs; see Chapter 25). It seems safe to argue that SDGs on climate, biodiversity, and oceans implicitly presupposes a strong notion of sustainability (Neumann et al. 2017). The metaphor of 'ocean health' also implies the idea that to reach and keep a good state of the ocean is a substance of global natural capital (Franke et al. 2020). Therefore, it is important to analyse how such notions of sustainability can be translated into the realm of law.

To summarise our argument so far: Bringing together the normative considerations of marine environmental ethics discussed above, we can substantiate the existence of *pro toto* commitments: If a community of discourse ('we') (a) gives weight to the many supporting, provisioning, regulating, and cultural services of the ocean (in the axiological dimension), and if it (b) adopts strong sustainability as a normative concept on a global scale out of intergenerational justice, and (c) adds some gradual inherent value for sentient ocean life), and if such community (d) recognizes that SDG 14 (see Chapter 25) is implicitly grounded in strong sustainability, and if it can and should interpret the metaphor of 'ocean health' in terms of productivity, resilience and diversity, and if such 'we' e) adds some principles of environmental justice (Ott 2020), then commonly shared premises are sufficient to ground a collective *pro toto* commitments with respect to the ocean. After having identified some crucial building blocks for an ocean ethics, we turn to the presumptive translation of these building blocks into law.

On the correlation of ethics and law in the context of the ocean

Ethics can make the dialectical relation between morals and law more explicit. On the one hand, the law should be independent from the many moral beliefs and moral-political doctrines which find support in pluralist societies. The law must *pretend* such independency. To Stanley Fish (1994), moral doctrines and partial interpretations are seen as two major threats to law's independency. A 'moralized' law would compromise its stability and credibility by favouring one of the many beliefs present in society. On the other hand, the law is never independent from common moral/legal sense, from the discipline of 'legal ethics' or from legislative policy making. Generally, there is a triangle of morals, law, and politics within which claims, demands, rights and obligations are oscillating. Rule of law should be based on democratic self-governance. The open question remains as to how such a triangle might operate on the trans- and international level. To put this question in other words: Might there be equivalents to democratic self-governance in international legal regimes?

Common moral/legal sense consists of a stabile set of convictions about lawfulness having been formed over generations within a cultural lifeworld. Such universal moral sense has been called 'common decency' within which the acceptance of legal orders is rooted. From moral and legal traditions, however, we lack a common decency in maritime affairs since the ocean has been regarded as a largely ungoverned space over centuries (Schmitt 1950). The normative orders of land and sea differ deeply since Grotius' (1609) doctrine of the freedom of the seas (*mare liberum*). In essence, Grotius claimed that the seas, other than land territory, were too vast to be controlled by single states and their resources too plentiful to be depleted, so that navigation and fishing should be free and unregulated. Despite efforts to change such perceptions, we have not yet reached sustainable marine governance.

The issue of enforceability

A key difference between morals and law lies in the enforcement of behaviour and sanctions. Morals are about blaming wrongful behaviour. They lack the power to enforce behaviour by means other than arguing, stimulating bad conscience, and withdrawing recognition. As opposed to morals, law has the force to punish, albeit to differing degrees. Although modern law uses a large toolbox of direct and indirect incentives instead of enforceable top-down obligations, the availability of punishment still sets law apart from morals. The law requires conformity in behaviour but not noble motives, while morality is also interested in motives and attitudes ('conscience'). While the law prefers clear-cut concepts of specific delicts (violations) and clarity with respect to its basic distinction of 'permitted versus forbidden' – and has to do so in accordance with the rule of law, e.g. if individuals are sentenced under criminal law – , morality also immerses into so-called 'thick concepts', as, for instance, 'nasty', 'repugnant', 'weird' and so forth.

In the context of global environmental change and international law, the question of enforceability arises. Public international law differs fundamentally from national legal systems. Instead of a hierarchical legal order and clear rules on enforceability of norms, it is organised horizontally on different levels and, due to the constant struggle to find compromise between sovereign states, it is overall more diffuse and vaguely worded. Despite obvious difficulties in ascertaining 'what the law is' within such fragmented legal order, the system with its leeway for specialisation may, however, be more responsive and flexible to change, if political will and consent by states allow. It is a particularity of public international law that the norm-makers are at the same time the creators and (often sole) addressees of norms. Moreover, the creation of legally binding obligations, including mechanisms for enforcement and dispute settlement, is based upon consent. Consent to a binding obligation does not automatically include consent to jurisdiction of an international court or tribunal. States remain the principal legal actors, although the relevance of international organisations and the individual has increased.

These particularities have raised the question whether public international law is 'law' at all. While this has been affirmed in the legal discourse of international law, comparisons to national legal orders often lead to the conclusion that international law is of a 'soft' character. However, 'soft law' as a term is used in international law to characterise norms that are not legally binding, such as recommendations by international organisations or codes of conduct.

Despite the lack of enforceability known from 'hard law', soft law may still reach the same effect to change behaviour if states consent to these norms and transform them into national law and policy. If states comply with the rules, they act as if there is 'hard' law. Since law in general only demands proper behaviour, states may overcome the hard-soft divide if they deliberately act in accordance with the rules. More importantly, they can enact enforceable hard law on the national level to sanction breaches by individuals. For example, by virtue of their sovereignty, they can create an enforceable national rule for fisheries under their jurisdiction that originally stems from the non-binding international Code of Conduct on Responsible Fisheries.

International environmental law

International environmental law is a prime example for the perception of being 'soft' in many respects, even if enshrined in binding treaties. International environmental law often contains obligations of conduct rather than result, which leaves ample room for interpretation. Clear standards or targets and timetables are the exception rather than the rule, and the search for compromise has led to formulations, such as in the Biodiversity Convention, whereby obligations shall be fulfilled 'as far as possible and appropriate'. The step from framework conventions to more specific protocols to implement the overall objectives is, for instance, often hampered by lack of ratification and accession by states to the latter, if protocols define stricter measures for enforcement.

The axiological-deontological divide resurfaces in international environmental law. The environment, and particularly the ocean, is acknowledged as a highly valuable entity, but how should it be regulated? Eudaimonic values, as in the beauty of nature, biophilic attitudes, transformative values, or place-making practices remain, despite some stipulations, rather alien to the law. Similarly, the demarcation problem also remains puzzling in the legal realm. Animal ethics can be, at least in part, incorporated into the law, but biocentrism and ecocentrism cannot. One may try to do so by giving natural entities the status of legal persons, as Stone proposed to do (Stone 2010). We will discuss below why considering nature or parts thereof – such as the ocean – as legal entities, is prone to conceptual and practical difficulties. Persons who are dealing with the law of the sea should recognise that the law as such might underrate and perhaps even misrepresent the ethical values of ocean life. If so, the law of the sea should be enabled to correct its own internal deficiencies with respect to the axiological dimension by means of law. This is a genuine political task for agency on behalf of the ocean within UNCLOS. We now turn to this question and ask how the translation from axiology to law can succeed.

From normative obligations to legal codes

In the previous section, we identified a gap between the ethical dimension and its legal implementation. In this section, we discuss potential remedies for this gap.

Human rights and the ocean

As Hannah Arendt argues, persons 'have' rights (in practical terms) if and only if there is a legal order which secures and fulfils such rights (1968). Having a right without structures to effectively guarantee this right is indeed an empty shell. Without protective legal orders, rights are nothing but moral entitlements. If so, there is a moral reason to institutionalise a set of fundamental rights and freedoms in constitutional law. One idea to integrate environmental ethics into a human-rights framework is to expand (a) the set of the human rights so as to include rights to environmental qualities and (b) apply such set also to future generations. The so-called second and third generations of the human rights beyond individual liberty rights and political rights are material endowment rights and collective rights. They include a right to food security, a right to freshwater, a right to education, medical services, and so forth but also rights to sustainable development for a people not for individuals as in the African Charter for Human and Peoples Rights. The set of material endowment rights might be expanded to a presumptive 'right to the ocean'. Such expansion looks, however, strange to our moral beliefs. Düwell and Bos see it as an open question which environmental goods 'are so important for human beings that we should take them to have a *right* to these goods' (2018: 18, emphasis in the original). Grounding sustainable management of the environment – and the ocean in particular – in the ability to exercise human rights leaves the extent of such sustainability demands severely undetermined: a rather deteriorated ocean might well suffice to exercise one's human rights.

We wish to cast doubt on the idea that an individual human right to the ocean, including its ecosystem services and amenities, deserves recognition. A warmer, polluted, eutrophicated, more acidic ocean will still be the ocean. Even if there might be more plastic than fish in the ocean, it remains doubtful from a legal perspective whether such an ocean violates or impairs a presumptive right. How can we qualify a presumptive right to the ocean? Should we speak of a right to a 'natural', 'healthy', or 'decent' ocean, for example? This question needs to be discussed separately from individual rights to damages if a private entity or the state, due to a violation of due diligence obligations, is liable for the loss of income from fisheries or tourism in case of a particular disaster with large-scale marine pollution.

Right-claiming is a specific discourse. If a right is claimed, one can either agree to or deny such a claim. A denial of specific rights (as a right to carry weapons in public) is perfectly compatible with a defence of another set of human rights. Is there a derivative right to affordable energy, paid labour, urban life and so forth? A right to whale watching for leisure or to whale hunting for subsistence? If a right to the ocean would include the abyss and the seafloor, a *reductio ad absurdum* looms. If there is a human right to food (that would also be linked to SDG 2), does that include a right to marine food from a sustainably managed ocean? We tend to argue that there is no right to marine food. All things considered, we do not see any moral and legal reasons to expand the set of the human rights with respect to the ocean.

Beyond these conceptual difficulties, a more practical issue remains. Human rights and fundamental rights in national legal orders give rights to a natural or, for some rights, juridical person against the state. The rights-bearer is an entity that is present in the sphere of jurisdiction of the state. A human right to the ocean as such, or to a healthy or sustainably managed ocean, does not fit well into this structure. One would need to construe that the national activities of a state in its marine realm or behaviour beyond its territorial national jurisdiction, such as on the high seas, which is attributable to the state, harms the human rights of this state's population. While human rights apply on the territory of states, including territorial waters of the ocean, a general extraterritorial application of human rights on or in regard to the high seas and deep seabed would thus break with the current structures of human rights' scope of application. Recognizing such limitation is not a denial of human rights. It rather takes human rights seriously by not applying and expanding them to any matter at stake. It maintains the necessary clarity of who is responsible for securing human rights in relation to individuals on what territory. Favouring a human rights approach to ocean protection would also reintroduce the issue of enforceability discussed above. A human rights approach that is not backed by strong institutions does not offer an effective toolbox that would overcome the deficiencies of state-cantered international law and policy on sustainability and marine environmental protection.

The ocean as a legal entity

The insufficiency of a human rights approach to global environmental protection is not mitigated by understanding such rights as obligations *erga omnes*. Obligations *erga omnes* in the context of human rights relate to the recognition that all states have a legal interest in their protection and owe respect to human rights vis-á-vis the international community as a whole. This community, however, is not a subject of international law and is not an equivalent to 'humanity' as a community of people. Instead, it translates into 'all other states'. This leads to a complex situation of rights and obligations on different levels. States owe compliance with specific human rights towards an individual but also towards all other states; other states have a right that human rights towards an individual in another state be complied with against that state (Dinstein 1992). Obligations *erga omnes* do not create new obligations in substance or new addressees with regard to communal interest of 'the international community' in sustainable oceans. The human rights approach remains focused upon what the state owes individuals with or without means for other states to claim compliance in this relationship between person and state via inter-state dispute settlement or retaliation and reprisal. The popular understanding of the international community as a community of states which is bound together by international law is severely limited and has been described as a circular definition (Simma and Paulus 1985): the community of states is seen as the legal community of international law, and this very legal community in turn refers to the states composing it. Even with contemporary, richer concepts of the international community, the question remains how a more communitarian, more diversely institutionalized international law is reflected by concepts for ocean governance to give effect to *pro toto* commitments. Public international law is still state-centred to the extent that neither the interests of the international community, which are formulated independently from a community of states, nor the interests of the ocean are yet enshrined in particular rights and obligations. Moreover, concepts of sustainability and interests of future generations would rely upon states to be actively introduced and translated into law.

At first glance, the concept of a 'common heritage of humankind' appears to capture such *pro toto* commitments. Designating the deep seabed and its resources as heritage at least by name bears resemblance to a shared interest in a common good. From a legal perspective, this term comes closest to 'the international community' or 'humanity' as a legally relevant entity. In its current form, however, this concept is limited to exploring and exploiting the deep seabed and a sharing of benefits from economic gain via states rather than the basis for stewardship for the sake of the ocean as a whole and humanity as a whole. It is far from being adapted to other resources in areas beyond national jurisdiction or the ocean as such, as originally envisaged by Arvid Pardo in his famous speech at the UN General Assembly (United Nations 1967). Currently, the law of the sea does not involve mechanisms to give effect to an international community that goes beyond the state. We will discuss a possible pathway below. (For more on provision of the law of the sea, see Chapter 2).

Instead of considering humanity as a potential rights-holder, one might instead suggest giving rights to the ocean as a legal person. The idea was outlined by Stone (2010). It is clearly based on legal positivism: The law is seen as a toolbox and one device within the box is 'status of a legal person'. Legal personhood for natural features has been granted in some national legal orders, for example for rivers in India and New Zealand, but not for the ocean as such. On the international level the consideration of legal personality for the ocean on the same level of states and international organizations would require settlement of the question of representation. If the ocean were to have a seat at international conferences or in international institutions, who would be entitled to formulate the interest and needs of the ocean, and by which standards? Could a human being be capable of formulating what is best for the ocean independently from anthropogenic considerations? Would ocean ethics and ocean justice require a particular stance in this respect? Analysing presumptive legal advocacy on behalf of the ocean expands the discourse-ethical framework beyond its current reach. In any case, conceptual difficulties are legion.

Civil society

Dropping the reliance on a binding legal framework and calling for a stronger role of non-state actors gives in to the problem even further (see Chapter 9). Admittedly, the UN Ocean Conference has made a significant contribution to such a pathway. Rather than investing in efforts to create further legally binding documents, it called for contributions from civil society globally. However, the actions of civil society are often neither concerted nor coherent. Although the legal path from codification, via implementation and compliance by states, to the level of regulation of individual behaviour is a lengthy one, it must not be given up so easily. One should, however, not overburden civil society and its organizations. If conferences of states fail to address problems of global environmental change, one cannot delegate problem-solving to nongovernmental organizations.

Ethical preamble

The problem of enforceability of international law between states in a human rights context similarly arises – and even in a more severe form – if we focus on enshrining environmental values in legal codifications explicitly. Codification might take the form of preamble-based clauses, as seen, for example, in the CBD. Contenting ourselves with such a solution would, however, commit us to non-enforceable legal codes since preambles are 'soft' parts of conventions, although they gain relevance to guide interpretation of a treaty. A preamble by which the axiological and deontological dimensions of ocean ethics are clearly outlined may serve as a common political grounding for how to address related global environmental change.

Putting obligations into practice

With the conceptual and practical difficulties of the approaches discussed so far, we now present a more promising alternative. To do so, we recall the existing international institutional and legal framework upon which we want to build. When the United Nations was founded toward the end of World War II, states did not add the protection of the environment to the goals to be achieved. The focus clearly was on peace, security, and human rights. Today, the Security Council and UN General Assembly acknowledge that environmental threats also threaten peace and security. In the meantime, the necessity of global environmental protection has resulted in hundreds of bilateral and multilateral treaties. During the last 30–50 years, multilateral treaty regimes of a global scale have shaped conglomerates of norms of binding and non-binding character and decisions by international institutions, such as CITES, UNFCCC and CBD. These regimes are cumbersome processes into the right direction to transform *pro toto* commitments with respect to endangered species, the climate system, and biodiversity into legal commitments. These regimes evolve over time but still remain largely based upon state consent and political will.

The legal framework, dynamic or evolutionary interpretation of norms in the light of new scientific insights and political agenda setting by institutions, together with incentives for implementation are not only theoretical chances but can gain practical relevance. The idea to upgrade UNEP from a programme to a more independent 'UN Environment Organization' might be a crucial step to create more effective institutionalised regimes that are potentially strengthened in regard to decision-making and implementation.

The UNEO-idea, however, has been blocked and it remains rather a long-term target. In the shorter run, one should improve the evolution of single regimes from the inside.

Therefore, we wish to consider an additional option that brings together ethically mandated actions and concern for political viability. It relies on states whose representatives and citizenry accept the building blocks of ocean ethics (from a moral point of view) and (from a political point of view) see no alternative than to remain within the state-centered system and the international regime of UNCLOS. Such states, be they small island states (see Chapter 14) or larger coastal states (see Chapter 13), could unite as a federal covenant within UNCLOS, promoting legal reforms in the overall spirit of marine ethics, environmental justice, and strong sustainability. UNCLOS as a framework convention explicitly allows for agreements among states that go beyond the convention as long as they do not breach its binding commitments or contradict leading principles of public international law, for example that states cannot establish binding treaties for non-consenting third states (the *pacta tertiis* rule). This political idea of a covenant of progressive states harbours a grain of truth within virtue ethics because representatives of those states may negotiate maritime affairs out of specific attitudes that stem from, say, the notion of 'ocean ethics in the making'. Such agents can rely on a large literature (see Franke et al. 2021, Böhm and Ott 2019, Neumann et al. 2017, Roberts 2012, Dallmeyer 2003), including many prudent pleas for ocean literacy, ocean stewardship, ocean sustainability, ocean recovery, ocean ethics, and ocean health (Blaustein 2016, Visbeck et al. 2014, Mengerink et al. 2014, Auster et al. 2008).

The United Nations' ocean decade is a window of opportunity for such a political covenant of forerunner states pushing an agenda of reforms, as part of the sub-targets of SDG-14 (see Chapter 25), closer to law. Such forerunner states could constitute, promote, and establish a future of common decency with respect to the ocean. The covenant of states within UNCLOS would be open to all states, including land-locked ones. We are confident that small island states would want to take a lead, as they did with respect to SDG 14. Recent policy processes in the European Union make it likely that it would also join the covenant.

Conclusion

This chapter first identified substantial building blocks of an 'ocean ethics': axiology, deontology, *pro toto* commitments, demarcation problems, principles of justice, and strong sustainability. Second, the divide between morals and international law has been addressed. Third, some ideas of how to bridge the gap between morals and law (e.g., human rights, ocean as legal person, civil society) have been discarded. Instead, we suggest an approach within the framework of UNCLOS, where a preamble explicates the common moral ground, and a covenant of progressive states is guided by ocean ethics. Such states (and other entities, such as the EU) can and should act on their marine and coastal territories, thus representing a 'best practice' of national ocean governance. They might also act within bilateral treaties with other states, as the German Advisory Council on Global Change (WGBU) (WGBU 2020: Ch. 4.5) has proposed. UNCLOS may become a global regime on behalf of future ocean sustainability. The next steps towards such an agenda are implicit in SDG-14. Even at the risk of stating the obvious at the end of our chapter, such an agenda should include (a) a global initiative for ocean recovery and restoration; (b) an aggressive reduction of pollutants, including carbon dioxide, from agriculture, urban areas, and industries; (c) enlargement of marine protected areas; (d) protection of coastal zones; and (e) sustainable regulation of fisheries and aquacultures. Moving in this direction would help to promote ethics, justice, and human rights in marine governance amidst global environmental change.

Acknowledgement

The authors wish to thank Lilo B. Rösch for her helpful comments.

References

Arendt, H. (1968) *The Origins of Totalitarianism*. Harcourt, San Diego.

Auster, J. et al. (2009) 'Developing an ocean ethic: Science, utility, aesthetics, self-interest, and different ways of knowing', *Conservation Biology*, vol 23, no 1, pp233–235.

Bennett, N. J. et al. (2021) 'Blue growth and blue justice: Ten risks and solutions for the ocean economy', *Marine Policy*, vol 125, p104387.

Blaustein, R. (2016) 'United Nations seeks to protect high-seas biodiversity', *BioScience*, vol 66, no 9, pp. 713–719.

Böhm, F., Ott, K. (2019) *Impacts of Ocean Acidification*. Metropolis, Marburg.

Callicott, J. B. (1980) 'Animal liberation: A triangular affair', *Environmental Ethics*, vol 2, no 4, pp311–338.

Carsons, R. (1951) *The Sea Around Us*. Oxford University Press, Oxford.

Chakrabarty, D. (2015) 'The Human Condition in the Anthropocene', *Tanner Lectures in Human Values*. Yale University Press, New Haven.

Dallmeyer, D. (Ed.) (2003) *Values at Sea*. University of Georgia Press, Athens.

Dinstein, Y. (1992) 'The erga omnes applicability of human rights', *Archiv des Völkerrechts*, vol 30, pp16–21.

Doorn, E. v. (2021) *Legal Implications of the »Common Heritage« Principle for Atlantic Bluefin Tuna*, Duncker & Humblot, Berlin.

Düwell, M., Bos, G. (2018): Why '*rights*' of future people? in Düwell, M., Bos, G., Steenbergen, N. van (eds) *Towards the Ethics of a Green Future*. Abingdon, Routledge, pp9–27.

Fish, S. (1994) 'The law wishes to have a formal existence', in S. Fish (ed) *There is No Such Thing as Free Speech*. Oxford University Press, Oxford, pp141–179

Franke, A. et al. (2020) 'Operationalizing ocean health: Toward integrated research on ocean health and recovery to achieve ocean sustainability', *One Earth*, vol 2, no 6, pp557–565.

Gorke, M. (2003) *The Death of Our Planet Species*. Island Press, Washington.

Grotius, H. (1609) *Mare Liberum*, Elzevir, Leiden.

Hannigan, J. (2016) *The Geopolitics of Deep Ocean*. Polity Press, Cambridge/Malden.

Mengerink, K. J. et al. (2014) 'A call for deep-ocean stewardship', *Science*, vol 344, no 6185, pp696–698.

Neumann, B., Ott, K. Kenchington, R (2017) 'Strong sustainability in coastal areas: A conceptual interpretation of SDG 14', *Sust Scie* vol 12, pp1019–1035.

Ott, K. (2012) 'Variants of de-growth and deliberative democracy: A Habermasian proposal', *Special Issue: Politics, Democracy and Degrowth*, vol 44, no 6, pp571–581.

Ott, K. (2016) 'On the meaning of eudemonic arguments for a deep anthropocentric environmental ethics', *New German Critique*, vol 43, no 2, pp105–126.

Ott, K. (2020) 'Grounding claims for environmental justice in the face of natural heterogeneities', *DIE ERDE – Journal of the Geographical Society of Berlin*, vol 151, no 2–3, pp90–103.

Ott, K., Reinmuth, K.C. (2021) 'Environmental evaluation between economics and ethics: An argument for integration', in Hobohm, C. (ed) *Perspectives for Biodiversity and Ecosystems. Environmental Challenges and Solutions*. Springer, Cham.

Parfit, D. (1984) *Reasons and Persons*. Oxford University Press, Oxford.

Roberts, C. (2012) *Ocean of Life*. Allen Lane, London.

Schmitt, C. (1950) *Der Nomos der Erde: im Völkerrecht des Jus Publicum Europaeum*. Greven, Köln.

Simma, B., Paulus, A. (1998), 'The 'international community': Facing the challenge of globalization', *European Journal of International Law*, vol 9, pp266–277.

Stone, C. (2010) *Should Trees Have Standing? Law, Morality, and the Environment*. Oxford University Press, New York.

United Nations (1967) 'UN General Assembly, 22nd session: 1st Committee, 1516th meeting' *UN Doc A/C.1/PV.1516*.

Visbeck, M. et al. (2014) 'Securing blue wealth: The need for a special sustainable development goal for the ocean and coasts', *Marine Policy*, vol 48, pp184–191.

von Negenborn, C. (2021) 'A fuzzy ontology: on the relevance of ecocentrism in marine environmental ethics', *WMU Journal of Maritime Affairs*. https://doi.org/10.1007/s13437-021-00254-y.

WBGU. (2020) *Landwende im Anthropozän*. WBGU, Berlin.

Widener, Patricia (2018) Coastal people dispute offshore oil exploration: toward a study of embedded seascapes, submersible knowledge, sacrifice, and marine justice. *Environmental Sociology*, vol 4, no 4, pp405–418.

PART 6

Conclusion

27

PROSPECTS FOR MARINE GOVERNANCE IN THE ANTHROPOCENE

Portents from the climate regime

Paul G. Harris

The preceding chapters help to establish the need for more effective marine environmental governance in the Anthropocene, the current geological epoch during which humanity has been causing and will continue to cause dramatic changes to the global environment (see Thomas, Williams, and Zalasiewicz, 2020). That need increases as the causes and impacts of global environmental change experienced across oceans and seas gather pace. Taken together, the chapters paint a picture of the increasingly desperate need for governments, industries, and civil society actors to change the behaviours that are causing adverse marine environmental changes. They also point to both the lack of mechanisms for coping with those changes if they worsen greatly, as they surely will, and potential avenues for negotiating and implementing revised and new arrangements that may enhance effectiveness in the future.

The chapters that comprise Part 2 of this book reveal how and why established and nascent international institutions for marine governance are seldom up to the task of addressing the implications of global environmental change. For example, the international law of the sea, while well-developed, has gaps that do not account for global environmental change, or it is sometimes little more than aspirational due to a shortage of resources for enforcement. It often lacks coordination with other regimes, not least the climate change regime, that are intended to govern global environmental change but not specifically in the marine context. This is not to say that marine environmental governance is doing nothing to address global environmental change; the chapters show that much is indeed being done. But the pace and scale of action and coordination do not match the pace and scale of the growing challenge of global environmental change at sea. Chapters in Part 3 reveal the extent to which non-state actors, ranging from experts and nongovernmental organisations to fishers and investors, have governed marine environments affected by the impacts of global environmental change. Those chapters also reveal the limitations of those efforts.

Chapters in Part 4 highlight the challenges of global change for marine governance across a variety of environments and regions, whether along or near shores, across the deep oceans, or in regional seas. The issues those chapters look into vary greatly, but they more or less consistently reinforce the emerging theme of new marine governance schemes responding to the manifestations of global environmental change even as the causes and consequences of the

DOI: 10.4324/9781315149745-33

change exacerbate faster than associated governance measures. Similarly, as the chapters in Part 5 survey and scrutinise specific emerging issues in marine environmental governance – the scourge of plastic pollution, the growth in marine transport, efforts to realise sustainable development in the ocean context, and important considerations of ethics, justice, and human rights – they expose the challenge: global environmental change at sea often outpaces even the most earnest efforts to realise effective marine environmental governance.

The chapters help to point policymakers and stakeholders in new directions. But what are the prospects for taking those new paths? In this chapter, I very briefly draw on the experience of the climate change regime – by far the most important environmental regime for addressing global environmental change, including in oceans and seas – to develop a perspective on the prospects for marine environmental governance in the future. What lessons does the climate change regime reveal, and what might those lessons tell us about the prospects for marine environmental governance in the escalating Anthropocene? If lessons from the climate change regime can be learned by governments, industries, and other actors, greater attention is likely to be given to the challenges of marine environmental governance. If repeating the past can be avoided, there are more effective ways for catching up with global environmental change and, because vanquishing it completely is not possible, at least coping with it into the future.

Some milestones toward climate governance

Many of the marine environmental challenges examined in the preceding chapters, such as ocean warming and acidification, are direct ramifications of global warming specifically and climate change more generally. Many others, such as changes to marine ecosystems that make them less able to provide food to communities that relied upon them in the past, are exacerbated by global warming and climate change. Whether, or more realistically to what extent, governments and other actors are able to govern climate change effectively will very largely determine the need for, and effectiveness of, new or more robust forms of marine environmental governance in the future. But the climate change regime – the collection of international agreements and developing norms for dealing with the climate crisis – is more than a cause of the growing challenges facing marine governance. The development of that regime, and its implementation around the world, offers potentially valuable guidance – or, perhaps more accurately, warning – about whether marine environmental governance will be up to the task of grappling with global environmental change.

As made evident in previous chapters (see especially Chapters 1, 3–5, 12–16, 19–20, 23–24), the most far-reaching challenge for marine environmental governance is climate change. International negotiations to address climate change, first to better understand it and later to reach agreement on action to mitigate its causes and adapt to its impacts, have been underway for more than three decades. Those negotiations initially resulted in the 1992 United Nations Framework Convention on Climate Change (UNFCCC), which ultimately aims for, among other things, the

> stabilization of greenhouse gas concentrations in the atmosphere at a level that would prevent dangerous anthropogenic interference with the climate system ... within a time frame sufficient to allow ecosystems to adapt naturally to climate change, to ensure that food production is not threatened and to enable economic development to proceed in a sustainable manner.
>
> *(United Nations 1992)*

Importantly, the UNFCCC called on economically developed countries to voluntarily and collectively reduce their greenhouse gas emissions to 1990 levels by no later than 2000. Those voluntary reductions were not realised.

Over subsequent years and decades, diplomats came together in more than two dozen annual conferences of the parties (COPs) to negotiate mechanisms for implementing the UNFCCC's objectives, in the process reaching a number of major agreements for mitigating greenhouse gas emissions, adapting to unavoidable impacts of climate change, and providing assistance to developing countries most affected by those impacts (for a brief summary of the COP process, see Harris 2021: 40–56). At the third conference in 1997, COP3, agreement was reached on the Kyoto Protocol. In the protocol, a number of developed countries agreed to reduce their greenhouse gas emissions by 5.2% by 2012 compared to what they were in 1990. While the protocol did not include major emitters, notably the United States (because it lacked support in the Senate) and China (because it was a developing country and thus not required to limit emissions), it was a step in the right direction. However, much as with the underlying UNFCCC, collectively the countries that promised to reduce their emissions again failed to do so. Another major milestone in development of the climate change regime came in 2009, at COP15, when countries acknowledged, in the Copenhagen Accord, the increasingly urgent need to stop the increase in global greenhouse gas emissions, agreed that global warming should be limited to 2°C, and promised to provide developing countries with $100 billion in annual aid by 2020 to help them mitigate and adapt to climate change. Despite apparent good intentions behind the Copenhagen Accord, greenhouse emissions continued to increase, as did global warming, while aid to developing countries did not reach the promised level.

In recognition that 'top-down' agreements such as the Kyoto Protocol and Copenhagen Accord were not achieving the objectives of the UNFCCC, climate negotiators shifted to a 'bottom-up' approach to climate governance (Harris 2021: 47–56). The result was the Paris Agreement on climate change, reached at COP21 in 2015. The overarching objectives of that agreement were to limit global warming to 'well below' 2°C, and preferably to no more than 1.5°C, and, once again, to provide aid to developing countries in the climate context (United Nations 2015). These objectives were to be met not by international agreement on each country's emissions, as in previous climate agreements, but through individual pledges – 'nationally determined contributions' – from each country to do what it could to achieve the Paris Agreement's objectives. Global greenhouse emissions and global warming continued to increase following COP21, with 2020 tied with 2016 for being the hottest year since record keeping began, and global warming reaching 1.2°C above the historical norm by the time of COP26 in 2021 (NASA 2021). COP26 resulted in the Glasgow Climate Pact, which essentially reaffirmed the Paris objectives while encouraging countries to implement policies to limit global warming to 1.5°C, including by moving away from coal – the first time that coal was mentioned in any climate agreement (although oil and natural gas were still not mentioned) – and promising to provide more funding to developing countries (despite such promises never having been realised up to that point in time) (United Nations Framework Convention on Climate Change, 2021).

It is perhaps a hopeful sign that the Glasgow Climate Pact mentions the importance of oceans in several of its clauses. Consistent with reporting from the Intergovernmental Panel on Climate Change (2019), the pact notes 'the importance of ensuring the integrity of all ecosystems, including ... the ocean' (preamble) and welcomes relatively new 'informal' scientific 'ocean and climate change dialogue' (Article 58). It invites relevant actors to 'integrate and strengthen ocean-based action' when addressing climate change (Article 60). An obvious question is whether this welcome recognition of the relationship between oceans and climate change will result in sufficient practical action to govern either issue more effectively.

The limits of climate governance

What has come from these and related efforts by governments to address climate change? At the international level, concern about climate change by most governments is now palpable, as demonstrated most recently in the Glasgow Climate Pact. While not all countries fully ascribe to the strengthening international mandate to address climate change, the message coming out of COP26 is that most of the world's governments now accept that climate change is an urgent problem – a climate crisis, or even a climate emergency – and that more action must be taken. That said, even if all of the nationally determined contributions that were pledged at COP26 toward achieving the objectives of the Paris Agreement were fully implemented, global warming would likely reach 2.4°C (Carbon Brief 2021), a global temperature that will bring devastation to ecosystems both on land and in oceans and seas (Intergovernmental Panel on Climate Change 2018).

All of these and other milestones in the evolution of the climate change regime occurred against the backdrop of growing greenhouse gas emissions and increasing impacts of climate change. Behaviours and activities that lead to greenhouse gas pollution are still expanding globally. Carbon dioxide and other greenhouse gas concentrations in Earth's atmosphere are increasing year by year, and both terrestrial and marine ecosystems are experiencing warming, in some regions dramatically so, over short periods (see Intergovernmental Panel on Climate Change 2021). In other words, despite decades of efforts to achieve the objectives of the UNFCCC and its successor agreements, the pollution causing climate change continues to *increase* globally and the 'dangerous' climate change that the convention seeks to avoid is already happening, with the only prospect in coming decades (at least) that things will become *more* dangerous. The climate change regime may have limited the increases in global greenhouse gas pollution somewhat compared to what they might have been without its associated agreements and actions, and full implementation of national pledges would certainly do so. However, the regime has failed utterly to achieve its intended objective, and experience over more than three decades suggests that many countries' pledges will not be fully implemented in time to achieve their stated purpose.

To be sure, in many countries around the world, much is being done to limit greenhouse gas emissions, and more developed countries and international agencies are providing assistance to developing countries to address climate change. A number of countries have pledged to become carbon neutral – to emit no more carbon dioxide than is absorbed or sequestered – although few have taken the concrete steps to do so as quickly as scientists say is necessary to avoid the worst effects of climate change. Around the world, many industries and businesses have made similar pledges, although how they will be accounted for raises questions about their seriousness in the long term. Growing numbers of individuals around the world are taking climate change seriously and changing their behaviours to limit their personal impacts on Earth's climate system. Yet, taken together, all of these actions, and many more around the world, have so far failed to reverse the upward global trend in greenhouse gas pollution. As some countries cut their greenhouse gas emissions, others increase them. As some businesses reduce their greenhouse gas pollution, others produce more. As some individuals live in more environmentally sustainable ways, many more join the global consumer classes to live and pollute in ways that make the problem of climate change much worse (Harris 2021: 130–157). At the same time, the resources needed by the least capable actors around the world – poor countries, communities, and people – to reduce their own emissions and, vitally, to adapt to the growing impacts of climate change, are severely lacking, in large part because developed countries' longstanding promises to provide aid have been only partially fulfilled (and even if all of those promises were made good today, they would scarcely meet the need).

The climate change regime has contributed to action around the world to combat climate change and its effects, but that action is not strong enough or happening fast enough to keep up with the causes and consequences of the problem for which the regime was crafted. With each passing year, the climate change regime falls further behind actual climate change. Does this portend similar failures for regimes that aim to address the impacts of climate change and other forms of global environmental change in and along the world's seas and oceans? To answer that question, it is helpful to consider why the climate change regime has lacked efficacy.

Portents from the climate change regime

As I have argued at length elsewhere (Harris 2021), efforts by governments and other actors to address climate change have suffered from major pathologies at all levels – international, national, and human – resulting in weak, slow-motion action even as the already fast-moving causes and impacts of climate change accelerate. At the international and national levels, countries are plagued by the key features of the system in which they interact: the international system is premised on non-interference and the individual, religiously observed sovereignty of each country to act as it wishes. By definition, each country aims to promote its own national interests as defined by its most powerful domestic actors. Those actors are still heavily vested in the fuels and behaviours that cause climate change and therefore resist making the sometimes-difficult changes needed to address the problem effectively. In short, international action on climate change is too often restrained by national inaction. This would not matter if it applied only to small countries producing limited greenhouse gas emissions. However, it applies to major countries – most notably China, the world's largest national source of greenhouse gas pollution by far (Ritchie and Roser 2021) – whose climate-changing pollution far exceeds the reductions in emissions implemented in other countries. Among nongovernmental actors, particularly businesses and individuals, there is an overriding focus on growth in income and material consumption. This growth is encouraged by – indeed it is the foundation of – the global economic system. While many people around the world need to consume more, sometimes much more, to meet their needs, too many who already consume and pollute far too much are consuming and polluting more and more. The climate change regime, even as it displays greater determination to tackle climate change, cannot keep up with these forces. Partly that is because the actors involved continue to believe that it is in their interests to continue with the status quo or at best free-ride on the actions of others.

The experience of the climate change regime may be a harbinger for global marine environmental governance. The question is to what extent regimes for marine environmental governance are afflicted with the same or similar pathologies. The chapters in this book suggest that many of them are so afflicted to varying degrees. The lack of action to reduce pollution or otherwise to effectively address threats to the marine environment is a feature of both the climate change regime and regimes for managing oceans and seas. The lack of funding and other resources that obtain in the climate change regime is likely to do so with respect to marine issues, particularly those in remote areas and those affecting the least capable – that is, the poorest – communities and countries. To be sure, it is possible to point to individual examples of successful climate governance in specific places around the world – cities, regions and even entire countries greatly reducing their greenhouse gas emissions – even as the problem of global climate change, and notably its impacts around the world, particularly in the world's least well-off countries, continues to grow worse. Similarly, there are examples of marine governance in particular locations around the world that have been successful, many of which have been recounted in preceding chapters. The future will see similar successes. But the climate change regime is a warning: those successes may be, or are likely to be, exceptions that prove the larger rule of global environmental change outpacing management schemes and leaving the marine environment, and those who depend upon it, increasingly worse off.

Two features of the world's efforts to address climate change bear on the question of marine environmental governance. First, the experience of the climate change regime has direct implications for marine governance: the most important and most damaging global environmental change affecting the world's oceans is climate change, as manifested in, among other impacts, ocean warming, sea level rise, and seawater acidification. All of the adverse manifestations of climate change felt by the oceans and those people and communities that utilise them *will definitely continue to increase* in coming decades (and perhaps coming centuries). This means that the challenges of global environmental change for marine governance will only increase. Second, the experience of the climate change regime points to flaws in the processes by which regimes for addressing global-scale environmental problems are formulated and implemented. If the climate change regime is any guide, not only will the environmental challenge for marine environmental governance be enormous, so will the challenge of devising and operationalising new schemes, and arguably updating extant ones, be extremely problematic. This does not bode well for marine environmental governance in the future, particularly if the weaknesses of the climate change regime are not taken extremely seriously and largely overcome by actors working to realise more effective marine governance amidst global environmental change.

Conclusion

Several decades of concerted international, national, local, and even individual efforts to address climate change demonstrate the limits of global environmental governance: even as efforts for responding to the causes and consequences of climate change (and global warming) have increased in pace and scale, the problem itself has grown increasingly worse. The same can be said of many manifestations of global environmental change across seas and oceans. Thus, much as action to address climate change must be scaled up greatly and rapidly if climate catastrophe is to be avoided in the coming decades (and centuries), so, too, must action to address the many forms of global environmental change afflicting oceans and seas. The status quo – continuing to do things as they have been done in the past and are being done in the present, in too many places and with respect to too many types of marine environmental change – is most definitely not a desirable option for the near future, least of all the long-term future. The experience of the climate change regime is a portent of what may happen – what is most likely to happen – with respect to marine environmental governance as the Anthropocene sees accelerating and deepening global environmental change. But portents have great value: they are warnings of bad things that are likely to happen in the future. With such warnings, assuming they are heeded, it is sometimes possible to take action to prevent them from becoming reality. That is the way that lessons from the climate change regime should be viewed: as signs of what to avoid and what to do differently in marine governance amidst global environmental change.

References

Carbon Brief. (2021) 'Analysis: Do COP26 promises keep global warming below 2C?', https://www.carbonbrief.org/analysis-do-cop26-promises-keep-global-warming-below-2c

Harris, P.G. (2021) *Pathologies of Climate Governance: International Relations, National Politics, and Human Nature*. Cambridge University Press, Cambridge.

Intergovernmental Panel on Climate Change. (2018) *Global Warming of 1.5°C*, https://www.ipcc.ch/sr15/

Intergovernmental Panel on Climate Change. (2019) *IPCC Special Report on the Ocean and Cryosphere in a Changing Climate*, https://www.ipcc.ch/srocc/download/#published

Intergovernmental Panel on Climate Change. (2021) *Climate Change 2021: The Physical Science Basis*, https://www.ipcc.ch/report/ar6/wg1/

NASA. (2021) '2020 Tied for Warmest Year on Record, NASA Analysis Shows', https://www.nasa.gov/press-release/2020-tied-for-warmest-year-on-record-nasa-analysis-shows

Ritchie, H. and Roser, M. (2021) 'China: CO_2 country profile', *Our World in Data*, https://ourworldindata.org/co2/country/china?country=~CHN

Thomas, J.A., Williams, M. and Zalasiewicz, J. (2020) *The Anthropocene: A Multidisciplinary Approach*. Polity, Cambridge.

United Nations. (1992) *United Nations Framework Convention on Climate Change*, https://unfccc.int/documents/36938

United Nations Framework Convention on Climate Change. (2015) *Paris Agreement*, https://unfccc.int/process-and-meetings/the-paris-agreement/the-paris-agreement

United Nations Framework Convention on Climate Change. (2021) *Glasgow Climate Pact*, https://unfccc.int/documents/310475

INDEX

Note: **Bold** page numbers refer to tables; *italic* page numbers refer to figures.

Made in the USA
Middletown, DE
20 September 2024